Integrated Watershed Management

Integrated Watershed Management

Principles and Practice

Isobel W. Heathcote
School of Engineering
University of Guelph

John Wiley & Sons, Inc.

New York / Chichester / Weinheim / Brisbane / Singapore / Toronto

This publication is designed to provide accurate and
authoritative information in regard to the subject
matter covered. It is sold with the understanding that
the publisher is not engaged in rendering legal, accounting,
or other professional services. If legal advice or other
expert assistance is required, the services of a competent
professional person should be sought.

Library of Congress Cataloging-in-Publication Data

Heathcote, Isobel W.
    Integrated watershed management : principles and practice / Isobel
W. Heathcote.
      p.   cm.
    Includes index.
    ISBN 0-471-18338-5 (cloth : alk. paper)
    1. Watershed Management. 2. Watershed management–Social aspects.
  I. Title
  TC409.H38   1998
  333.91'15–dc21   97-30290

Printed in the United States of America

10 9 8 7 6 5

Contents

Preface

Conservation is a state of harmony between men and land.

Aldo Leopold, *A Sand County Almanac*

To waste, to destroy, our natural resources, to skin and exhaust the land instead of using it so as to increase its usefulness, will result in undermining in the days of our children the very prosperity which we ought by right to hand down to them amplified and developed.

Theodore Roosevelt, Message to Congress, December 3, 1907

Water is possibly our most precious natural resource. The abundance and quality of water drives all human systems and those of most other organisms as well. Yet until the 1970s, most water management practices sought to solve single, localized problems without taking account of the impacts of those actions on the biophysical, economic, and social elements of the larger watershed system. Over the past twenty years, a strong global consensus has begun to develop around the notion that the watershed is, in fact, the best unit for the management of water resources. Now, countries in every part of the world try to place water management actions in the context of natural and human systems: watersheds, and the human communities in them.

This book builds on the experience of many individuals and agencies around the world as they have grappled with the challenge of integrated watershed management. Almost without exception, meeting that challenge has required collaboration among specialists in widely varying disciplines—engineering, biological sciences, economics, sociology, law, and ethics—and among government agencies, private industries, nongovernment organizations, and the public. This level of collaboration marks a dramatic change from the technocratic approaches to water management that were typical of earlier generations. It also marks the beginning of a new, and integrated, style of water management, embracing the contribution of a variety of disciplines and viewpoints in the development of strong water management strategies.

This book aims to present an integrated approach to watershed management to those who are now, or in future will be, responsible for the management of precious water resources. It is intended for use by senior undergraduate and graduate students and by water management professionals. It has several key objectives:

1. To present a rational framework for the development of water resources management strategies.
2. To provide an introduction to the technical elements and tools of water management.
3. To illustrate the interplay of disciplines in the management of water resources problems.
4. To demonstrate that water resources planning is not a clear-cut scientific or engineering technique with a single "right" answer, but rather a continuous social process intended to move a community closer to its goals for environmental quality.

Many agencies and individuals have contributed to the development of this book. In particular, I would like to recognize the advice, comments, and technical support provided by past and present staff of the Ontario Ministry of Environment and Energy's watershed management group; the Grand River Conservation Authority (Cambridge, Ontario); the Institute for Water Resources, U.S. Army Corps of Engineers; the U.S. Environmental Protection Agency water management and hydrologic simulation groups; and the Canadian Institute for Environmental Law and Policy (Toronto). I would also like to acknowledge the special support and advice I have received from many individuals, in particular Walter Lyon, Harold "Jack" Day, David Eaton, Mark Killgore, Tony Smith, Bill James, Trevor Dickinson, Ramesh Rudra, Hugh Whiteley, Doug Joy, Don Weatherbe, George Zukovs, and Mike Fortin. On a personal level, I would like to thank my husband, Alan Belk, and my children, Elspeth Evans, Zoë Belk, and Edward Belk, and my parents, Blake and Barbara Heathcote, for their continued support and encouragement through the long process of research and writing. Finally, warmest thanks go to the editorial and production staff at John Wiley and Sons, particularly Dan Sayre, Neil Levine, Ira Brodsky, and Millie Torres-Matias, who have cheerfully and enthusiastically seen this book through its development and publication.

ISOBEL HEATHCOTE

University of Guelph
November 1997

1

Introduction

Water is one of our most precious resources. In moist, temperate regions, water is the fundamental mechanism in chemical flux and cycling. In arid regions, access to water lies at the heart of much conflict. Every living organism on this planet requires water in some form. Water, therefore, regulates population growth, influences world health and living conditions, and determines biodiversity (Newson 1992).

For thousands of years people have tried to control the flow and quality of water. McDonald and Kay (1988) document water disputes of 4,500 years ago in the Mesopotamian cities of Lagash and Umma. Engineering works related to military and urban development, drainage works, irrigation projects, and water diversions can all be documented over thousands of years. Bonnin (1988) notes that the year 1989 was the 2,000th anniversary of a Roman decree (*senatus-consultus*) to the effect that:

Ne quis aquam oletato dolo malo ubi publice saliet si quis oletarit sestertiorum X mila multa esto.

That is,

It is forbidden to pollute the public water supply: any deliberate offender shall be punished by a fine of 10,000 sesterces.

Water provided resources and a means of transportation for development in North America—and placed limits on that development in some areas. Even today the presence or absence of water is critical in determining the uses to which land can be put.

Yet despite this long experience in water use and water management, humans have failed to manage water well. Through the nineteenth century and much of the twentieth century, economic development in many countries was rapid, and often at the expense of sound water management. Frequently, optimism about the applications of technology—whether dam-building, wastewater treatment, or irrigation measures—vastly exceeded concerns or even interest in their environmental shortcomings. Pollution was viewed as the inevitable consequence of development, the price that must be paid if economic progress was to be achieved.

Rachel Carson's publication of *Silent Spring* in 1962 was a turning point in public views about the environment in general and about water in particular. The book drew attention to the rapid deterioration of water quality and the role of industrial polluters in that decline. Over the next decade governments around the world strove first to understand, and then to limit, misuse of water, establishing stronger environmental-protection legislation, more efficient administrative structures, and better oversight of public and private water users.

In March 1977, the United Nations (UN) sponsored a conference on water at Mar del Plata, Argentina. The conference is viewed by many researchers (e.g., Lee 1992; Koudstaal et al. 1992; Biswas 1992) as a landmark event in water management. The conference resulted in an "action plan," including recommendations targeted at meeting the goal of safe drinking water and sanitation for all human settlements by 1990. The Mar del Plata conference made specific reference to the problem that water resources would be increasingly under siege as the need for economic development came in conflict with the desire for protection of the environment. The Mar del Plata recommendations for water management policy can be summarized as follows:

1. Each country should formulate and keep under review a general statement of policy relating to the use, management, and conservation of water as a framework for planning and implementation. National development plans and policies should specify the main objectives of water-use policy, which in turn should be translated into guidelines, strategies, and programs.

2. Institutional arrangements adopted by each country should ensure that the development and management of water resources take place within the context of national planning, and that there be real coordination among all bodies responsible for the investigation, development, and management of water resources.

3. Each country should examine and keep under review existing legislative and administrative structures concerning water management and, where appropriate, should enact comprehensive legislation for a coordinated approach to water planning. It may be desirable that provisions concerning water resources management, conservation, and protection against pollution be combined in a unitary legal instrument. Legislation should define the rules of public ownership of water and of large water engineer-

ing works, as well as the provisions governing land ownership problems and any litigation that may result from them. This legislation should be flexible enough to accommodate future changes in priorities and perspectives.

4. Countries should make necessary efforts to adopt measures for obtaining effective participation in the planning and decision-making process involving users and public authorities. This participation can constructively influence choices between alternative plans and policies. If necessary, legislation should provide for such participation as an integral part of the planning, programming, implementation, and evaluation process.

This "action plan" emphasizes a strong, centralized, and national commitment to water management. Yet even 20 years later, the problems it was intended to solve remain significant. Lee (1992) lists the following difficulties as continuing to exist, despite a growing global consensus, confirmed at Mar del Plata, about the need for careful, strategic water management:

1. The dominance of unregulated water uses
2. Inadequate and ineffective water resources management
3. A high degree of inefficiency in many water-related public utilities
4. A failure to retain trained staff of all types
5. Overcentralization and bureaucratization of decision-making authority
6. Inappropriate and inadequate water legislation

The significance of Mar del Plata probably lies in the fact that it recognized, formally and globally, that existing water management policy was failing to reach its goals. The disappointing progress in the years since the conference has encouraged many authors to reexamine the Mar del Plata action plan and the reasons for continued inaction. Many of these papers were written between 1990 and 1992, possibly in preparation for the UN Conference on Environment and Development held in Rio de Janeiro in June 1992.

The Rio meeting provided an important forum for the discussion of global environmental issues and reinforced the need for continued action, including the protection of biological diversity. At the meeting, 156 nations signed a Convention on Biological Diversity, which aims to protect biodiversity and restore damaged ecosystems. Many countries have begun to develop formal policy and programmatic responses to the Convention, and a Global Environment Facility is under development by the UN and the World Bank to provide loans for projects that have environmental benefits in preserving biodiversity and maintaining natural habitats, which reduce the emission of greenhouse gases, stop pollution of international waters, and protect the ozone layer.

These actions clearly speak to the need for management of systems, not system components. The level of consensus on this notion is now almost unprecedented. Lee (1992) argued that overcentralization of water management, like overcentralization of social and economic systems, has failed and must be replaced with locally responsive systems at the watershed level. Koudstaal et al. (1992) reaffirm this idea, noting that there is no single, clear water management "problem," so it is difficult to focus public attention on water and to develop a single centralized approach to water management. Increasingly, authors are calling for an emphasis on "achieving rational, efficient use of water locally" (Lee 1992), including water management institutions that are "appropriate to local conditions and not centrally, inflexibly, imposed."

This perspective, widely endorsed, clearly supports the notion of water management on a watershed, not state or national, basis. The *Oxford English Dictionary* defines *watershed* as "a narrow elevated tract of land separating two drainage basins," or "the thin line dividing the waters flowing into two different rivers." A watershed, therefore, is the boundary of a drainage basin. In the two decades since Mar del Plata, however, the term *watershed* has come to mean also the drainage basin itself, or "catchment" of the river system. This book employs the common usage of *watershed* as a drainage basin: an area of land within which all waters flow to a single river system.

As early as 1980 authors such as Schramm were observing that a watershed is an integrated system, "holistic in nature ... [with the] whole ... greater than the sum of its parts" (Schramm 1980). Today there is a clear global consensus that the watershed is the most appropriate unit for water management (cf. Newson 1992; Lee 1992; Koudstaal et al. 1992; Goodman and Edwards 1992; Nickum and Easter 1990; McDonald and Kay 1988). Figure 1.1 illustrates the interplay of forces affecting integrated watershed management.

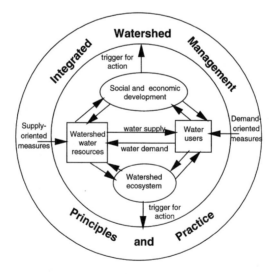

Figure 1.1 Forces affecting integrated watershed management (after Koudstaal et al. 1992).

This book responds to that call for water management on a watershed basis, providing a range of techniques and approaches that can be used to investigate the biophysical, social, and economic forces affecting water and its use. This chapter examines some of the fundamental issues currently facing water resources and proposes a general framework for integrated water management.

1.1 CURRENT ISSUES IN WATER MANAGEMENT

In the wide debate leading up to the Rio meeting, a number of authors analyzed the forces affecting water management. There is remarkable consensus among these authors—who come from countries around the world—about the current issues confronted by water managers (Viessmann 1990; Goodman and Edwards 1992; Nickum and Easter 1990):

Water Availability, Requirements, and Use

- Protection of aquatic and wetland habitat
- Management of extreme events (droughts, floods, etc.)
- Excessive extractions from surface and ground waters
- Global climate change
- Safe drinking water supply
- Waterborne commerce

Water Quality

- Coastal and ocean water quality
- Lake and reservoir protection and restoration
- Water quality protection, including effective enforcement of legislation
- Management of point- and nonpoint-source pollution
- Impacts on land/water/air relationships
- Health risks.

Water Management and Institutions

- Coordination and consistency
- Capturing a regional perspective
- The respective roles of federal and state/provincial agencies
- The respective roles of projects and programs
- The economic development philosophy that should guide planning
- Financing and cost sharing
- Information and education
- Appropriate levels of regulation and deregulation

- Water rights and permits
- Infrastructure
- Population growth
- Water resources planning, including

    Consideration of the watershed as an integrated system

    Planning as a foundation for, not a reaction to, decision making

    Establishment of dynamic planning processes incorporating periodic review and redirection

    Sustainability of projects beyond construction and early operation

    A more interactive interface between planners and the public

    Identification of sources of conflict as an integral part of planning

    Fairness, equity, and reciprocity between affected parties

There is a marked trend from the early 1980s to the present time in several aspects of water management. Lee (1992) suggests that these trends may derive from shifts in economic forces, from aggressive economic growth in the early period to more fragmented, less successful growth in recent years. Certainly, a global economic recession through the early to mid-1990s has made governments more cost conscious and businesses more willing to evaluate the impacts of their actions in advance of implementation. The 1980s saw significant growth in public concerns about the environment, and these have not abated even in the face of economic downturns. The trends that can be observed include:

1. A move from end-of-pipe (reactive) pollution control measures toward pollution prevention.

2. Increasing concern about chronic effects and "invisible" threats—for instance, to human health—as compared with acute effects and visible problems.

3. Increasing awareness that point-source controls are generally well in hand, and that urban and agricultural nonpoint sources of pollution are now major contributors to surface and groundwater impairment.

4. A shift from local action (e.g., abatement of a single point source) to global management strategies (e.g., the Rio Convention on Biodiversity). This trend is also reflected in the growing consensus about the value of watershed management, as compared with management by political boundaries such as those of municipalities.

5. Growing mistrust of, or perhaps understanding of the limits of, technology, and increased reliance on education and extension activities to change consumer behavior.

6. Increasing consensus that the user or exploiter of resources should pay for any damage done by that use: the "user pay" or "polluter pay" principle.

These principles are inherent in an effective integrated water management strategy, as discussed in Section 1.2.

1.2 CHARACTERISTICS OF EFFECTIVE WATERSHED MANAGEMENT

Generally speaking, water management can be considered effective when it:

1. Allows an adequate supply of water that is sustainable over many years
2. Maintains water quality at levels that meet government standards and other societal water quality objectives
3. Allows sustainable economic development over the short and long term

We may, in fact, have reached a point—perhaps signaled by recent environmental disasters like Love Canal and the Cuyahoga River and by water supply crises in many communities—at which it is clear that future water use must be sustainable or development in some regions will halt. Sustainability implies closer cooperation between water users than has typically been experienced in the past. It also implies consideration of the needs of the community, not just the individual—a difficult proposition for many water users.

Goodman and Edwards (1992) state that the in this context word *plan* can mean any one of the following:

- A single-purpose, single-unit plan to serve a specific need, such as a demand for water or abatement of a water-related problem
- A multipurpose and multiproject plan
- A regional plan for water resources development, preservation, or enhancement, staged over a period of time with one or more planning horizons
- A national plan for water resources development, preservation, or enhancement

Planning may proceed in many ways. Bishop (1970) notes that clear goals may be set or a process may simply proceed without goals. One agency may lead and control the process, perhaps even forbidding participation by other groups or agencies. Or the process may be clearly a multiparty and multiperspective one, with community consensus established at every step. Some planning processes consist of a rigid schedule of meetings, formally chaired and run; others employ a flexible workshop or "kitchen table" approach in which discussion is open and unstructured. There are advantages and disadvantages to each method, but the point that must be stressed is that the choice of a planning process is often highly context-dependent; that is, its success will depend very much on the characteristics of the planning area, the water management issues in the area, and the interests and needs of the community of water users. It is increasingly clear, however, that unilateral planning processes that seek to exclude the public will fail—if not in the planning stage, then in implementation. The rapid rise of public interest in, and knowledge about, environmental issues through the 1980s has created a climate in which public participation is expected and, indeed, required in almost every planning situation.

Schramm (1980) offers the following general guidelines for successful river basin planning:

1. The institutional framework for the project must allow consideration of a wide range of alternatives to solve observed problems, including those that may be outside the specific responsibilities of the planning bodies.
2. The planning agencies must have the expertise needed for multiple-objective planning and evaluation procedures, especially in economic, social, and environmental areas.
3. The institutional framework must facilitate adaptation of the plan to meet changing national, regional, and local priorities.
4. The institutional framework must seek representation of all parties affected by the specific development plans and management.
5. The institutional framework must reward initiative and innovation among the members of the technical team and within cooperating agencies.
6. The technical team must be sufficiently free from day-to-day responsibilities so that they can concentrate on long-range planning and anticipation of future problems.
7. The institutions must have the capacity for learning and improving over time, including sufficient continuity over time and the ability to evaluate past programs.
8. There must be sufficient authority within the institutional framework to enforce conformity of execution with construction and operating plans.
9. The institutional framework must be capable of guaranteeing an acceptable minimum level of professional performance by the technical team.
10. The plan implementation stage must include provisions for the timely and qualitatively and quantitatively sufficient supply of needed services by other agencies, as well as provisions to assure continued functioning—i.e., operation, repair, and maintenance of the facilities and services provided.

Schramm emphasizes the need for coordination and cooperation at local, regional, and national levels, noting that:

Planning and plan implementation do not proceed in a rarified vacuum derived from lofty, immutable principles that are a law unto themselves. Planning is done for people and people have different and often competing wants, desires, and hopes; political institutions should be designed to meet those wants. One of the best ways to condemn planning efforts to oblivion or failure is to turn the task over to a self-contained, isolated team of experts who fail to communicate with one another, the people their plans are to serve, and those with political, decision-making authority. Within this dynamic, competing world of human wants and values there is no ultimate reality or single-dimensioned optimum that can be determined by scientific methods alone.

1.3 WHY "INTEGRATED" MANAGEMENT?

The idea of trans-media environmental management—management using the "ecosystem" concept—is a relatively new one. In large part, it was born of experience showing that single-medium or single-source management was not successful in meeting short- or long-term goals. Until the mid-1970s, for instance, almost all pollution control effort was directed at controlling point sources like sewage treatment plants and industrial discharges. The International Joint Commission's Pollution from Land Use Activities Reference Group (PLUARG) (PLUARG 1983) examined the reasons that phosphorus-reduction efforts in the Great Lakes Basin had stalled. Its research showed that remediation of the lakes would require integrated management plans for both point and nonpoint sources throughout the entire Great Lakes Basin. In some areas, point-source controls would be most cost-effective; in others, the focus would have to be on nonpoint-source controls. Without the overview provided by an integrated strategy, costly management efforts would continue to fail.

Sometimes water management efforts have been unsuccessful because they have focused on a single medium (water) rather than on other environmental components such as sediment, air, or biological tissue. Mercury poisoning at Minamata, Japan, and the Wabigoon-English River system, northwestern Ontario, Canada, are excellent examples. In each case, an industrial facility had discharged large volumes of wastewater containing inorganic mercury into receiving waters. In each case the inorganic mercury (which is relatively non-toxic to humans and other organisms) was converted in the water column and sediments to methyl mercury, which is highly bioaccumulative and persistent in body tissues. The methyl mercury was readily taken up by invertebrates in the waters, which in turn were eaten by larger species such as fish, and these larger animals were consumed by humans. The humans, at the top of this particular food chain, received concentrated doses of methyl mercury, which accumulated in their own body tissues, causing a wide range of nervous system impacts. Subsequent abatement efforts aimed at eliminating mercury-using technologies in the Wabigoon facility (a pulp and paper mill) were successful. Nevertheless, more than 20 years after the technology change, mercury continues to be released into the water from river and reservoir sediments and, possibly, from residual deposits within the mill. Consideration only of effluent quality from the mill might suggest that the problem was "solved" 20 years ago; in fact, methyl mercury continues to cycle through the Wabigoon-English River system as a result of trans-media environmental phenomena.

Very often, water management strategies have failed because they neglected to incorporate the full range of values and perspectives present among water users or agencies with an interest in water management. Wilkes (1975) noted that the provision of adequate water supplies in the Rhine River watershed is hampered because different agencies are responsible for water supply and for water quality, and the two are not always effectively coordinated. Wilkes also observed that watershed management requires the involvement of regional, state, national, and international agencies—a measure that was unnecessary at the level of local water management and pollution control. The transition from local to watershed management can be difficult, because interested agencies may not have the necessary authority to take on new management roles, may encounter varying political influences, or may simply not work very well together in managing water resources. In multilingual countries or watersheds, or in less developed countries where external agencies like the World Bank may be involved in planning initiatives, cooperation across agencies and disciplines may be more difficult still.

"Integrated" watershed management, although a strategy that is increasingly advocated in the literature, is therefore still a relatively new concept. McDonald and Kay (1988) observe that there have been "few real attempts to provide integrated management information and even fewer evaluative studies of the policy and management of integration within the water resources field." More and more agencies are now establishing administrative frameworks that permit and even encourage management of water on a watershed basis. Less frequently are water management activities integrated with other resource management activities affecting or affected by water. Heathcote (1993) notes that these may include, at minimum, the intensity and nature of agricultural activities, forestry, and commercial fisheries. Although integration at the watershed level is increasingly possible, integration at larger scales is, in the words of McDonald and Kay (1988), "conspicuously absent," although there are clear advantages to integrated water management at the international scale (especially in international river basins) and even at the global scale.

In the mid-1980s, the Canadian federal government established a formal inquiry on federal water policy, in response to a growing awareness that Canadian water resources were potentially at risk of overuse and underprotection. The inquiry called for "visionary policies" for the management of water resources in Canada (Pearse et al. 1985).

The inquiry drew attention to rising water consumption rates in Canada, conflicting water uses in many areas, and deteriorating water quality, especially in the heavily populated Canada-U.S. border region. Throughout their reports, the inquiry panel stressed the need for caution and prevention, rather than careless use and reaction. Among their recommendations were several relating to the administrative structures of water management and the need for what they termed "comprehensive management." In particular, they called for (Pearse et al. 1985):

- A watershed plan sufficiently comprehensive to take into account all uses of the water system and other activities that affect water flow and quality

- Information about the watershed's full hydrological regime

- An analytical system, or model, capable of revealing the full range of impacts that would be produced by particular uses and developments in the watershed

- Specified management objectives for the watershed, with criteria for assessing management alternatives in an objective and unbiased way

- Participation of all relevant regulatory agencies

- Provisions for public participation in determining objectives and in management decisions

These recommendations are particularly notable because they come from a seasoned group of experts in a country that has long considered itself to have infinite water resources. In the decade following the release of the inquiry's report, many of the panel's admonitions about excessive water use and deteriorating water quality have been proven correct, and the need for integrated watershed management is now seen as urgent. In less-water-rich countries, including the United States and many European countries, population density and a limited resource base make integrated watershed management essential for sustainable water use. As global consensus about the need for integrated management grows, it may now be social and economic forces, rather than technical considerations, that determine the success of an integrated watershed planning effort. In this regard, Thompson (1982) has remarked:

Without systematic methods for taking account of uncertainty, the tendency of the regulator is to obscure the fact that scientific controversy exists. This tendency in turn reinforces the public's unrealistic expectations that science and technology can supply the answers. Closely related to the uncertainty problem is the fact that scientific conclusions, when they are applied to solving human problems, invariably incorporate a range of value judgments. If these are acknowledged, the tendency of the pragmatic professional is to say that, since the issue involves value questions, it might as well be confronted as a political choice without the need for an expensive and time-consuming scientific analysis. Alternatively, the regulator may ignore the value question and carry forward the pretense that his decision is a purely technical one. In this case it is better not to pursue the scientific inquiry too far!

1.4   A RECOMMENDED PLANNING AND MANAGEMENT APPROACH

This book presents a planning approach that is rooted in rational decision making—that is, systematic development and comparison of management alternatives. But this general approach is placed within the context of an informed public and a rapidly changing external environment. It endorses Bishop's (1970) observation that water management planning is a process of achieving social change. In that sense, it is a consensus-building process, not a unidimensional scientific exercise. This theme is stressed throughout the book. Chapter 4 offers a range of techniques that have been proven to be effective in developing social consensus about water management planning.

The general approach presented in this book can be summarized as follows:

1. Develop an understanding of watershed components and processes, and of water uses, water users, and their needs (Chapter 2).
2. Identify and rank problems to be solved, or beneficial uses to be restored (Chapter 3).
3. Set clear and specific goals (Chapter 3).
4. Develop a set of planning constraints and decision criteria, including any weights that may be assigned to criteria (Chapter 5).
5. Identify an appropriate method of comparing management alternatives (Chapters 6 and 7).
6. Develop a list of management options (Chapters 5, 6, and 7).
7. Eliminate options that are not feasible because of time, cost, space, or other constraints (Chapters 5, 6, and 7).
8. Test the effectiveness of remaining feasible options using the method identified in (5) and the decision criteria and weights identified in (4) (Chapters 6 and 7).
9. Determine the economic impacts and legal implications of the various feasible management options (Chapters 8 and 9) and their environmental impacts (Chapter 10).
10. Develop several good management strategies, each encompassing one or more options, for the consideration of decision makers (Chapter 11).
11. Develop clear and comprehensive implementation procedures for the plan that is preferred by decision makers (Chapter 12).

Planners (and authors) find it helpful to divide the planning process into these discrete steps. In reality, however, the planning process is dynamic and continuous. Several tasks or steps may be under way simultaneously. Planning direction may change radically if new information comes to light, if political forces change dramatically, or if community consensus is redirected for other reasons. Above all, integrated watershed planning and management must be responsive and adaptive to changing conditions. This means that a good watershed plan is not a single product, such as a document that sits on an agency bookshelf. Instead, it is a framework for continued dialogue about water and the watershed. Ideas endorsed by the planning team must be revisited and reviewed periodically to determine whether they are still acceptable or could be improved. New technologies and management thinking must be incorporated into the evolving plan. Most of all, the watershed management plan must reflect the current societal consensus about the value of water as a resource, about responsibilities and social attitudes, and about the community's vision of an ideal watershed state. Integrated watershed management is, therefore, a journey, not a destination.

REFERENCES

Bishop, Bruce. 1970. *Public Participation in Planning: A Multi-Media Course.* IWR Report 70-7. Fort Belvoir, Va.: U.S. Army Engineers Institute for Water Resources.

Biswas, Asit K. 1992. Sustainable water development: A global perspective. *Water International* 17 (1992): 68–80.

Bonnin, Jacques. 1988. Were urban water systems improved over the last 20 centuries? *Water International* 13 (1988): 10–16.

Goodman, A. S., and K. A. Edwards. 1992. Integrated water resources planning. *Natural Resources Forum* 16(1): 65–70.

Heathcote, Isobel W. 1993. An integrated water management strategy for Ontario: Conservation and protection for sustainable use. In *Environmental Pollution: Science, Policy and Engineering* , edited by B. Nath, L. Candela, L. Hens, and J. P. Robinson. London: European Centre for Pollution Research, University of London.

Koudstaal, Rob, Frank R. Rijsberman, and Hubert Savenije. 1992. Water and sustainable development. *Natural Resources Forum* 16(4): 277–290.

Lee, Terence. 1992. Water management since the adoption of the Mar del Plata Action Plan: Lessons for the 1990s. *Natural Resources Forum* 16(3): 202–211.

McDonald, Adrian T., and David Kay. 1988. *Water Resources Issues and Strategies.* New York: Longman Scientific and Technical and John Wiley & Sons.

Newson, Malcolm. 1992. Water and sustainable development: The "turn-around decade"? *J. Envir. Planning and Management* 25(2): 175–183.

Nickum, James E., and K. William Easter. 1990. Institutional arrangements for managing water conflicts in lake basins. *Natural Resources Forum* 14(3): 210–220.

Pearse, P. H., F. Bertrand, and J. W. MacLaren. 1985. *Currents of Change: Final Report of the Inquiry on Federal Water Policy.* Ottawa: Queen's Printer.

Pollution from Land Use Activities Reference Group (PLUARG). 1983. *Nonpoint Source Pollution Abatement in the Great Lakes Basin: An overview of Post-PLUARG Developments.* Report to the Great Lakes Water Quality Board of the International Joint Commission. Windsor, Ont.: International Joint Commission.

Schramm, Gunter. 1980. Integrated river basin planning in a holistic universe. *Natural Resources Journal* 20: 787–805.

Thompson, Andrew R. 1982. Water law—The limits of the management concept. In

Environmental Law in the 1980s: A New Beginning, edited by P. Z. R. Finkle and A. R. Lucas. Proceedings of a colloquium convened by the Canadian Institution of Resources Law, Faculty of Law, University of Calgary, Calgary, Alberta, November 27–29, 1981.

Viessman, Warren, Jr. 1990. Water management issues for the nineties. *Water Resources Bulletin* 26(6): 883–891.

Wilkes, Daniel. 1975. Water supply regulation. In *Regional Management of the Rhine*, edited by Chatham House Study Group. London: Chatham House.

2

The Watershed Inventory

The lifeblood of a watershed ecosystem is water. Water movement in the system is affected by many physical, chemical, and biological features and processes. An understanding of these features and processes is an essential first step in assessing the condition of a watershed ecosystem and the impacts of management actions on it. A watershed inventory also provides the building blocks with which to create predictive models of the system for application in evaluating the impacts of proposed management actions.

The following sections provide an overview of the elements contained in a typical watershed inventory.

2.1 PHYSICAL FEATURES AND LANDFORMS

2.1.1 Bedrock Geology

Rock underlies all watershed ecosystems. The nature of this rock determines the character of overlying soils and influences the ultimate movement of water draining through those soils.

Rocks are formed in one of three ways: by an igneous, sedimentary, or metamorphic process. Each type has its own special physical qualities and chemistry.

Igneous rocks are formed when molten rock (magma) is pushed up through the cooler regions of the earth's crust from deeper zones. As the magma cools, the minerals within crystallize and bond in ways that create characteristic patterns. Igneous rocks typically show a texture of finely to coarsely interlocking grains, sometimes containing larger crystals (usually of quartz or feldspar) in a finer-grained matrix. The chemical properties of these rocks become important through the process of rock weathering, in which the quality of water flowing over the rock surface is affected by the chemistry of the rock itself. Rock weathering can occur either through physical processes, such as fracturing and freeze-thaw forces, or through chemical processes such as dissolution and absorption.

Sedimentary rocks are composed either of fragments of mineral and rock eroded from preexisting rocks and then deposited by natural forces, or of materials precipitated from aqueous solutions. Thus, sandstones come from sandy sediments, limestones and dolomites come from finer sediments with a high proportion of shell and carbonate, and shales come from the finest clay sediments. Salt and gypsum are sedimentary rocks formed from the precipitation of materials in solution. As with igneous rocks, the chemistry of sedimentary rock becomes a significant influence on watershed soil and water chemistry. This is particularly true for rocks containing high concentrations of calcium carbonate, an important buffering agent especially in freshwater systems. Sedimentary rocks usually exhibit a layered or stratified structure, reflecting the depositional processes involved in their creation.

Metamorphic rocks are created when igneous or sedimentary rocks are subjected to further heat and pressure, thus modifying their crystal structure (but not their chemistry) and their physical properties. Slate, for example, is the metamorphosed form of shale; marble is the metamorphosed form of limestone. Metamorphic rocks show a more pronounced alignment of crystals than is seen in igneous or sedimentary rocks, a result of the action of heat and directional pressure on preexisting crystal structures.

In the context of watershed management, the geological processes that create these minerals are probably of less interest than their physical structure and chemistry. Important aspects of physical structure include fractures, folds, and faults, all of which (among other processes) are important in groundwater flow and, thus, in the distribution of water within the watershed ecosystem.

The United States and Canada, and many other countries worldwide, maintain extensive geological databases. Often, this information is available in map or report form specific to a given area. The agencies involved may also have electronic bulletin board facilities and/or World Wide Web sites summarizing investigative activities and available publications. The following are the primary Web sites currently operated by these agencies:

U.S. Geological Survey   www.usgs.gov

Geological Survey of Canada   www.emr.ca/gsc

2.1.2 Surficial Geology and Landforms

Overlying bedrock in most regions is a layer of softer sediments, or soils. These materials develop gradually from the action of plants and physical forces on bedrock and contain a mixture of eroded rock and mineral fragments and organic debris; the precise mixture determines the soil chemistry, physical properties such as drainage characteristics, and fertility.

These soft sediments can be shaped and realigned in many ways by the action of ice, water, and wind. In northern regions the force of advancing and retreating glaciers tens of thousands of years ago had an important and lasting impact on landforms. A variety of glacial features can be seen in these areas, including moraines, created by the pushing effect of massive ice sheets; till plains, mixtures of soil and rock left behind by the plowing glacier; and sorted beds of sand and gravel formed by meltwater drainage.

Glacial action was also very important in carving the landscape in these northern regions, including, for instance, the shaping of lake and river basins. It is often possible to see vestiges of ancient beaches and shorelines near the margins of modern water bodies. These structures can provide clues about the water levels and flow patterns that prevailed in glacial times and thus give insight into modern drainage phenomena rooted in glacial structures.

In warmer regions the action of wind and water can be as important in sculpting landforms as glacial action in the north. Arroyos, for example, reveal where water flows seasonally or less often, or perhaps where it flowed in the distant past. In the prairie, wind forces can be important in sorting and redistributing soils, as was the case during the dust storms of the 1930s.

State and national geological surveys should also be able to provide information on watershed topography and the historical forces that led to the development of present landforms. Detailed topographic maps, available for most regions, are an excellent source of information about the degree, shape, and length of slopes (important in controlling runoff and erosion). This information, in conjunction with a knowledge of bedrock geology, lays the foundation for a more detailed inventory of the natural and built features of a watershed. It can give a general indication of the historical and modern forces affecting water flow and, thus, the distribution of surface waters, indigenous and domesticated flora and fauna, and the human systems based on these factors.

2.2 CLIMATE

Climate, including temperature, wind force and direction, and precipitation, is an extremely important influence on water resources and biological processes in a watershed.

2.2.1 Temperature, Evaporation, and Transpiration

The national weather services in most countries collect information about ambient solar radiation, maximum and minimum air temperature, and related measurements across a wide network of monitoring stations. These data are usually available to the researcher through public and university libraries or directly from the agencies involved. Increasingly, such information is available in electronic format, vastly simplifying its use in predictive computer models or other analytical methods. As with geological data, the World Wide Web is an excellent starting point for information about the availability of meteorological data. Current Web sites for key U.S. and Canadian agencies involved in meteorological data collection are:

National Weather Service (NWS) (links to NWS offices throughout the United States)   vortex.atms.unca.edu/~staylor/nws

Atmospheric Environment Branch, Environment Canada   www.ns.doe.ca/ aeb/aebhome

As with precipitation measurements and other climatological data, it is important to remember that climate can vary dramatically with the local environment, so a general rule of thumb is that climatological data must be collected as close as possible to the watershed area(s) of interest. Where available data are inadequate or too distant to be relevant, it may be appropriate to collect additional measurements using recording gauges at key locations in the watershed.

For water managers, air temperature is of interest because of its influence on evaporation and transpiration, and on the growth of plants in the watershed. Evaporation is the transfer of water from the liquid state to the vapor form. It occurs over the surface of open water bodies through the escape of water molecules (which are always in motion) from the liquid into the air. Evaporation is said to occur when the transfer from liquid to air exceeds the transfer from air to liquid; when the reverse is true, precipitation is said to occur. Dalton's law gives the formula for evaporation from free-water surfaces:

$$E = C(e_s - e_d) \qquad (2.1)$$

where: E = evaporation rate
C = a constant
e_s = the saturated vapor pressure at the temperature of the water surface in mm Hg
e_d = actual vapor pressure of the air (e_s times relative humidity) in mm Hg

The value of the constant has been estimated by Rohwer (1931) as:

$$C = (0.44 + 0.073W)(1.465 - 0.00073p) \qquad (2.2)$$

where: W = average wind velocity in km/hr at a height of 0.15 m
p = atmospheric pressure in mm Hg at 0∘C

The value of E calculated using this constant has units of mm/day. For reservoirs, the calculated value of E should be multiplied by 0.77 (Schwab et al. 1993). Meyer (1942) calculated the value of C for pans and shallow ponds using the relationship:

$$C = 15 + 0.93W \qquad (2.3)$$

where: W = average wind velocity in km/hr at a height of 7.6 m

(Note that the value of E calculated in this fashion has units of mm/month, that vapor pressure should also be measured at 7.6 m height, and that air temperature is calculated as the average of the daily minimum and daily maximum temperatures.) Meyer's calculation of C for small lakes and reservoirs is similar:

$$C = 11 + 0.68W \qquad (2.4)$$

Evaporation also occurs from land surfaces. Where soils are saturated and the water table is high, evaporation rates may approach those from free-water surfaces. As saturation decreases, evaporation rates will diminish rapidly. Artificial cover on a soil surface (for instance, by mulching) or a natural cover (for instance, by densely growing vegetation) further restricts evaporation from the land.

Transpiration is the transfer of water molecules into the air through the tissues of living plants. So in an area such as a rice paddy, evaporation (from the water surface) and transpiration (from the growing plants) may occur simultaneously. Transpiration rates are usually measured in the field—a complex process that can involve "bagging" plants or trees to capture all moisture. Experimental data of this type is sometimes subject to criticism in that the process of enveloping a growing plant with plastic or other film may alter tissue temperatures and thus affect the rate of natural processes.

Transpiration is sometimes expressed as a ratio of the weight of water transpired to the weight of dry plant tissue. Schwab et al. (1993) quote typical transpiration ratios for common plants, as shown in Table 2.1.

Table 2.1   Transpiration Ratios for Common Agricultural Plants

Plant	Transpiration Ratio
Sorghum	250
Corn	350
Red clover	450
Wheat	500
Potatoes	640
Alfalfa	900

(*Source:* Schwab et al. 1993)

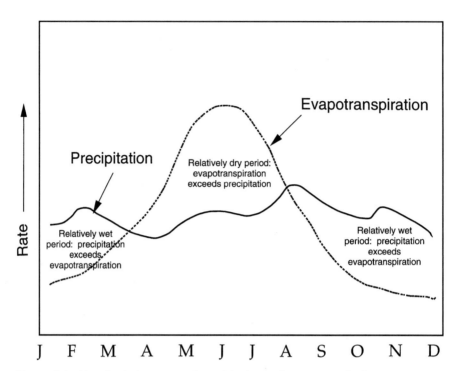

Figure 2.1 *Hypothetical patterns of precipitation and evapotranspiration over a watershed.*

Evaporation and transpiration rates are often combined into a single "evapotranspiration" rate, which can be determined from field measurements using either controlled evaporation techniques and/or mass balance approaches (calculation of evapotranspiration losses by difference, given known inputs and outputs from a system).

Across most of eastern Canada and the United States, approximately to the Mississippi River valley, average annual precipitation exceeds evaporation. This regime also applies in the Pacific Northwest. For the rest of the country west of the Mississippi, evaporation rates generally exceed precipitation rates, with the result that there is a net annual loss of water from the regions of this area. Figure 2.1 illustrates these concepts.

2.2.2 Precipitation

Ultimately, all the water in a watershed comes from precipitation—rain, snow, or dew. Water vapor is always present in the atmosphere, even over the driest regions. The deposition of this moisture as precipitation depends on complex physical phenomena, both at the local and at the regional—or larger—scale. The most casual observer knows that rainfall is seldom uniformly distributed over a given area: it may be raining in one part of a city, for example, but dry in another. It is virtually impossible to model atmospheric processes, and thus to predict rainfall intensity, with certainty. A sound understanding of precipitation phenomena must therefore be based on observation, preferably over a period of time long enough to capture the full range of natural variability.

The following discussion focuses on the collection of rainfall data as the most general case; a brief discussion of snowfall/snowmelt data is included at the end of this section.

Rainfall measurement aims to collect information about the duration of a rainfall event, as well as its intensity, or rate. There are two challenges in collecting this information. The first is to ensure that the rain gauge collects a representative sample (in other words, accurately reflects the quantity of rain that would have fallen on open ground, as if the gauge had not been present). The second is to ensure that the number and placement of rain gauges is sufficient to give an accurate picture of rainfall conditions over the entire area of interest—for instance, a watershed.

The first challenge, that of collecting a representative rainfall sample, relates primarily to the design of the rain gauge and its placement in the study area. Careless placement of gauges (for instance, under trees or close to sheltering buildings), can reduce the amount of rain reaching the gauges and give an unrepresentative sample. Less obvious errors include problems arising from water splashing into, or out of, the gauge container, errors related to improperly leveled gauges, and instances in which the gauge itself absorbs water and measured rain is less than actual rain entering the gauge. A correctly placed gauge is located close to the ground but sufficiently elevated to exclude water splashed in from the surrounding land surface. Usually, this position is somewhere between 30 cm and 1 m from the ground; the decision may relate to routine practice in the agency responsible for rain gauging, to topographic features, to expected snow cover, or to some combination of these factors.

The simplest rain gauge is simply a graduated cylinder. A funnel placed over the opening can allow collection of rain over a larger area (thus improving collection efficiency) but requires that the gauge be calibrated to reflect the larger area of capture. Simple gauges of this type are generally limited to sites with good access, where regular visits for reading and maintenance are possible. In more remote areas, gauges of larger capacity ("storage" or "totalizer" gauges) are often used instead.

Most modern rain gauges incorporate capability for continuous recording of data, whether by mechanical (e.g., clockwork) or digital means. Three types of gauges are commonly used: tipping bucket gauges, weighing gauges, and gauges that record water level by means of a float device. Tipping bucket gauges use a small two-compartment tipping container that, when full, tips to one side, closing an electric circuit and recording a specified (calibrated) quantity of rain. As the container empties, it brings the other side under the filling funnel and the process begins again. Tipping bucket gauges are durable and reliable and are widely used. In heavy rains they may, however, underrepresent actual rainfall through spillage during filling/tipping. They also generate rainfall records that are stepped rather than continuous (smooth) and are usually considered unsuitable for measuring snowfall.

Weighing devices, as the name suggests, measure precipitation by weight. Rain or snow falling into the container is continuously weighed on a balance, and the result is transmitted to a recording device. This type of gauge is better than tipping bucket devices in areas with frequent snow, but its capacity is, of course, limited, so it cannot be left unattended for long periods in areas with heavy precipitation. Weighing devices are also sensitive to wind effects, which can cause the balance to tremble and create an error in the recorded data.

Float-type gauges employ a float that rises with accumulating rainwater. The position of the float is then continuously recorded on a strip chart or other recording device. Without heating devices (which have sometimes been employed in cold climates), these gauges are unsuitable for use in freezing weather.

Newer gauges employ digital recording techniques in place of older strip chart recording and are capable of transmitting data in electronic form over dedicated telephone lines or by similar means.

The second challenge in monitoring precipitation is that of collecting a reasonable sample across a diverse area. Weather radar technology is now widely used for the prediction of rainfall events and for assessment of the spatial distribution and intensity of rainfall over the area of interest. (Increasingly, this technology is also used in flood prediction and other hydrological forecasting requiring distributed rainfall data as input; see, for example, Kouwen and Soulis (1994).

Radar data may be available more or less continuously over the period preceding, during, and following a storm, so the analyst can piece together a time series of precipitation phenomena. Although methods are available to convert radar data into quantitative estimates of precipitation volume, these methods can be difficult to apply accurately. Radar depends on echo intensity from rain droplets, and because storms vary tremendously in the size and distribution of these droplets, it has been difficult to develop a consistent relationship between echo response and actual rainfall rate.

The simplest method of estimating precipitation over an area (as opposed to a single point) is probably to calculate the arithmetic average of several point measurements throughout the area of interest. Although useful for small, topographically homogeneous watersheds, this method is flawed in that precipitation seldom falls uniformly over a given area (as a result of topography and other factors) and gauges are seldom uniformly distributed throughout the area of interest. A simple arithmetic average may therefore yield an unrepresentative estimate of "average" precipitation. More complex methods, such as the isohyetal method and the Thiessen method, attempt to interpolate precipitation between gauges, the former method using weighted average precipitation between isohyets (lines of equal rainfall depth) and the latter using polygonal subareas, the center of each being a rainfall gauge. The Thiessen method is relatively straightforward to employ but is not suitable for areas of high elevation. The isohyetal method is considered the most accurate but requires considerable skill and careful application in areas of varied topography. Figure 2.2 illustrates the disposition of precipitation falling over a drainage basin.

In colder regions snow may comprise 50% or more of total precipitation. It

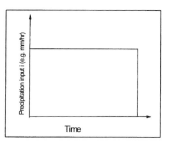

Rain falls continuously over a
drainage basin, causing . . .

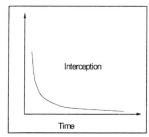

(1) Interception of initial flows by
trees and other plants, then. . .

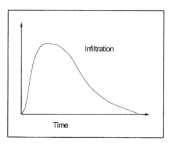

(2) Infiltration into soils and
groundwater, and . . .

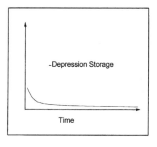

(3) Filling of small surface
depressions, and finally . . .

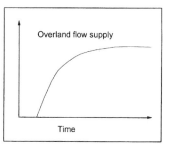

(4) Overland flow of
remaining precipitation.

Figure 2.2 *Disposition of precipitation falling over a drainage basin (adapted from Viessman and Lewis 1996).*

is therefore an important element in the watershed water budget, and essential to include in a watershed inventory.

Snow sampling presents special challenges in comparison with measurement of rainfall or streamflow. The most important difference, of course, arises from the need to accommodate freezing conditions in the collection, recording, and transmission of data.

A simple and obvious means of measuring snow is to insert a ruler in the snow pack to the full depth of the snow. Yet measurement of snow depth may be of little use in estimating water content, which differs depending on the age and consistency of the snow. To obtain information about water equivalency, snow packs may be cored and the cored material either weighed or melted and measured in a graduated cylinder. Radioisotope methods involve burial of a gamma emitter (e.g., cobalt 60) under the ice pack. The thickness of the accumulating snow pack is then proportional to the decrease in gamma particle emissions. Weighing snow gauges have also been developed, but they are likely to be less accurate because of the tendency of adjacent snow to support the weighing plate and thus cause an underestimate of snow accumulation.

Snow cover (i.e., proportion of area covered by snow) is also of interest, particularly during snowmelt periods. Even aerial photographs of snow cover can give an indication of watershed areas that are probably still frozen and storing water as snow, versus areas that are melting and thus in a position to release their stored moisture.

2.2.3 Wind

Wind conditions affect many land and water phenomena in a watershed, particularly evapotranspiration and wind erosion. Wind, like precipitation and ambient temperature, is commonly measured by regulatory agencies through a network of meteorologic stations, in part because of its primary importance in aviation. Also like those parameters, wind conditions are typically highly variable, even over small areas, as a result of natural topography and the influence of structures. Wind speed and direction are usually measured with an anemometer, which may be either mechanical (e.g., rotating cup type) or electronic (e.g., "hot wire" or "hot film" anemometers, which measure wind speed as a function of heat loss over a surface). Wind records are widely available in printed and electronic formats from regulatory agencies such as the United States National Weather Service and Environment Canada's Atmospheric Environment Branch.

2.3 SOILS, INFILTRATION, AND RUNOFF

2.3.1 Soil Classification and Permeability

In effect, soil is a porous medium that blankets bedrock. Its ability to transport water depends on the size and condition of channels through the porous medium. These factors in turn depend on the size of soil particles, the degree to which individual soil particles are aggregated into larger masses, and the arrangement of individual particles and aggregates. Where soil channels are large and relatively permanent, the soil provides an effective transport mechanism for water—in other words, it is highly permeable.

Soil permeability can change over time, particularly through soil compaction or the development of a thin, impermeable layer at the surface, such as may be caused by repeated action of raindrops on soil particles. The energy in a raindrop dislodges soil particles and moves them, with the flowing water, over the soil surface. Smaller particles eventually settle into the gaps between larger fragments, so over time the porous surface can become clogged and compact.

Soils are classified into various groups depending on their permeability, infiltration rate, and surface runoff potential. Permeability refers to the rate of water movement through a saturated soil (when gravitational forces dominate) and is defined by Darcy's law:

$$q = KhA/L \qquad (2.5)$$

where: q = flow rate (volume/time)
  K = hydraulic conductivity of the flow medium (distance/time)
  h = head or potential causing flow (distance)
  A = cross-sectional area of flow (distance^2)
  L = length of the flow path (distance)

Soil permeability is affected by a range of factors, including particle size distribution (texture), degree of compaction, and chemical composition, especially salt concentration. Infiltration rate is a broader concept, referring to the ability of a soil to "draw up" water through a combination of gravitational forces and capillary action. Infiltration rate is expressed in units of volume per unit time per unit area or, roughly, depth per unit time. Infiltration is the mechanism by which plants receive water and dissolved nutrients and by which groundwater supplies are maintained and replenished. It is affected by the physical structure of the soil and any cover on the soil surface (such as mulch or crops) and by soil moisture content, temperature, and rainfall intensity. Figure 2.3 illustrates the relationship between soil type, and soil porosity and moisture content.

The United States Soil Conservation Service uses four categories to define soil infiltration characteristics (see Table 2.2). *Infiltration* refers to the entry of water into the soil surface (Schwab et al. 1993) as opposed to other drainage phenomena such as overland flow. Soil classification maps are available from the SCS for most regions of the U.S.

2.3.2 Infiltration

Infiltration rate (that is, water transport through an unsaturated soil) can be predicted by Horton's (essentially empirical) formula (Horton 1939):

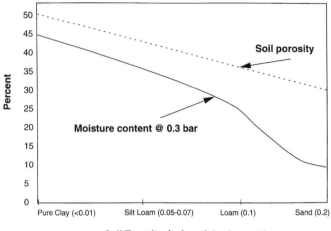

Soil Type (typical particle size, mm)

Figure 2.3 Typical drainage characteristics of soils (adapted from Novotny and Olem 1994).

$$f = f_c + (f_0 - f_c)e^{-kt} \qquad (2.6)$$

where: f = infiltration capacity (maximum rate at which a soil can take water through its surface) (length/time)

f_c = the constant infiltration capacity as t approaches infinity

f_0 = infiltration capacity at time of onset of infiltration

k = a positive constant for a given soil and initial condition

t = time

Table 2.2   U.S. Soil Conservation Service Soil Infiltration Categories

Soil Group	Infiltration Rate	Surface Runoff Potential
A: Well-drained beds of sand and gravel	High (>25 cm/hr) even when thoroughly wetted	Low
B: Moderately fine to moderately coarse soils such as sandy loams	Low to moderate (6.3 to 25 cm/hr)	Low to moderate
C: Fine-textured soils and soils with underlying layer impeding drainage	Low (0.5 to 6.3 cm/hr)	High
D: Fine clay soils and soils with underlying impermeable layer; soils in areas of permanently high water table	Very low (<0.5 cm/hr)	High

Other methods for predicting infiltration, such as that proposed by Green and Ampt (1911), are essentially variations on Darcy's equation using assumptions about uniformity of vertical flow, water content, and soil hydraulic conductivity. Holtan (1961) proposed a method that reflects the gradual exhaustion of soil moisture storage as a soil becomes saturated. Simplified versions of more complex models, such as that proposed by Philip (1983), appear in several major hydrological models. Philip's model calculates infiltration rate from soil sorptivity and conductivity of the wetting front and takes the following form:

$$f(t) = \frac{1}{2} St^{-1/2} + K \tag{2.7}$$

where: $f(t)$ = infiltration rate at time t (distance per unit time)
        S = sorptivity of the soil (a function of soil "suction")
        K = conductivity of the wetting front
        t = time

In the field, infiltration is frequently measured using a device called an infiltrometer. An infiltrometer is essentially an enclosed plot of land to which moisture is applied using sprinklers. Infiltration is simply computed as the difference between the volume of water applied and the volume collected as surface runoff. Infiltration rates can vary significantly even over small areas of soil (tens of square meters or less), so this method can be expected to yield only an approximate estimate. Infiltrometer results should therefore be evaluated relative to other infiltrometer experiments, and should not be considered to be accurate representations of natural phenomena.

Infiltration data can also be obtained by comparing direct measurements of rainfall and measured runoff from gauged streams. These "infiltration indices," as they are called, give a general picture of infiltration over a watershed (or subwatershed) area rather than the infiltration characteristics of a particular plot of land. See Figure 2.4.

2.3.3 Runoff

Precipitation falling over land that does not evaporate or infiltrate into the soil must, by default, run off the land surface. Runoff estimation is important, for instance, in the design of urban stormwater structures such as storm sewers and treatment wetlands. Runoff rates can be calculated in several ways. An older method, now often regarded as unacceptably simplistic by many authors, is termed the rational method or the rational formula. It is expressed as:

$$q = 0.002CiA \tag{2.8}$$

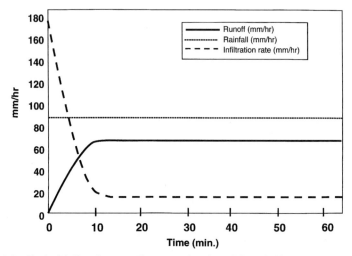

Figure 2.4 *Typical infiltration-runoff curves developed from infiltrometer data (adapted from Schwab et al. 1993).*

where: q = design peak runoff rate
  C = runoff coefficient (dimensionless ratio of peak runoff rate to rainfall intensity)
  i = rainfall intensity
  A = watershed area in hectares (ha)

This formula requires that a design return period (a period of time over which performance will be judged) be selected. For example, large structures may be designed to contain major runoff events that may occur only once every 20 to 50 years, or more, whereas constructed wetlands or nonstructural measures may be designed to contain runoff that occurs once every 10 years, or even less. Clearly, the more extreme the event (the "once-in-200-year storm," for example), the larger the structure must be to contain it.

The formula also requires selection of a rainfall intensity (for the design return period), which is equal to the "time of concentration" of the watershed—the time required for water to flow from the most remote parts of the watershed to the outlet. Time of concentration assumes saturated soil conditions and all minor depressions filled with water. When the duration of a storm exceeds the time of concentration, the analyst can assume that all parts of the watershed (not just those closest to the outlet) are contributing to observed flow patterns. Time of concentration can also be estimated in various ways, and indeed there is a long-standing debate in the literature about the best way to calculate this parameter. One simple, if crude, approach is to divide the length of flow by the estimated velocity of flow to obtain travel time. The sum of channel travel time(s) and those for overland flow equals the time of concentration.

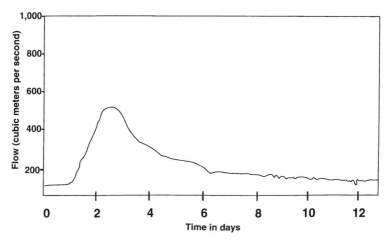

Figure 2.5 *Measured hydrograph for the Grand River at Galt, Ontario, January 3–10, 19?? (adapted from Kouwen and Soulis 1994).*

The rational method makes several important simplifying assumptions:

1. That the frequencies of rainfall and runoff are similar;
2. That rainfall occurs at uniform intensity over the time of concentration; and
3. That rainfall intensity is uniform over the watershed area.

These assumptions are usually acceptable in smaller, homogeneous basins but quickly break down in larger systems. The rational method should therefore be used cautiously or replaced with another runoff estimation approach such as the Soil Conservation Service method (United States Soil Conservation Service 1990).

A hydrograph plots runoff rate versus time and is a useful representation of watershed response to a given storm (see, for example, Figure 2.5). The SCS method of estimating runoff employs a simplified triangular hydrograph approach and was developed from hundreds of measurements compiled over a variety of real systems and storms. In the SCS method, the peak runoff rate (the top of the hydrograph) is calculated as:

$$q = q_u A Q \tag{2.9}$$

where: q = peak runoff rate (m3/sec)
q_u = unit peak flow rate (m3/sec per ha/mm of rainfall)
A = watershed area in ha
Q = runoff depth in mm

Q is calculated separately as:

$$Q = \frac{(I - 0.2S)^2}{I + 0.8S}$$ (2.10)

where: I = storm rainfall in mm
  S = maximum potential difference between rainfall and runoff
  in mm, starting when the storm begins
  (usually obtained empirically)

In the SCS method, the time to peak flow (top of the hydrograph) is:

$$T_p = \frac{D}{2} + T_L = \frac{D}{2} + 0.6T_c$$ (2.11)

where: T_p = time to peak
  T_L = time of lag
  T_c = time of concentration
  D = duration of excess rainfall (units of time)

T_c is the longest travel time in the basin and can be calculated as:

$$T_c = \frac{L^{0.8}[(1000/N) - 9]^{0.7}}{4407S_g^{0.5}}$$ (2.12)

where: T_c = time of concentration
  L = longest flow length
  N = runoff curve number
  S_g = average watershed slope in m/m

Like the rational method, the SCS method is best used on small, homogeneous watersheds without steep slopes or major areas of ponding.

Chow (1988) discusses a variety of other methods for the prediction of runoff. Typically, these relate peak runoff to watershed area and surface characteristics using an exponential relationship of the general form:

$$q = KA^x$$ (2.13)

where: q = predicted peak runoff rate
  K = coefficient intended to represent the physical characteristics
  of the watershed
  A = watershed area
  x = a constant for a given location

2.4 STREAMFLOW

Streamflow is an important component of water resources in a watershed ecosystem. Certainly, it is among the most visible and perhaps also the most easily measured. For these reasons, most jurisdictions maintain networks for the sampling, often continuous, of streamflow. This section describes some basic approaches to these measurements, along with their advantages and disadvantages.

Streamflow is the rate of flow of water and is expressed in units of volume per unit time (e.g., m^3/sec). To determine streamflow, the analyst must obtain two pieces of information: the cross-sectional area of the stream at the location of interest and the velocity of that flow. The first involves units of area, for instance, square meters; the second involves units of distance per unit time—for instance, meters per second. Multiplying the two yields units of volume (e.g., cubic meters) per unit time.

Measurement of streamflow therefore comprises two activities: estimation of stream cross-sectional area and measurement of flow velocity. There is often a large quantity of historical data available for major rivers and streams. As a result, it is often possible to discern for each location in a given system a characteristic relationship between water level ("stage") and the flow (or "discharge") passing that location. This relationship is of considerable value in the collection of streamflow data, because if the stage-discharge relationship is well understood, simple measurements of water level can yield reasonably accurate information about flow.

Water level, or stage, is measured by reference to a fixed point of known position. The simplest water level measurement is obtained with a vertical rod or stick clearly marked with appropriate units (inches, feet, or centimeters). This gauge is attached to some permanent vertical structure, like a bridge or piling, in such a way that it is visible at all possible water levels. Water level is then simply read off the gauge by an observer standing at the side of the river.

Another simple device for measuring water level is the wire-weight gauge. With this method, a reel of wire or metal tape is attached to a fixed vertical structure (again, a bridge or piling is suitable) and a weight attached to the end of the wire. Either the wire (or tape) can be marked with appropriate graduated units or the gauge can be calibrated so that each revolution of the reel allows a known length of wire to be unwound. Water level is then measured by lowering the weight from the reel until it touches the surface of the water.

These manual gauges require regular observation by trained personnel and thus are subject to human error both in measurement and in recording. The need for on-site staff usually reduces the frequency at which manual measurements are taken; most manual gauges are read only once a day, yet the stage may change dramatically over a period of an hour or less. It is therefore both costly and onerous to collect detailed streamflow data using manual gauges.

More convenient, particularly in remote locations, are automated recording stream gauges. These may be of several types. Float-type gauges, as the name implies, use a float on the water surface to track water level changes. Stage is then automatically recorded on a strip chart or digital device and can also be telemetered to remote locations. Float gauges are easily upset by river-borne debris and ice, and by vandalism, so they are usually housed in an enclosure with a stilling basin linked to the stream by pipes. Usually, more than one pipe is used to ensure continuous recording even when one pipe becomes clogged—a not infrequent occurrence at periods of high flow and high sediment transport. Pressure-activated gauges do not require a stilling well; however, they are also usually enclosed in a housing to protect against damage. Pressure-activated gauges employ a sensor that tracks changes in a pressure diaphragm submerged in the river. Pressure is transmitted to a manometer located in the gauge house, where it is recorded on a continuous recorder and possibly also telemetered to a remote location.

Recording gauges are typically accurate to within a few millimeters and may be operated for periods of up to a year or more without major maintenance. More usually, these types of gauges are serviced on a monthly basis to ensure smooth functioning and guard against major gaps in the streamflow record that could be caused by malfunction.

As with precipitation gauges, the placement of water level gauges is critical to their accuracy. Because the stage-discharge relationship is often used to generate streamflow data, water level gauges should be located where water level and flow are strongly correlated (e.g., scatter less than 5% around a plotted linear relationship) and where the stage-discharge relationship is relatively independent of flow conditions. This condition may exist upstream of a small dam or weir, for example, provided that the location is far enough upstream to avoid backwater effects from the control structure. If downstream channel characteristics change through significant erosion or deposition, the stage-discharge relationship will probably also change. In such a case, it will be necessary to redefine the stage-discharge relationship by collecting additional data on stream cross sections and flow velocity. Figure 2.6 illustrates a typical stage-discharge relationship.

Discharge, or flow, measurement requires an understanding of both stream cross-sectional area and average flow velocity across the stream. Estimation of average velocity requires a number of velocity measurements across the cross section, which are then averaged to give an estimate for the full channel width. Channel depth is relatively easily measured using graduated lines or wires with a weighted end. Depth measurements are typically obtained at regular intervals across the channel. Twenty or more measurements may be taken across a single stream cross section. At each measurement point, stream velocity is also measured, usually with the aid of a mechanical current meter. In shallow streams these measurements can be obtained on foot using hip- or chest-waders. In deeper waters measurements are usually taken from a bridge or a moored boat. Velocity is measured just above the stream bed (at 80% of total depth) and just below the surface (at 20% of total depth) in each segment, and the two values are averaged. This approach produces a more accurate velocity reading

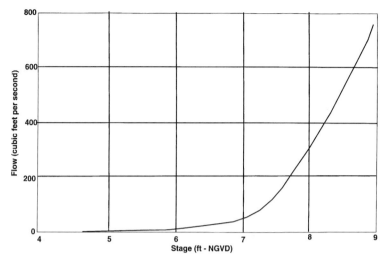

Figure 2.6 *Computed stage-discharge curve for Little Sixmile Creek, Jacksonville, Florida (from Schmidt et al. 1996, reproduced with permission from Computational Hydraulics International).*

because it avoids the effects of drag at the sediment-water and water-air interfaces. In very shallow water velocity is measured at mid-depth. This approach is illustrated in Figure 2.7. Under some circumstances, especially where flow dynamics are complex, it may be appropriate to collect a more complete velocity profile at certain stations, with measurements at 10 cm intervals from surface to bottom.

When all necessary data have been collected for a given segment, the flow for that segment is calculated as:

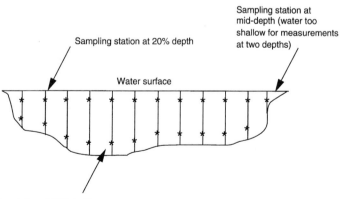

Figure 2.7 *Hypothetical stream-gauging scheme showing location of cross-sectional segments and velocity measurements.*

$$q = \frac{a(n_{0.2} + n_{0.8})}{2} \qquad (2.14)$$

where: q = flow through individual segment of stream cross section
(volume per unit time)
a = area of segment (= measured depth × width,
expressed in length2)
$n_{0.2}$ = measured velocity at 20% of total depth
$n_{0.8}$ = measured velocity at 80% of total depth

Total flow through the full cross section is then calculated as the sum of all individual segment flows, as follows:

$$Q = q_1 + q_2 + q_3 \ldots + q_n \qquad (2.15)$$

where: Q = total section flow
$q_1, q_2, q_3 \ldots q_n$ = individual segment flows

Viessman and Lewis (1996) and other basic hydrology texts provide a full discussion of current streamflow measurement practice.

From these basic measurements, it is possible to develop a plot relating water level (stage) to flow (discharge). This plot is called by several names, most commonly "stage-discharge curve" or simply "rating curve." Once the descriptive data have been collected, the stage-discharge curve should remain applicable unless major changes occur in the stream cross section or in the stability of downstream controls. If these changes occur, a new stage-discharge curve must be developed. In some rivers this will occur several times a year; in others, perhaps only once every few years.

Streamflow data are collected and published by a variety of government agencies. Reported values usually are mean daily flows, but average annual and average monthly flows, as well as annual maximum and minimum values, may also be reported. If the data have been collected on a continuous basis (as is likely the case for most major river systems), continuous flow data may also be available, often in electronic format. Increasingly, such data (like those for the well-known Hubbard Brook research program) are available freely over the World Wide Web or on Internet bulletin boards.

2.5 GROUNDWATER

Water that has passed through the land surface into underlying soils and rocks is called groundwater. This subsurface region contains both saturated and unsaturated zones. In the very deepest regions of the earth, water in liquid form is unavailable and all water molecules are chemically combined with other substances. Above this layer lies a deep, saturated zone through which liquid water flows, confined by impervious rock strata. Novotny and Olem (1994) define an aquifer as a saturated permeable geologic underground stratum that can transmit significant quantities of water, and an aquitard as a less permeable layer that may be significant in the regional transport of water but which has a permeability insufficient to permit economical development. (The term *aquiclude* refers to a geologic formation that is entirely impermeable—a rare occurrence, and consequently a rarely used term.) Most aquifers are formed of unconsolidated materials including soil and gravel. Some sedimentary rocks also permit the movement of large quantities of water, but igneous and metamorphic rocks, which are essentially impermeable, form aquifers only when they are fractured or otherwise mechanically rendered permeable.

The top of the saturated zone forms a "water table," which can be located with the aid of drilling apparatus. Closer to the land surface, smaller basin-shaped rock formations may capture water, creating locally saturated zones and "perched" water tables. The transition from saturated to unsaturated zones is not an abrupt one. Rather, capillary action at the interface acts to draw water up from saturated zones into less saturated areas, creating a "fringe" effect at the saturated-unsaturated boundary.

Groundwater is measured by injecting and tracking tracer materials in the groundwater, by the use of field permeameters, or by drilling into the aquifer and determining the pressure of the flowing water. Normally, water in the drill hole will rise to the surface of the water table. When the water surface in the well is higher than the groundwater table, the aquifer is said to have artesian conditions.

When groundwater is extracted—for instance, pumped out for well water extraction—the water surface in and near the drill hole or well will be lower than the original groundwater table. The distance between the well and the point where no significant influence on water level is observed is called the "zone of influence" of the well; the zone of lowered water level is referred to as the "drawdown," which is located in a "cone of depression." These terms are illustrated graphically in Figure 2.8.

The one-dimensional flow of water in a saturated zone or aquifer is described by Darcy's law:

$$q = KhA/L \qquad (2.16)$$

where: q = flow rate (volume per unit time)
K = hydraulic conductivity of the flow medium
    (distance/unit time)
h = hydraulic head or potential causing flow (distance)
A = cross-sectional area of flow (distance^2)
L = length of the flow path (distance)

This equation is valid for most conditions of one-dimensional flow, whether

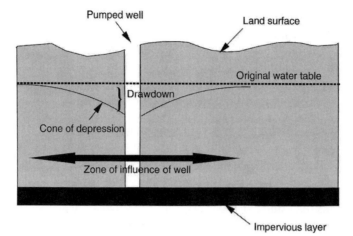

Figure 2.8 *Influence of well water extraction on ground water table (after Viessman and Lewis 1996).*

vertical (as in downward infiltration), horizontal, or upward. Groundwater is discharged into surface waters (lakes, streams, and wetlands) at the land-water interface. Groundwater is therefore a major (perhaps *the* major) contributor to the base flow for most surface waters. Table 2.3 gives typical hydraulic conductivities for a variety of soil and rock materials.

The chemistry of water moving through an aquifer is altered by reaction with chemicals in the soil, adsorption onto solids in the aquifer, bacterial action, and similar processes. (Note that this is, nevertheless, a less complicated system than in surface waters, which are also subject to the influence of light and volatilization.) Important in these chemical transformations are acid-base reactions—for example, those that occur when acidic precipitation falls on neutral or alkaline soils. As rainwater passes through soils and over rocks, it dissolves minerals

Table 2.3 Typical Hydraulic Conductivities of Soil and Rock Materials

Soil or Rock Type	Typical Porosity (%)	Typical Hydraulic Conductivity (cm/s)
Mixed clay, sand, and gravel (till)	50–60	10^{-6}–10^{-4}
Mixed sand and gravel	20–35	10^{-3}–10^{-1}
Gravel	25–40	0.1–1
Coarse sand	35–40	0.01–0.1
Silty sand	25–35	10^{-5}–10^{-3}
Clay	45	10^{-2}–10^{-1}
Dense solid rock	<2	<10^{-8}
Shale	5–10	10^{-10}
Sandstone	10–20	10^{-6}–10^{-3}

(which minerals, and how much is dissolved, depends on the relative characteristics of the rainwater and the soil or rock). Carbonate chemistry is often important in such a system. Bicarbonate (HCO_3^-) is created when limestone and dolomite are dissolved. The dissolution of limestone is expressed as:

$$H_2O + CO_2 \rightleftharpoons H^+ + HCO_3^-$$

That of dolomite is expressed as:

$$CaCO_3 + H_2CO_3 \rightleftharpoons Ca^{2+} + 2HCO_3^-$$

These sedimentary rocks are common throughout much of the world, and thus their influence on surface water and groundwater chemistry is significant and well documented. When acidic precipitation (rich in the hydrogen ion, H^+) flows through limestone or dolomite zones, the bicarbonate content of the rock buffers, or absorbs, excess hydrogen ion and thus reduces the acidity of the flowing water, yielding water and carbon dioxide:

$$CaCO_3 + 2H^+ \rightleftharpoons Ca^{2+} + H_2O + CO_2$$

This reaction has the effect of increasing the content of polyvalent ions in the water, particularly calcium (Ca^{2+}) and magnesium (Mg^{2+}). Water supplies in limestone/dolomite areas are typically high in these ions and are termed "hard."

A variety of other minerals are also dissolved in groundwater flow. Among these are sulfides of zinc, lead, and other heavy metals, which, when dissolved, release those metals into solution and thus into the domain of drinking water, irrigation water, and livestock water supplies. For example:

$$ZnS \rightleftharpoons Zn^{2+} + S^{2-}$$

and

$$PbS \rightleftharpoons Pb^{2+} + S^{2-}$$

Other chemical phenomena that can affect the quality of groundwater include redox reactions, in which oxidized forms such as sulfates are converted to reduced forms such as sulfites (and vice versa). Adsorption-desorption phenomena are also important in transportation of certain nutrients and priority pollutants such as heavy metals and trace organic compounds. In this process, adsorption sites are provided by the electrically charged surfaces of clay minerals and organic particles and certain other chemical forms. Phosphorus, ammonia, many heavy metals, and hydrophobic organic compounds, such as some pesticides, are attracted by these charged sites and bind to the particle. When the particle is transported—for instance, in spring runoff—the attached pollutant is transported with it. (This phenomenon is partly responsible for the thorny problem of remediating large volumes of polluted sediments in urbanized streams, lakes, and harbors. To remove the target pollutants, it is often necessary to remove the entire polluted sediment mass. Removal by dredging or suction can disturb settled sediments and cause desorption to occur, with the result that older, buried pollutants are once more released into surface waters.)

Groundwater quality is also strongly affected by local sources such as malfunctioning septic systems, leaking below-ground manure storage systems, land application of sewage sludges and manures, and similar processes. These activities can convey bacteria, viruses, parasites, and cysts into groundwater along with nutrients and solids. Aging or ruptured underground storage tanks formerly used for volatile or nonvolatile wastes are often sources of various industrial chemicals introduced into groundwater. Indeed, the range of wastes, coupled with the difficulty of reversing groundwater contamination, make this one of the most serious environmental challenges currently faced by our society. Even sanitary (municipal) landfill sites, if not properly designed and operated, can be a significant source of numerous materials including solids, phenols, nitrogen compounds, and various toxins into groundwater.

A common soil quality impact in warmer climates occurs when land is overirrigated. Water applied to fields contains some level of dissolved minerals, but water evaporated from those fields is, of course, mineral-free. Thus, with time, there is an accumulation of minerals, or salts, in irrigated land. This problem can be addressed by applying excess irrigation water or by growing salt-tolerant crops. Excess application of water can in turn leach other materials from the soil and into underlying groundwaters.

A direct impact on human health can occur when high levels of nitrate are leached into groundwater from agricultural fields where nitrogen fertilizers or manures have been applied. The disease methemoglobinemia is particularly troublesome in infants and older adults, who tend to produce less gastric acid and have higher digestive system pH levels than the rest of the population. At these higher pH levels, nitrate ions (NO_3^-), which are common in food and drinking water, are readily converted into nitrite ions (NO_2^-). The nitrite ions then react with hemoglobin to create an inactive methemoglobin form, reducing its oxygen-carrying capacity and causing oxygen deficits in the affected individual. (Healthy adults are usually able to convert methemoglobin back to oxyhemoglobin and thus retain sufficient blood oxygen, even at relatively high levels of nitrate/nitrite.)

2.6 WATER QUALITY

2.6.1 Influences on Water Quality

Water quality directly affects virtually all water uses. Fish survival, diversity, and growth; recreational activities such as swimming and boating; municipal, industrial, and private water supplies; agricultural uses such as irrigation and livestock watering; waste disposal; and general aesthetics—all are affected by the physical, chemical, biological, and microbiological conditions that exist in watercourses and in subsurface aquifers.

Many factors influence water quality; some of these have been discussed in previous sections. The chemistry of bedrock and surficial geology and the drainage characteristics of soils can determine whether natural waters are acid or alkaline, sediment-laden or clear, high in heavy metals and dissolved salts or relatively free of those constituents. Physical processes like erosion can add large quantities of suspended sediment to surface waters.

Biological processes draw nutrients from surface waters and soil moisture, and decaying tissues release nutrients into the watershed ecosystem. Photosynthesis uses sunlight and carbon dioxide to create plant sugars and oxygen, so where aquatic plant growth is dense, daytime dissolved oxygen levels are likely to be high. At night, however, respiration—which uses oxygen for tissue maintenance and repair—dominates over photosynthesis, so dense aquatic plant growth can also mean nighttime dissolved oxygen depletion in a stream.

Although pristine waters are suitable for most purposes, the minimum acceptable quality of water depends very much on the water use. Water for irrigation, for example, should be low in dissolved salts, but water intended for livestock should be low in bacteria. Water for industrial processes should usually be of much higher quality than water for industrial cooling. Water for municipal drinking water supply must not only be safe to drink, but should ideally contain low concentrations of materials such as calcium carbonate, iron, and similar materials that can cause costly infrastructure damage or add unpleasant characteristics to the finished water. Human health considerations demand low levels of a range of contaminants, yet salmon and trout populations may be even more sensitive than humans to many water quality constituents.

Water quality impairment is often a trigger for conflict in a watershed, simply because degraded water quality means that desired uses are not possible or are not safe. Chapter 3 discusses approaches to "scoping" a watershed management plan, including compilation of a list of desired uses and the water quality targets appropriate for each. This simple matching process provides an early evaluation of where use impairments may exist and, thus, where remedial action may be warranted.

Water quality indicators are of several main types. The following sections give a brief overview of the principal indicators in current use. A more complete discussion is available in Tchobanoglous and Schroeder (1987).

2.6.2 Physical Indicators of Water Quality

Water Clarity   The most basic indicator of water quality, and an excellent measure of aesthetic potential, is water clarity. Clarity is measured in different ways. One method often used for lakes, and occasionally for rivers, is Secchi disk depth. The Secchi disk is a flat disk about 30 cm in diameter, with the top surface painted in quarters alternating black and white. The disk is lowered on a line until it disappears from view, and a note is made of the depth of disappearance. The disk is then lowered farther and raised again, with a note made of the depth at which it is once again visible. The average of the two depths-of-visibility is recorded as the "Secchi disk depth." This measure is inexpensive and easy to use and provides results that are intuitive for the lay person. It does not, however, yield information as to the *causes* of reduced clarity, which may be of diverse origin.

Water clarity is also measured with mechanical devices, including the turbidity meter and the transmissometer. These devices, which measure light transmission through the water column, are more costly and more difficult to operate (and calibrate) than a Secchi disk, and the data they generate is harder for the lay person to understand. They are, however, less subject to human error and interpretation than the Secchi disk and have largely replaced that device in modern water quality monitoring.

Suspended Sediment Suspended sediments (particulates) arise from eroded silts and clays, organic detritus and plankton. Waters high in suspended sediment have a turbid or "muddy" appearance that makes them unattractive for swimming and bathing. High solids content can also demand additional treatment for municipal water supplies and can limit the utility of water for industrial purposes. Sediment settled out of turbid water can foul fish spawning habitat, clog fish gills, and otherwise disrupt a habitat for aquatic organisms. Although natural processes like erosion contribute a high proportion of suspended sediment in some systems, human activities such as agriculture can increase the rate of these processes and can add new sources of sediments such as municipal and industrial wastewaters. Urban storm drainage is a major source of suspended solids in many cities with storm sewer systems discharging to surface waters. Suspended sediments can also indicate the presence of other pollutants such as phosphorus, heavy metals, and some pesticides, which readily adsorb to the surface of clay and silt particles. Table 2.4 gives particle size ranges for a variety of inorganic and organic solids.

Table 2.4 Particle Size Ranges for Organic and Inorganic Solids Found in Water

Particle Size (mm)	Particle Type
10^{-8}–10^{-6}	Dissolved solids
10^{-6}–10^{-3}	Colloidal solids, clays
10^{-5}–10^{-4}	Viruses
10^{-4}–10^{-1}	Bacteria
10^{-3}–10^{-1}	Algae
10^{-3}–10^{0}	Suspended inorganic solids
10^{-2}–10^{0}	Settleable solids

Conductivity   Conductivity is a measure of the electrical conductance of water—its ability to conduct an electric current. Electricity is conducted by ions in a solution, so conductivity tends to increase with increased ionic strength (quantity of dissolved salts). Although a gross measure of water quality, conductivity can be a valuable surrogate for other water quality constituents that may be both more costly and more time-consuming to measure. Continuous conductivity measurement is common in industrial operations (as is continuous pH measurement) to provide an early warning of problems ("upset") in the process or other nonroutine circumstances.

Hardness   Dissolution of calcium and magnesium minerals can add high levels of these multivalent cations to natural waters. This "hard" water can cause problems with scaling and clogging in pipes and other infrastructure, reduce the life of water-using appliances such as water heaters and electric kettles, and reduce the effectiveness of household washing agents. Pipe clogging due to hard-water "scale" is a major contributor to reduced water pressure in residences and institutions.

Water Temperature   Water temperature is often a useful measure of water quality, although it is understandably influenced by many factors, including local vegetative cover (which may shade a stream from incident sunlight and thus reduce its temperature), meteorological conditions such as cloud cover and wind, and industrial or other effluents—for instance, cooling waters (which by definition have received excess heat from heat-generating processes such as thermal power generation). Temperature is a critical factor in habitat choice for aquatic organisms. Too high or too low an ambient temperature will deter certain organisms from entering an area; extreme high or low temperatures can be lethal. Temperature is often a central factor in the decline of desired "pan fish" species such as salmon and trout, especially in urban areas. These species, sometimes termed "cold-water" species, prefer cool temperatures and, although they may pass through warm water zones, will not reproduce successfully in conditions where ambient temperatures are too high. In heat-polluted surface waters, it is not uncommon to see less desirable species such as carp and alewife come to dominate the fish community. Thermal pollution—elevated temperature in receiving waters—is common near the effluents of thermal generating stations and where urban stormwater runoff has picked up heat from sun-warmed roofs and parking lots. The role of temperature in determining species distribution and reproductive success probably relates to the fact that temperature affects the rate of many biochemical processes. Temperature can, of course, also affect the rate of chemical reactions in the abiotic environment and is a central factor in evaporative processes.

Aesthetics   The term *aesthetics* encompasses a broad, if subjective, range of concerns about water quality. In many systems, surface waters are valued not just for their utility in drinking, recreation, agriculture, and industrial use, but

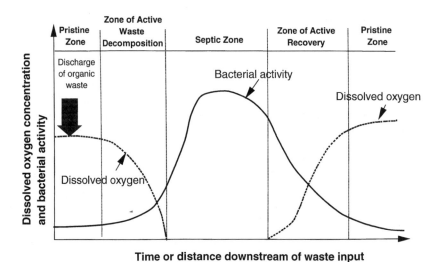

Figure 2.9 Influence of organic wastes on oxygen concentrations in a stream (after Ontario Ministry of Environment 1987).

also simply for their beauty. Degraded aesthetics may not have direct health effects but can create major economic impacts—for instance, through diminished revenues from tourism and recreation. Waters of high aesthetic quality are typically free of materials that will settle and form objectionable deposits, contain little or no floating debris, oil, or scum, and do not have an objectionable color, odor, taste, or turbidity. Aesthetics may also be taken to include the presence of desirable species, such as trout and salmon, or the absence of undesirable species, such as carp or excessive insect fauna.

2.6.3 Chemical Indicators of Water Quality

Dissolved Oxygen   Of all chemical indicators of water quality, dissolved oxygen may, in many ways, be the most important. Most aquatic animal species require abundant oxygen for life and reproduction; some will tolerate lower oxygen conditions, and it is often the presence of these in the aquatic community that signals water quality impairment. High in-stream oxygen levels are also helpful in promoting the biological assimilation of organic and nitrogenous wastes—an important consideration downstream of effluent discharges from sewage treatment plants and some industrial facilities. High levels of dissolved oxygen are not usually of concern, but low levels, highly variable levels, or a complete absence of oxygen can create conditions that are inhospitable for aquatic life. See Figure 2.9.

Nutrients   Several common water pollutants are often grouped together as "nutrients" because they are essential for the growth and maintenance of biological tissue. The three pollutants most commonly described as nutrients are phosphorus, nitrogen, and silica; the latter is in such abundant supply relative to demand that it seldom controls the growth of nuisance plants or algae, and it is not discussed further in this section. Phosphorus and nitrogen are, however, considered to be important parameters for pollution control and are therefore often the focus of concerted efforts in watershed cleanup. Both of these elements are commonly found in nature. Phosphorus is used in many biological systems, most fundamentally in adenosine diphosphate and triphosphate, which are key players in glucose metabolism and energy conversion. Nitrogen is probably most familiar for its role in building protein, and in the decay products of protein such as ammonia and urea. In most temperate freshwater systems nitrogen is abundant, and it is phosphorus that is the so-called limiting nutrient—the nutrient whose concentration determines whether additional biomass can or cannot be produced. In marine systems, the reverse is often true: phosphorus is abundant and nitrogen is the limiting nutrient.

There are many sources of nutrients in most watersheds, including atmospheric processes and rock weathering, but in most developed watersheds the most significant sources are often municipal and industrial waste discharges and agricultural activities. Phosphorus and nitrogen compounds are considered "nonconservative" in nature, because they are likely to react with other substances and with biological tissue. They are unlike suspended sediment (a "conservative" pollutant) in that it can be difficult to develop a mass balance for phosphorus and nitrogen compounds in a watershed—that is, to account for all the sources and sinks of these materials within the system and fully explain their behavior. These nutrients are therefore challenging to simulate accurately, and it can be very difficult to develop accurate predictions of ecosystem response to nutrient management actions.

In converting from one form to another, nitrogen compounds can cause impacts on other water quality parameters, especially dissolved oxygen. For example, the chemical conversion of ammonia to nitrite, and then nitrate, uses oxygen. High ammonia levels, as may be found downstream of a sewage treatment plant outfall or a poorly managed manure storage facility, can therefore also have a significant impact on instream dissolved oxygen resources (the unionized fraction of ammonia is also acutely toxic to aquatic life). High concentrations of nitrate in drinking water can cause serious health effects in human infants through methemoglobinemia (see Section 2.5).

Heavy Metals Our society uses many metals in many forms, and with time these have found their way into soils and receiving waters and, in some cases, into biological tissue as well. Heavy metals such as copper, nickel, zinc, cadmium, chromium, lead, and mercury are common in industrial use. In urban centers most small industries and many larger ones have for decades used the municipal sewer system to convey liquid effluents to a central sewage treatment plant. This has been a good arrangement for many public utilities, which can charge sewer-use fees and generate revenues while providing a convenient treatment option for industry. The difficulty has been that sewage treatment plants are not designed to remove metals. In some cases wastewater concentrations of a given metal have been high enough to affect treatment processes at the plant. In many, perhaps most, others, metals simply attach to particulate matter in the wastewater stream and are removed with "biosolids," the mixture of dead bacterial cells, inorganic particles, and other materials that comprise sludge. Up to 70% "removal" of metals may occur in publicly owned treatment plants. These metals are, however, not truly removed but simply transferred to the solid phase. Some metals do remain in solution and are discharged with treated effluents into receiving waters. Metals may therefore reenter the ecosystem in several ways: in aqueous solution, in leachate from landfills or agricultural lands where sewage sludges have been spread, or in air emissions from incinerators used to dispose of sewage sludges.

Metals are of concern as a group because they can impair enzyme function. The degree of impairment depends very much on the metal and the organism affected. In humans, heavy metal poisoning characteristically involves symptoms of gastric upset and central nervous system impairment, including altered perception (e.g., peripheral vision) and gait. Several metals, including lead and mercury, have been linked to smaller birth weights and delayed cognitive development in young children. High concentrations of heavy metals also cause acute and chronic toxicity in nonhuman organisms and (depending on the metal) may also have mutagenic, teratogenic, or carcinogenic effects. Some metals (mercury, lead, and tin, for example) are known to convert from relatively nontoxic inorganic forms to highly toxic, bioaccumulative, and persistent organic (methyl or ethyl) forms, probably through bacterial mediation.

Trace Organic Compounds   Almost unknown 20 years ago, industrial organic compounds, pesticides, and herbicides are rapidly gaining prominence in the field of water management. Until recently laboratory analytical protocols were unable to measure these substances at very low levels, and it was assumed that levels so low could not be important to human or ecosystem health. In recent years, however, improved analytical techniques and better knowledge of the toxicology of these agents have led to increased concern about them.

There are at least 60,000 chemicals in regular use in the Great Lakes Basin alone, and many of them have been developed in the last 50 years. Some substances, which may be almost undetectable in water, have been labeled "hydrophobic" because of their affinity for particulate matter or, in some cases, for fats and oils. Many are toxic (that is, cause impaired function in one or more organs), and some are highly persistent and bioaccumulative. A substance is considered bioaccumulative when plant or animal enzyme systems cannot effectively detoxify and excrete it. Over time, an organism's "body burden" of the substance increases, simply because intake rates vastly exceed elimination rates. A corollary of this finding is that animals higher in the food chain—second- or third-level carnivores (animals that eat other carnivores)—will themselves receive a highly concentrated "dose" of the substance with each meal. When humans consume a top carnivore species like lake trout, they too ingest all of that organism's accumulated body burden of the contaminant—and cannot themselves readily excrete it.

(Many of the most toxic and persistent organic (carbon-based) substances, including dioxin, DDT, and PCBs, contain chlorine and are thus termed "chlorinated organics." Residual chlorine from water and wastewater disinfection or bleaching processes is not itself in this category—in fact, it is highly volatile—but it is acutely toxic to fish and other aquatic life forms even at relatively low concentrations.)

2.6.4   Biological Indicators of Water Quality

Bacteria   Thousands of bacteria strains can be detected in even the most routine household items; few of these will actually cause disease in humans or other organisms. For decades, fecal coliform and fecal streptococcus bacteria (common in the feces of warm-blooded animals) have been used as indicators of the presence of fecal pollution. These groups are easy to identify, primarily by appearance; laboratory analyses for their presence are quick and inexpensive. The groups do, however, encompass a wide range of bacteria, both pathogenic and nonpathogenic. There is now a growing body of epidemiological evidence showing that fecal coliform and fecal strep counts do not correlate well with the incidence of gastrointestinal illness, so these indicators are gradually being replaced by more specific measures. Assays for individual organisms, or smaller groups, has become more routine with the development of improved laboratory techniques. Sometimes fecal coliform, total coliform, or fecal streptococcus assays will be used with measurements of one or more specific strains to obtain a more precise picture of the likelihood of illness occurring.

One organism now commonly used in water quality assessment is *Escherischia coli*, a common gut bacteria in warm-blooded animals, including humans. Enterococci (species of the fecal streptococcal group, which includes *Streptococcus faecium* and *S. faecalis*, which are always found in the human gut) and *Pseudomonas aeruginosa*, which causes eye, ear, and skin infections in humans, give a more reliable indication of health risk than traditional indicator groups. Other pathogens, for instance *Salmonella, Shigella, Staphylococcus aureus, Campylobacter jejuni, Legionella* spp., and viruses can also be measured in water, but their analysis is much more costly and time-consuming than tests for indicator groups, so they are seldom used unless a specific problem is under investigation.

The sources of bacteria and viruses primarily involve the feces of warm-blooded animals, including humans. In a watershed, this can mean sewage treatment plant effluent that is incompletely disinfected, stormwater runoff (carrying animal feces), combined sewer overflows (carrying a mixture of human sewage and stormwater and diverted to a watercourse before treatment), and animal wastes such as manure storage and feedlots. These sources are directly affected by precipitation events. As a result, levels of bacteria and viruses in receiving waters are typically highly variable, difficult to correlate with other biological indicators or bacteriological species, and almost impossible to model with accuracy. Simulation is even more difficult because these organisms, as living entities, grow, reproduce, and die and can even overwinter in lake and river sediments if conditions are suitable.

Parasites A number of parasitic organisms occur in freshwater systems; some of them have serious health implications for humans and other animals. Among these are *Schistosoma* sp., parasitic worms causing schistosomiasis (bilharziasis). In temperate climates some species of schistosomes do not actually live in the human body but are carried by freshwater molluscs and birds. These species cause dermatitis, or "swimmer's itch," in humans exposed to them. In warmer regions most schistosome species live in the human body and can cause serious gastrointestinal illness and even death.

Giardia lamblia, a waterborne parasitic protozoan carried by the beaver, is of increasing concern in recreational areas, where it may be ingested in drinking water or in water used for swimming or bathing. *Giardia* causes mild to serious gastrointestinal illness in the affected person and is readily transmitted by oral contact with contaminated materials. About 7% of the parasitologic samples submitted in the United States contain this organism. *Giardia* is more resistant to chlorination than are bacteria and, in some cases, may also be more resistant to medical intervention than other parasitic infections.

Cryptosporidium is a more recently identified pathogenic protozoan that may be as important as *Giardia* in causing diarrheal illness. Outbreaks of cryptosporidiosis are associated primarily with ingestion of inadequately treated drinking water. It is now believed that *Cryptosporidium* is widely present in the feces of warm-blooded animals, including domestic farm animals, so major sources of this organism are thought to be improperly stored manure and, in some cases, faulty septic systems.

2.7 PLANT AND ANIMAL COMMUNITIES

In addition to geologic and meteorologic forces, a watershed is affected by a variety of biological processes. An inventory of watershed biota therefore aims to do three things:

1. Determine the number and types of plant and animal species present in the watershed area;
2. Estimate the number of individuals of each species; and
3. Investigate the interrelationships between the species and their abiotic environment.

Depending on the desired detail of the watershed inventory and the complexity of uses and issues in the basin, these objectives may be satisfied by a review of existing data (for instance, in the scientific or government literature) or new data may be required. The latter is a costly and time-consuming prospect and one that may be unnecessary except at a research level. In either case, the analyst is usually dealing with problems of sampling from a vastly diverse and complex natural system, with all the pitfalls such sampling implies.

At a cursory assessment level it is usually possible to obtain relative or qualitative assessments of the plant and animal species present in the watershed, along with their probable responses to management actions of various types. It is, however, very difficult to obtain a detailed and comprehensive understanding of these systems and, thus, to predict ecosystemic responses accurately and quantitatively. For major management initiatives likely to have impact on these systems, there may be value in pilot scale implementation to test outcomes "in nature" before full, and costly, implementation.

2.7.1 Using Existing Data

The literature on field ecology provides extensive guidance on estimating the number of species present in an area and the number of individuals per species. Most state and provincial natural resources agencies have available detailed information about regional biota, including the biotic region or ecotone within which the watershed falls, the common and uncommon species in the area of interest, and the nature and location of fragile or valued habitats. Information about rare or endangered species can also be obtained from these agencies or from regional or national biodiversity protection organizations.

Detailed information about the distribution of individual species can be found in one of the many readily available field guides such as those produced by Peterson or Audubon.* These guides may be most helpful for larger species of mammals, birds, and reptiles and for trees and shrubs. For these organisms, field guides usually provide maps of geographic distribution accompanied by detailed commentary—for instance, about species habitat preference and diet.

Information of this kind is, however, seldom as available for invertebrate species such as insects and their larvae. Yet these organisms can be very important in assessing the health of aquatic systems or soil quality, especially at a micro-habitat level. Surveys of the nature and diversity of macroinvertebrate populations have been used for many years to identify areas where water quality impairment has had an impact on stream ecology. These surveys are relatively inexpensive (as compared with laboratory analysis of individual chemical substances) and can be helpful in identifying areas requiring more detailed assessment. State and provincial agencies responsible for natural resources or envi-

---

*The Peterson Field Guide Series is published by the Houghton Mifflin Company, Boston; there are about 30 guidebooks in this series. The dozen or so books in the Audubon Society Field Guide Series are illustrated with photographs rather than sketches; they are published by Alfred A. Knopf, New York.

ronmental protection may have historical data on invertebrate fauna. University departments of biology, agriculture, or soil science are also often engaged in research in these areas. Theses reporting this research are usually available on microfilm or microfiche through university or public reference libraries.

Occasionally it may be necessary to conduct field surveys within the watershed to update existing information or fill data gaps. The following sections provide a brief overview of techniques for this purpose.

2.7.2   Indices of Biodiversity and Similarity

Classical ecological wisdom holds that ecosystems that are diverse and complex are also likely to be more stable and more able to recover following disruption. Although there is not universal agreement on these concepts, it is often true that a healthy, robust ecosystem is characterized by a large number of species but with no species heavily dominant or overrepresented in terms of individuals.

Environmental biologists use a variety of indices to describe the species diversity of a habitat and the similarity of that habitat to another. Some of these indices are quite old and now well entrenched in the ecological literature; few, if any, of them are wholly objective or accurate. They may, however, provide useful tools with which to assess the condition of a watershed system of interest.

Diversity indices usually employ measures of the number of species present and the number of individuals per species. Since these measures also imply the probability of encountering an individual of a given species (or reencountering one, in the case of mark-recapture methods, as discussed in Section 2.7.3), some indices express number of species present in terms of probability.

Among the oldest and most famous of diversity indices is Simpson's Index of Diversity (Simpson 1949), which is expressed as:

$$D = 1 - \Sigma(p_i)^2 \qquad (2.17)$$

where: D = Simpson's Index of Diversity (values from 0 to 1.0)
   p_i = proportion of individuals of species i in the community

Simpson's Index reflects the simple principle that the probability of randomly picking two organisms of different species is equal to 1 minus the probability of picking two organisms of the *same* species. The probability of picking an organism of any given species is simply its proportion in the community, p_i. So the probability of picking *two* organisms of the same species is simply the joint probability $(p_i \times p_i)$, or $(p_i)^2$. The difficulty with Simpson's Index is that it tends to give little weight to rare species and more weight to common species.

The Shannon-Weiner Index attempts to add information about the evenness of individuals among species. This function was developed independently by Shannon and Weiner and bears both their names. It is sometimes mistakenly referred to as simply "Shannon's Index" or as the "Shannon-Weaver Index." It is expressed as:

$$H = \Sigma(p_i)(\ln p_i) \qquad (2.18)$$

where: H = Shannon-Weiner Index of Diversity
p_i = proportion of individuals in species i

It is possible to calculate a theoretical maximum equitability (evenness) of individuals among species using the Shannon-Weiner approach, as follows:

$$H_{max} = -S\left(\frac{1}{S} \ln_2 \frac{1}{S}\right) = \ln S \qquad (2.19)$$

where: H_{max} = species diversity under conditions of maximum equitability
S = number of species in the community

Equitability, sometimes written as J, can then be calculated as the ratio of observed diversity (H) to H_{max}:

$$J = \frac{H}{H_{max}} \qquad (2.20)$$

Similarly, indices typically employ measurements of the number of species occurring in each of two communities to be compared, and the number of species occurring in *both* communities. A good example is Jaccard's Similarity Coefficient, I:

$$I = \frac{j}{a+b-j} \times 100 \qquad (2.21)$$

where: I = Jaccard's Similarity Coefficient (units of percent)
j = the number of species occurring in both communities
a = the number of species occurring in community a
b = the number of species occurring in community b

2.7.3 Field Data Collection Techniques

Field data collection is an expensive prospect, so if the effort is made to undertake a field survey, it is worth taking care to make sure that the results are both quantitative and representative. Usually, this means careful attention to the data requirements of desired analytical techniques, such as statistical hypothesis tests. Ideally, the analyst should have a clear question in mind in designing the field survey, such as "How many species are present in the sample plot?" or "What is the average number of individuals species in the sample plot?" Clear formulation of the hypothesis to be tested simplifies selection of an analytical technique and provides valuable guidance in sampling design.

Frequently, sampling of biological populations and communities (whether aquatic or terrestrial) employs random placement of plots of known area and a complete inventory of the biota within the test plot. Plots of a square-meter area are typically used for sampling small organisms such as insects and aquatic invertebrates. Larger plots are necessary for larger organisms, and plots of several hundred square meters or even kilometers may be required to survey very large or uncommon species or those that routinely move across large distances.

Field sampling of animal populations is usually much more difficult than sampling of plant populations, for the simple reason that animals are mobile. Sampling of animal populations can employ mark-recapture methods, which involve capturing and "marking" (for instance, tagging) a number of individuals of the target species, releasing those individuals, and then recapturing animals of the same species after some period of time. The recaptured sample will include both marked and unmarked individuals, the proportions of which give an estimate of total population size.

Aerial photography can be helpful in assessing large tracts of vegetation, such as forest cover or crop cover, and in tracking certain kinds of large animals, especially those that are very mobile but quite visible. The harp seal population in the Gulf of St. Lawrence has been estimated in this way with some success. The disadvantage of aerial methods is that it is not always possible to distinguish the sex or age of an animal from the air, yet these characteristics can be vital in establishing the health and breeding status of a population. Similarly, it can be difficult to identify plants to the species or subspecies level, especially if the canopy is dense or the vegetation diverse. Although sometimes less labor-intensive to mount, aerial techniques may therefore have less value than traditional on-ground methods in locations where there is heavy forest cover, where access by air is dangerous, or where the presence of aircraft can interfere with normal behavior patterns and, thus, population distribution.

Air photo techniques and satellite imagery have been used successfully in inventories of managed forests and other terrestrial plant systems and may have potential for the remote evaluation of some water quality characteristics. The advent of Global Positioning System technology has made it possible to collect field data with considerable accuracy as to geographic location and thus simplifies the problem of locating a precise sampling station when sampling is to be repeated.

All field sampling techniques, whether on-ground or remote, involve some error in estimation. Depending on the sampling method and the way it is applied, this error can be very large and thus difficult to interpret in the formation of public policy. It is therefore essential that estimates of population size and distribution (including sex and age distribution as well as density) be reported with an accompanying estimate of sampling error or uncertainty.

A more complete description of current ecological field data collection techniques is provided in Magurran (1988).

2.7.4 Assessing Interrelationships

Although counting the number of species and the number of individuals per species is a relatively straightforward matter, it is much more difficult to evaluate the interrelationships between species and (perhaps of most concern to watershed managers) the implications of management actions that might change those relationships.

A common method of representing the interrelationships between species is the food web diagram. The food chain is a related, but simpler, concept. A food chain shows only a single species at each tropic level: one species of primary producer (plant), one species of primary consumer (herbivore) that feeds on that plant, one species of secondary consumer (carnivore) that feeds on the herbivore, and so on to the top of the food chain, arriving at the species that has no major predators, the "top carnivore." For the purposes of watershed management, the food web offers a more useful approach, showing the interlinkages of individual food chains and, thus, usually a range of habitats from aquatic to terrestrial, a range of species encompassing herbaceous and woody plants, and a range of vertebrate and invertebrate animals. Figure 2.10 illustrates a typical food web.

Conceptual representations like a food web diagram are helpful in elucidating relationships that may not be apparent on first glance but that may be vital to key watershed functions. For example, heavy grazing pressure in rangeland tends to favor low-growing plants like *Polygonum* and discourage reproduction of taller grasses and herbs. Release of that grazing pressure on a grassland (by fencing to exclude cattle access) could (depending on other climatic and soil conditions) allow grass species to recover and even dominate the system. As a result, surface roughness is significantly increased, runoff is diminished, and changes in local stream hydrology may be observed. It is likely that these changes would also be accompanied by shifts in the composition of small mammal communities, especially granivores like rats and mice, and larger animals who feed on those species. These changes may in turn affect recreational opportunities, aesthetics, and agricultural activities.

Although schematics like the food web diagram can help identify relationships that may be in place, they are not as helpful in evaluating—or predicting—the rates of mass- or energy-transfer phenomena. These are strongly affected by local physical, chemical, and biological conditions; an accurate evaluation therefore requires detailed investigation. There is a steadily expanding body of literature on ecosystem processes and the factors that affect them. This literature may provide sufficient information for an initial estimate of system rates and insight into what may control them in the watershed of interest. An example of qualitative food web analysis is provided in Bodini et al. (1994).

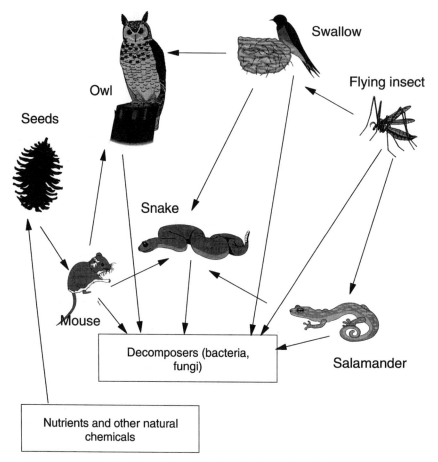

Figure 2.10 Example of a simple food web.

2.8 LAND USE

Up to this point, we have primarily been concerned with taking inventory of natural features in a watershed and (to a certain extent) understanding their behavior in space and time. Equally important in most systems are what may be termed the "built" components of the environment—the structures, machines, and systems constructed by humans and the social and economic systems that have encouraged those changes in the natural watershed system.

A critical first step in assessing this built environment is to understand current and anticipated land use within the watershed. The term *land use* implies use by humans: the ways in which humans formerly, presently, or may in the future change the landscape for the purposes of resource extraction and processing, housing, and transportation.

Assessment of land use demands three elements: an understanding of the

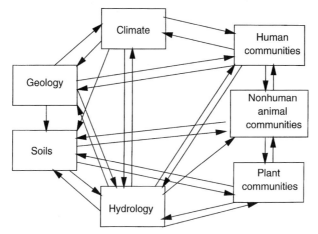

Figure 2.11 *Interaction of natural processes in a watershed ecosystem (after Lang and Armour 1980).*

nature of human activities currently practiced in the basin, an estimate of the areal extent of each activity (and perhaps its proximity to valued watershed features), and the ways in which these aspects are likely to change in the near and far future. An analysis of land use yields insights into many facets of the watershed system, including altered drainage regimes, pollutant sources, valued natural and built features, and, indirectly, regional economic forces and community priorities. These in turn become important in anticipating the impact of possible management actions within the basin.

Figure 2.11 gives a conceptual representation of the interrelationships between major systems in a watershed. Even at this simple scale, it is easy to see that minor changes in one system (for instance, climate) can have important implications for the structure and function of other systems (for instance, soils, hydrologic processes, human and nonhuman animal communities, plant communities, and so on). Similarly, a change in human activities, especially land use, can have a dramatic impact on natural processes.

Figure 2.12 illustrates the effect of increasing urbanization on a hypothetical stream system. With the building of more roads, parking lots, housing, and major industrial, commercial, and institutional structures, more of the land surface becomes impermeable. Rain falling over the area no longer infiltrates as easily or as quickly, so much more is "excess" flow, or runoff, over the land surface. This runoff reenters streams much more quickly now than in the pre-urban state, so streamflow response to rainfall, and to the cessation of rain, is much faster in an urbanized watershed.

The element of time is also important in understanding land use. Past land use practices have an influence on the present watershed and the ways in which current land uses may be constrained. For example, past open-pit mining activities may have created physical changes (pits, altered drainage, and so on)

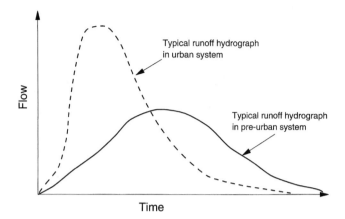

Figure 2.12 *Effect of urbanization on rate and volume of runoff (after Lang and Armour 1980).*

that limit the usefulness of the site for residential development. The same will clearly hold true for the influence of current land use on future development. For this reason, many urban centers have devoted time and thought to the development of long-term land use forecasts to guide present development. These forecasts may include estimates of projected population growth and reflect community consensus about intended areas for future residential, industrial, and commercial growth. Long-term planning for transportation corridors will likely also be a part of the plan. Many communities employ a system of land use zoning, so that individual parcels of land are designated for use only for certain activities.

Compiling an inventory of current land use need not be a tedious task. An obvious place to start is local and regional governments with an interest in planning and zoning. Many such governments have official land-use plans incorporating current and projected land use maps. If maps for a particular watershed are not readily available, topographic maps or even road maps can provide a starting point for a land use inventory. Agricultural extension professionals usually have quite detailed information about regional farming practices and even individual farm operations. Information held by public agencies is increasingly available in digital format, such as Geographic Information Systems, and thus readily extracted for use in simulation models or for other analysis.

Remote sensing technology is also well developed in this area. Satellite or high-altitude aerial photography is an excellent source of accurate information and, indeed, forms the basis for most modern watershed cartography. The frequency of imagery may not be sufficient to identify small changes in the watershed with time (changes that may, for example, be important in assessment of biological communities) but will likely be more than adequate for a general inventory of land use activities in the basin.

Land use is often divided into categories, such as open space (rural, non-

Table 2.5 Pollutant Generation Potential of Typical Land Use Categories

Land Use Category	Typical Pollutant Contribution
Rural nonagricultural	Low
Agricultural—cropland	Moderate to high, depending on activity and season
Agricultural—rangeland	Low to moderate, depending on grazing pressure
Forests and woodlots	Low
Parkland	Low, but may contain pesticide/herbicide residues depending on management practices
Low-density residential	Low to moderate
Medium-density residential	Moderate
High-density residential	Moderate to high
Parking lots	Moderate to high
Commercial	Moderate to high
Light industrial	Moderate to high
Heavy industrial	High
Roads and highways	High

agricultural, undeveloped land), forested/woodlot, parkland, agriculture (which may be subdivided into rangeland and cropland or even into more specific categories reflecting production type), residential, commercial, and industrial. Again, depending on the interests of the analyst, these broad categories may be further subdivided. Table 2.5 gives an overview of typical land use categories and the pollutant generation potential of each.

These different land uses alter the land in different ways. In terms of the movement of water in the basin, land use often changes the permeability of the land surface, thus affecting surface drainage systems and natural hydrology. The most dramatic example of this is the widespread paving of roads, highways, and parking lots, and the high proportion of roof area to total watershed area, in heavily urbanized watersheds. Water diverted off these impermeable surfaces is routed overland or to stormwater collection systems and thence to receiving waters, moving faster and carrying more heat and pollutants than would be the case in an undeveloped watershed.

Land use also affects pollutant yield, both in terms of the type of pollutant generated by a particular land use and in terms of the total mass of pollutant released. This phenomenon has been recognized for many years, and there is now an excellent literature on the unit-area pollutant loads that can be expected from various land use types (see Table 2.6).

Land use may also be important in determining the applicable legislation (Chapter 9), oversight agency, and interested nongovernment organizations. Agricultural activities, for example, are subject to different legislation than are manufacturing facilities, have available different loan and grant opportunities, are overseen by government agencies with an interest in protecting and promoting agriculture and its products, and have well-developed farmer and soil

Table 2.6 Typical Pollutant Unit-Area Loads From Various Land Uses (kg/ha/yr)

Land Use	BOD_5	Suspended Solids	Total Kjeldahl Nitrogen	Total Phosphorus
Commercial	43	825	11.0	0.18
Industrial	42	1082	14.0	0.92
Residential	33	619	5.0	0.42
Rural	18	569	11.7	0.90
Wooded/Idle	5	40.7	5.15	0.08

(*Source:* Waller and Novak 1981; Wanielista 1978)

conservation outreach networks. Different land uses clearly *look* different to an observer, implying landscape aesthetics that are altered to a greater or lesser degree. Although appearance may be a minor concern in some communities or landscapes, it can also be an important contributor to community economics, for instance, through tourism and recreational opportunities available, or unavailable, in the basin.

Systematic decision making regarding appropriate land uses for different parts of a watershed system has a central role in protecting or impacting valued landscapes. Often this entails zoning—a system of land use categories with allowable uses designated for every part of the basin. Without a clear land use planning system, development can proceed both unattractively and, arguably, unsustainably. Bangkok, Thailand, experiences staggering urban traffic congestion, forcing most commuters to spend four to six hours a day in their cars. This in turn creates serious urban air pollution problems and directly affects the health of hundreds of thousands of Bangkok residents every year. Yet Bangkok's traffic problems may have their roots in the simple absence of a comprehensive land use planning framework. Without such a plan, landowners have been able to build houses, dig roads and laneways, and subdivide property without regard for the cumulative result on the community. Some parcels of land have now been subdivided to the point where they are too small to build on and cannot be adequately serviced, frequently deteriorating into slums. An impressive system of urban highways has been built and more are under construction, but the city lacks an adequate system of "feeder" roads to service these highways. Drivers exiting a freeway may enter directly into a small laneway system—and grind to a halt. In Bangkok the jurisdictional authority for transportation and land use planning is not at all clear, and it will require a major effort on the part of Thailand's federal and state governments to clarify these roles before the problem of land use, and transportation, planning can be resolved.

2.9 SOCIAL AND ECONOMIC SYSTEMS

The natural and built environments of a watershed are usually overlain by human social systems. Beyond the obvious systems of urban centers and farmsteads, human systems can include the social and economic infrastructures necessary to support thousands or millions of people.

In the past these social and economic systems were often ignored in water management, viewed, perhaps, as separate from and irrelevant to surface or ground water hydrology and the structures that might be built to manage those systems. Today, however, it is widely recognized that social and economic systems are an integral part of the watershed ecosystem, affecting not only the physics and chemistry of natural waters but also the attitudinal and economic forces so central to successful implementation of water management actions. Indeed, under many environmental assessment laws, "the environment" is defined as including human social and economic systems. Yet these systems can be difficult to describe—and to inventory.

What is meant by "social and economic systems"? The term can encompass the obvious components of commercial and industrial activity, major institutions such as hospitals and universities, and residential development. Less obviously, it can be assumed to include religious systems, because they entail socially significant structures (e.g., churches, temples, synagogues; graveyards; sites of aboriginal religious value) and the belief systems of the religion(s), which have at least the potential to conflict with watershed activities. Considerations such as social and economic stability may also be important elements within a community, as well as what might be termed "quality of life"—that intangible quality of a safe, satisfactory living environment, with all essential services available and affordable.

Social and economic systems have the potential to influence a watershed ecosystem in three main ways beyond those described earlier in regard to land use:

1. By influencing the attitudes and priorities of watershed residents and decision makers
2. By affecting the value that may be placed on individual watershed features and activities, and thus the importance they are given in watershed planning
3. By constraining the financial resources available to resolve watershed issues

These forces must not be underestimated in watershed management. A company that is profitable and progressive can provide important leadership in the basin through its use of effective waste management practices, water reuse and recycling, and responsible ultimate disposal of wastes. On the other hand, a major employer that is struggling financially may be hard pressed to pay its routine bills, let alone install costly new pollution control devices. In the latter case, community residents may support the company even when it is a known polluter, simply because of their desire to retain jobs and economic stability in the community. Community attitudes are especially influential in actions relating

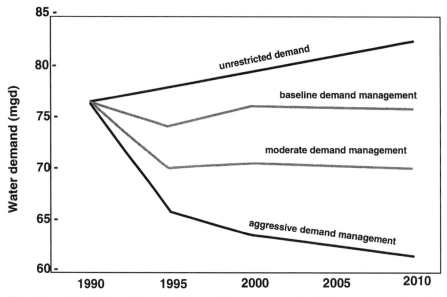

Figure 2.13 *Projected influence of demand management strategies on community water consumption patterns for the City of Providence, Rhode Island (after Deb et al. 1995).*

to water conservation. Figure 2.13 shows the influence of demand management strategies (measures to encourage consumers to use less water) on projected water consumption patterns in the city of Providence, Rhode Island. The figure illustrates four possible water consumption scenarios, depending on whether demand is unconstrained (no water conservation measures employed), or "baseline," "moderate," or "aggressive" demand management is undertaken. It is not necessary to understand the details of these schemes to realize that community support is an essential factor in successful implementation. That support will depend on community attitudes and social priorities: whether a green lawn is highly prized, for example, or whether residents would favor a brown lawn to conserve limited community water supplies.

Information about social and economic systems is often qualitative rather than quantitative, although census surveys can provide a good overview of average household income, age and sex distribution, spending patterns, religious preference, and similar information related to these concerns. Most valuable of all are sources within the basin, including local Chambers of Commerce, environmental or other (for instance, labor) public interest groups, and service clubs such as Rotary or Kiwanis Clubs. Community leaders usually have an excellent understanding of the economic and attitudinal forces at play in their areas and can confirm impressions gleaned from published sources. Even community "yellow pages" listings can give a sense of the major industrial, commercial, and institutional activities in the area and may yield insights into community attitudes and priorities.

2.10 VALUED FEATURES AND ACTIVITIES

Each watershed ecosystem contains certain valued features or activities. Often these are unremarkable to the casual observer, yet fiercely defended by local stakeholders. It is almost impossible to give a guide to such features, except by example.

The city of Wellington, New Zealand, is located on the South Coast of that country's North Island. Just east of the city lies a rock formation called Red Rocks—of little interest to European settlers but prominent in local Maori myths and legends. In recent years quarrying activities near the site have been closed down and relocated farther inland to protect the appearance and cultural importance of this site.

Another example illustrates the very local nature and small scale of some valued features. Asked to suggest an example of a valued feature, a young student from a farming background cited his grandfather's horseradish patch, which had been cultivated by the family for more than a hundred years. The family was adamantly opposed to any management intervention such as tile drainage or grassed waterways that might affect the quality and yield of horseradish from this plot.

Other examples of valued features may be historical sites such as a standing rock circle or a battleground, a vista or feature of great natural beauty, or simply a geological oddity such as a rock shaped like a man's head. These features may have little obvious importance to the casual observer, and indeed their value may be invisible (as in the case of the battleground), but to ignore them is to invite antagonism and public displays of concern.

An inventory of valued features is best compiled with the assistance of local residents, including interest groups, historical societies, community leaders, and similar individuals. Widespread advertising in local media can invite submissions by the wider population and ensure a good coverage of issues of concern.

REFERENCES

Bodini, Antonio, Giovanni Giavelli, and Orazio Rossi. 1994. The qualitative analysis of community food webs: Implications for wildlife management and conservation. *J. Envir. Management* 41: 49–65.

Bruce, J. P., and R. H. Clark. 1966. *Introduction to Hydrometeorology.* Oxford: Pergamon Press.

Chow, Ven Te, David R. Maidment, and Larry W. Mays. 1988. *Applied Hydrology.* New York: McGraw-Hill.

Deb, A. K., F. Grablutz, and P. Gadoury. 1995. Demand management strategies for Providence Water Supply Board. In *Integrated Water Resources Planning for the 21st Century*, edited by M. F. Domenica. Proceedings of the 22nd Annual Conference, Cambridge Massachusetts, May 7–11, 1995. New York: American Society of Civil Engineers, Water Resources Planning and Management Division.

Green, W. H., and G. A. Ampt. 1911. Studies in soil physics. The flow of air and water through soils. *J. Agr. Sci.* 4: 1–24.

Holtan, H. N. 1961. *A Concept for Infiltration Estimates in Watershed Engineering.* Washington, D.C.: USDA Agricultural Research Service.

Horton, R. E. 1939. Approach toward a physical interpretation of infiltration capacity. *Proc. Soil Sci. Soc. Am.* 5: 399–417.

Kouwen, N., and E. D. Soulis. 1994. Weather radar and flood forecasting. In *Current Practices in Modeling the Management of Stormwater Impacts*, edited by W. James. Boca Raton, Fla.: Lewis Publishers.

Lang, R., and A. Armour. 1980. *Environmental Planning Resourcebook.* Ottawa: Environment Canada, Lands Directorate.

Magurran, A. E. 1988. *Ecological Diversity and Its Measurement.* London: Croon Helm.

Meyer, A. F. 1942. *Evaporation from Lakes and Reservoirs.* St. Paul: Minnesota Resources Commission.

Novotny, V., and H. Olem. 1994. *Water Quality: Prevention, Identification, and Management of Diffuse Pollution.* New York: Van Nostrand Reinhold.

Ontario Ministry of the Environment. 1987. *Technical Guidelines for Preparing a Pollution Control Plan.* Report from Urban Drainage Policy Implementation Committee, Technical Sub-Committee No. 2. Toronto: Ontario Ministry of the Environment.

Philip, J. R. 1957. The theory of infiltration: 1: The infiltration equation and its solution. *Soil Sci.* 83(5): 345–357.

———. 1969. Theory of infiltration. In *Advances in Hydroscience*, Vol. 5, edited by V. T. Chow. New York: Academic Press.

———. 1983. Infiltration in one, two, and three dimensions. In *Advances in Infiltration.* ASAE Publi. No. 11-83. St. Joseph, Mich.: American Society of Agricultural Engineers.

Rohwer, C. 1931. *Evaporation from Free Water Surfaces.* USDA Tech. Bull. 271. Washington, D.C.: U.S. Government Printing Office.

Schmidt, M. F., M. J. Bergman, D. R. Smith, and B. A. Cunningham. 1996. Calibration of SWMM-EXTRAN using short-term continuous simulation. In *Advances in Modeling the Management of Stormwater Impacts*, edited by W. James. Guelph, Ont.: Computational Hydraulics International.

Schwab, Glenn O., Delmar D. Fangmeier, William J. Elliot, and Richard K. Frevert. 1993. *Soil and Water Conservation Engineering.* 4th ed. New York: John Wiley & Sons.

Simpson, E. H. 1949. Measurement of diversity. *Nature* 163: 688.

Tchobanoglous, G., and E. D. Schroeder. 1987. *Water Quality: Characteristics, Modeling, Modification.* Reading, Mass.: Addison-Wesley.

United States Soil Conservation Service. 1990. *Engineering Field Manual.* Chapter 2. Washington, D.C.: US SCS.

Viessman, W., Jr., and G. L. Lewis. 1996. *Introduction to Hydrology.* 4th ed. New York: HarperCollins College Publishers.

Waller, D. H., and Z. Novak. 1981. Pollution loading to the Great Lakes from municipal sources in Ontario. *J. Water Pollution Control Fed.* 53(3): 387.

Wanielista, M. P. 1978. *Stormwater Management—Quantity and Quality.* Ann Arbor, Mich.: Ann Arbor Science.

3

Problem Definition and Scoping

The watershed inventory discussed in Chapter 2 gives an overview of physical, chemical, and biological processes operating in a watershed ecosystem. It seldom, however, points the way to a clearly defined "problem" to be solved by management actions. Yet disagreements about what problem is to be solved can create significant obstacles to effective watershed management, even if stakeholders agree on most issues and conditions in the system.

The process of problem definition begins with a vision of what the watershed should be like, in terms of beneficial water uses and the quantity and quality of water required for them. It then continues to an evaluation of the disparity between existing and ideal conditions. The vision of an "ideal" watershed will vary according to the basin and the people living and working in it. But whatever the ultimate goals, this process is fundamental in determining where conditions are less than ideal and where restorative actions should be begun.

Often, watershed residents are acutely aware of degraded conditions, simply because degradation often constrains beneficial uses. Section 3.1 examines some approaches to identifying water uses and use impairments in a watershed system.

3.1 IDENTIFYING CURRENT WATER USES AND USE IMPAIRMENTS

Virtually every watershed is of some use, or value, to human communities, even if that use is simply one of aesthetic enjoyment or preservation of the natural condition. A helpful list of potential water uses developed by Great Britain's National Rivers Authority (NRA) is presented in Table 3.1.

Table 3.1 Water Use Categories Employed by the National Rivers Authority in Catchment Planning

Water Use Categories	Typical Uses
Potable (Drinking) Water Supply	Municipal water supply (surface or groundwater sources)
	Residential water supply (private wells)
Industrial Water Supply	Process water supply
	Cooling waters
Agriculture	Irrigation waters
	Livestock watering
	Milkhouse wash water
	Livestock housing wash water
Flood Control	Impoundment of high flows for delayed release
	Construction of dams, reservoirs, levees, and channel protection
Thermal Electric Power Generation	Cooling waters
	Settling pond waters
	Water for pipe flushing and maintenance
Hydroelectric Power Generation	Impoundment of water for power generation
	Construction of dams and reservoirs
	Pumping and drawdown of water levels
Navigation	Recreational boating (e.g., sailing, canoeing, motor boat traffic)
	Commercial shipping
	Commercial navigation for tourism purposes (e.g., sightseeing)
Water-based Recreation	Recreational fishing
	Recreational boating and windsurfing
	Swimming
	Hiking
	Picnicking
	Nature enjoyment activities (e.g., bird-watching)
	Aesthetic enjoyment
Fish and Wildlife Habitat	Aquatic and riparian habitats
	Protection of community structure
	Protection of rare and endangered species
Water Quality Management	Protection of minimum flows for water quality preservation
	Low-flow augmentation from reservoirs
	Assimilation of waste discharges from municipalities and industries
	Assimilation of storm- and combined-sewer discharges

(*Source:* NRA 1933a)

Although not obvious "water users," a variety of land-based activities are also important in affecting water resources in a watershed. Most important among these are probably forestry, including the logging and replanting of trees, agricultural practices such as tillage, planting, harvesting, and drainage works, and construction activities. Each of these activity classes has potential to affect local hydrologic processes and soil quality and permeability. As a result, each can be important in altering sedimentation regimes and changing the ability of upland areas to retain moisture, thereby affecting the volume and patterns of surface and groundwater flows. These processes, if altered, can in turn have an impact on valued watershed features and activities such as potable water quality, fisheries, and wildlife habitat. Their influence cannot, therefore, be ignored in an evaluation of water uses and water-using interests in the basin.

3.1.1 Estimating Population Size

A preliminary list of water uses is an essential foundation for more detailed analysis. The next step, that of calculating current and future demand for those uses, is more complex.

Most estimates of water demand require an estimation of population size and potential growth over the planning horizon. Population size is a key variable in determining not only water demand, but demand for other water-using activities such as electric power generation, manufacturing, agriculture, and recreation, and the intensity of these activities within the watershed. Forecasted population size is also a useful measure of the potential market for water-related goods and services.

Since many water structures such as sewer systems and dams have long useful lives, it is appropriate that estimates of population size be made for the present, the near future, and the distant future—the time at which major structures are likely to require replacement. Goodman (1984) observes that our record of accuracy in long-term forecasting is not good, in part because of changing societal patterns (such as an increasing number of women in the work-place, leading to a higher frequency of delayed parenting and smaller average family size). Goodman reviews several authors' methods of predicting population growth for the city of Kingston, Jamaica (see Figure 3.1), concluding that projections of population growth within two to five years can vary by as much as 10%. Ten years into the future, estimates differ by 20% to 25%. Fifteen years into the future, the spread in estimates can be 50% or more, depending on the estimation method used. (Simple extension of historical growth trends yielded an estimate that was in the low to moderate range; this projection was ultimately adopted as "most likely.")

Despite difficulties in achieving an accurate projection, long-term population forecasting is an important element in watershed management and one that can be continuously updated as more current information becomes available.

The size of the current population is most easily estimated from census records, usually compiled every 10 years, and/or municipal (e.g., tax) records and similar documents. The simplest population growth model takes the form:

$$P_t = P_0 + B - D + I - O \qquad (3.1)$$

where: P_t = population at end of time t
P_0 = population at present time

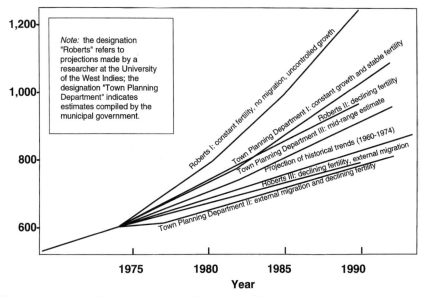

Figure 3.1 *Population projections for Kingston and St. Andrew, Jamaica (adapted from Goodman 1984).*

B = number of births in planning period
D = number of deaths in planning period
I = number of individuals immigrating into area during planning period
O = number of individuals emigrating from area during planning period

Census information, often available for many decades, is an excellent source of data for the individual terms in this equation. Generally speaking, projections are more accurate for larger areas (e.g., a nation) than for smaller ones (e.g., a county) and for overall population size rather than for the specific characteristics (e.g., average income) of a given population. Forecast accuracy is also affected by the age of the data used, so it is always best to begin with the most recent data available.

Although complex models incorporating detailed county-level data on age, sex, fertility rates, and similar details are available, these are difficult to apply and increase the number of possible error sources in the final population projection. This type of model may have application in some specialized watershed planning situations but is likely to be less accurate, and, hence, less useful to the watershed manager, than aggregated (lumped) models of total population size.

Whether based on aggregated or disaggregated models of population growth, estimates of population size can be used to generate forecasts about work force availability, markets for goods and services, demand for schools and hospitals, and similar elements important in watershed planning. Information about the existing distribution of industry and other activities in the basin can be used with population forecasts to predict the intensity of land-based activities such as agriculture and construction in the future. These predictions will, however, be only as accurate as the information on which they are based, and they will be most useful if regularly updated to reflect changing watershed conditions.

Despite all precautions, population forecasts are always fraught with error. This problem may be dealt with in two ways. First, the data used in developing population growth forecasts should, where possible, be corroborated by other sources or methods. For example, data obtained from census information should be checked using municipal tax records or similar sources. And forecasts can often be compared with national or regional trends prepared by government agencies.

A second way of guarding against error in population forecasting is to avoid using a single, probably uncertain, estimate. Instead, the analyst can employ a sort of sensitivity analysis, in which high, moderate, and low estimates are made to generate a range of projected population sizes. Although this renders the analysis somewhat more complex, there is immense value in these best case/typical case/worst case estimates, particularly in forecasting long-term (and therefore very uncertain) population growth and associated water demand. As an example, the full range of forecasts presented in Figure 3.1 could be used to generate a range of domestic water demand estimates. A "best case" (lowest demand) estimate could be compiled using the lowest population growth estimates. A "worst case" estimate, using the highest population growth estimates, would give an idea of the highest likely demand. The two together bracket a range of possibilities within which the true condition will almost certainly fall.

3.1.2 Estimating Water Demand

Estimates of population size and the intensity of various human activities form the basis for estimates of water demand. Water demand is usually expressed in two categories: *water withdrawals*, the total amount of water that is removed from natural systems for human use, and *consumptive demand*, the volume of water that is removed by humans but which is "consumed" in the removal and thus unavailable for return to natural systems.

Pearse et al. (1985) estimated that, of 38 billion m^3 of water withdrawn from Canadian sources in 1981, 54 billion m^3 were used (recirculation allowed these supplies to be "stretched" 1.4 times) and 4 billion m^3 (about 10% of the total) were consumed. For consumptive uses, thermoelectric power generation and agriculture were by far the largest consumers.

Water demand varies widely, depending on geographic location and activity (see, for example, Van der Leeden et al. 1990; *World Water* 1989). In general,

Table 3.2   Estimated Per Capita Water Use in Selected Countries

	Water Use gal/cap/day	Water Use L/cap/day	Rank
Australia	476	1802	1
United States	159	602	2
Canada	125	473	3
Sweden	120	454	4
Kuwait	117	443	5
New Zealand	116	439	6
Portugal	116	439	7
Japan	107	405	8
Spain	101	382	9
Italy	99	375	10
Malaysia	83	314	11
France	81	307	12
Bulgaria	80	303	13
Switzerland	78	295	14
Iceland	78	295	15
Saudi Arabia	74	280	16
Belgium	72	273	17
United Kingdom	71	269	18
Brazil	71	269	19
Norway	68	257	20
Finland	64	242	21
Venezuela	63	238	22
Romania	63	238	23
Libya	63	238	24
Denmark	61	231	25
Czechoslovakia	61	231	26
Philippines	59	223	27
Austria	55	208	28
Argentina	54	204	29
Chile	54	204	30
Egypt	52	197	31
Poland	51	193	32
Korea, Dem. People's Rep.	50	189	33
Cuba	49	185	34
Guyana	48	182	35
Iraq	48	182	36
Colombia	47	178	37
Turkey	46	174	38
Suriname	46	174	39
Panama	45	170	40
Israel	43	163	41
United Arab Emirates	43	163	42
Yugoslavia	42	159	43
Nicaragua	41	155	44
Greece	40	151	45
Mauritius	38	144	46
Peru	37	140	47

(*Source:* World Resources Institute 1993)

Table 3.2 (*Continued*)

	Water Use gal/cap/day	Water Use L/cap/day	Rank
Netherlands	35	132	48
Hungary	33	125	49
Mauritania	30	114	50
South Africa	27	102	51
Mexico	27	102	52
Ireland	26	98	53
Tunisia	26	98	54
Ecuador	25	95	55
Gabon	25	95	56
Trinidad and Tobago	23	87	57
Singapore	22	83	58
Jordan	22	83	59
Iran, Islamic Rep.	21	79	60
Lebanon	21	79	61
Korea, Rep.	19	72	62
Zambia	19	72	63
Mongolia	19	72	64
Morocco	18	68	65
Algeria	18	68	66
China	17	64	67
Thailand	16	61	68
Dominican Rep.	14	53	69
Honduras	14	53	70
Swaziland	13	49	71
Lao, People's Dem. Rep.	13	49	72
Syrian Arab Rep.	13	49	73
Costa Rica	12	45	74
Bolivia	12	45	75
Zimbabwe	12	45	76
Cameroon	11	42	77
Togo	11	42	78
Paraguay	10	38	79
Afghanistan	10	38	80
El Salvador	9	34	81
India	9	34	82
Liberia	9	34	83
Madagascar	9	34	84
Uruguay	9	34	85
Pakistan	9	34	86
Mozambique	9	34	87
Guinea	9	34	88
Cote d'Ivoire	9	34	89
Kenya	8	30	90
Indonesia	8	30	91
Equatorial Guinea	8	30	92
Congo	8	30	93
Zaire	7	26	94
Nigeria	7	26	95

Table 3.2 *(Continued)*

	Water Use gal/cap/day	Water Use L/cap/day	Rank
Belize	7	26	96
Vietnam	7	26	97
Jamaica	7	26	98
Senegal	6	23	99
Oman	6	23	100
Fiji	6	23	101
Nigeria	5	19	102
Papua New Guinea	5	19	103
Sri Lanka	5	19	104
Sudan	5	19	105
Angola	5	19	106
Guatemala	5	19	107
Ghana	5	19	108
Myanmar	5	19	109
Benin	5	19	110
Burundi	4	15	111
Lesotho	4	15	112
Sierra Leone	4	15	113
Bangladesh	4	15	114
Namibia	4	15	115
Malawi	4	15	116
Nepal	4	15	117
Chad	4	15	118
Rwanda	3	11	119
Central African Rep.	3	11	120
Ethiopia	3	11	121
Bhutan	3	11	122
Burkina Faso	3	11	123
Tanzania	3	11	124
Albania	3	11	125
Djibouti	3	11	126
Botswana	2	8	127
Uganda	2	8	128
Guinea-Bissau	2	8	129
Cambodia	2	8	130
Mali	2	8	131
Somalia	2	8	132
Gambia, The	1	4	133
Haiti	1	4	134
Solomon Islands	0	0	135

Canada and the United States rank among the heaviest domestic water users in the world (see Table 3.2). By contrast, some water-poor countries such as Botswana, with its Kalahari Desert, may exhibit domestic water consumption rates of only 8 L/person/day.

Table 3.3 illustrates the fact that different countries use water differently. Certainly, climate has much to do with these differences. Warm, dry countries

Table 3.3 Domestic Water Use in Selected Countries

Country	Total Withdrawal (km^3)	Per Capita Withdrawal (m^3)	% Public Use (includes firefighting)	% Industrial Use	% Electric Power Generation	% Agriculture/ Irrigation
United States	472.000	1,986	10	11	38	41
Canada	30.000	1,172	13	39	39	10
Egypt	45.000	962	1	0	0	98
Belgium	8.260	836	6	37	47	10
Former USSR	226.000	812	8	15	14	63
India	380.000	499	3	1	3	93
China	460.000	460	6	7	0	87
Poland	15.900	423	14	21	40	25
Oman	0.043	350	2	0	0	98
South Africa	9.200	284	17	0	0	83
Barbados	0.027	102	45	35	0	20
Malta	0.023	60	100	0	0	0

(*Source:* McDonald and Kay 1988)

such as Egypt and India use a high proportion of extracted water for irrigation to replace or supplement natural rainfall. By contrast, cool, wet countries like Canada and Poland use a higher proportion of water for electric power generation (especially cooling waters) and for industrial purposes.

Water use also implies nonwithdrawal uses such as fishing, swimming, and navigation. Although these uses do not extract water, they may require a certain volume of water to be present and thus must be included in any consideration of present or future water demands.

Western European countries such as Belgium, France, Germany, and the Netherlands have a limited resource base and a much higher population density than much of Canada or the United States. These factors may have influenced their modest water use (consumption rates in those countries are typically 150–200 L/person/day) and high water prices. (Note that not all water-poor countries are careful in their water use. In some wealthy but water-poor Gulf nations, high water use is considered a sign of affluence. These cultural attitudes have been serious obstacles to water conservation efforts.)

Generally speaking, water demand is influenced by population size and density, annual per capita income, quality of supply, and annual rainfall. These variables can be used to make reasonably accurate predictions of water demand for a given area. Clearly, the larger the population, the more water consumed within the area. Warmer climates usually require higher water use, for drinking, irrigation, livestock watering, building foundation watering (to avoid cracking), and similar uses. Annual rainfall affects the volume of water available for withdrawal and thus may impose limits on allowable water use.

Much water resources planning is, however, still based on assumptions of steadily increasing per capita demand. Certainly, as growth proceeds in a watershed area, total water demand may also be expected to increase. But per capita demand may also increase through social forces and life-style expectations. A recent residential subdivision application in the city of Guelph, Ontario, included a proposal to require all new houses to install in-ground sprinkler systems. The houses were to be costly "executive" homes, and the developer's vision included uniformly green lawns across the development. In response to citizen outcry against this *requirement* to use water, Guelph City Council required the deletion of this provision in approving the plan of subdivision. By contrast, the neighboring regional municipality of Kitchener-Waterloo, which faces increasingly urgent water supply shortages over the next decade, has significantly reduced water demand by aggressive public education programs and municipal subsidies of water-conserving devices. In Kitchener-Waterloo, a brown lawn is seen as prestigious, a symbol of responsible citizenship and concern for the environment.

Water consumption also varies widely in industrial, commercial, and institutional settings. Even within the food processing industry, for example, production of canned fruit may use only one-fifth to one-tenth of the water required for the same quantity of beer or soft drinks. Pulp and paper manufacturing and mineral mining and refining are among the heaviest industrial water users,

Table 3.4 Typical Industrial Wastewater Discharge Rates for the United States and Canada

Brewery wastes	900–1,440 L/bbl
Apple canning wastes	110–174/case
Baked beans processing wastes	151 L/case
Fresh green beans processing wastes	540–800 L/case
Fresh lima beans processing wastes	216–1,110 L/case
Dried lima beans processing wastes	76–125 L/case
Milk processing wastes	900–4,180 L/100 kg
Cheese processing wastes	10,780–19,300 L/100 kg
Poultry processing wastes (conventional technology)	17,600 L/1,000 kg live weight
Poultry processing wastes (advanced technology)	9,200 L/1,000 kg live weight
Petroleum refining wastes	76–227 L/bbl
Pulp and paper (bleached kraft) processing wastes	120–220 cu m/tonne
Steel processing wastes	57 m^3/tonne
Textiles—wool processing wastes	516 m^3/tonne of finished cloth
Textiles—rayon processing wastes	25–58 m^3/tonne of finished cloth

(*Source:* Eckenfelder 1970; Metcalf and Eddy 1991)

with withdrawal rates approaching 2,000 times those of some food processing industries. Table 3.4 gives some typical wastewater discharge rates (indirectly reflecting water withdrawal rates) for industries in the United States and Canada.

Estimating current water demand, like current population size, is a relatively straightforward task. The simplest approach is to estimate daily per-capita water demand (municipal water supply records are a good source of this information) and multiply per-capita use by the population served. This estimate can be extended into the future by using projected population figures and assuming the same usage rate as currently exists.

This crude estimate can be improved if separate information is available for different user types, especially residential versus industrial, commercial, and institutional users. Many municipalities employ water metering to track water use by individual residences and industries so that users can be billed for water use on a unit basis. (Both increasing block pricing—the more you use, the more you pay per unit used—and decreasing block pricing—the more you use, the less you pay per unit—are used in charging metered consumers for water service; the former is clearly preferable in terms of encouraging efficient water use.) These metered records provide the most current and accurate basis on which to calculate current basin water uses.

In some areas (many older residential areas in Canada, for instance) residential water use is unlimited and unmetered, and billed on a flat-rate basis. In England less than half of 1% of the residential population has water meters (Heathcote et al. 1996). Water metering is much more common in water-poor areas, in newer development, and in industrial facilities.

Month	Water Production (L/month)	Water Sales (L/month)
May	5,352,793	3,912,486
June	5,378,911	4,180,107
July	5,328,188	4,111,971
Total	16,059,892	12,204,564
Average (L/day)	174,563	132,659

Metcalf and Eddy (1991) offer an example of a water demand calculation based on the following water supply data:

Solution: Assuming a serviced area of 147 households and an average household ("service") size of 2.43 people based on local planning data, daily water consumption can be calculated from the given water supply data above:

$$\text{Consumption} = \frac{132,659\text{L/d}}{(147 \text{ services})(2.43 \text{ people/service})} = 371\text{L/capita.day}$$

Metcalf and Eddy's hypothetical data illustrate another important point about water supply and demand calculations. In almost every system a certain proportion of water is "lost" through pipe leakage and other unaccounted system losses. In a well-maintained system, this amount will be a small proportion of the total, perhaps less than 15%. In most older cities the proportion is somewhat higher, 30% or more. A very high loss rate points to an urgent need for infrastructure repair. In the preceding example unaccounted-for system losses amount to about 24% of the total, calculated on a daily basis as follows:

$$\text{Losses} = \text{production} - \text{sales}$$

or

$$\text{Losses} = \frac{174,563 - 132,660}{174,563} * 100\% = 24\%$$

It is important to use data on water *sales* (consumption) rather than water *production* because of the problem of unaccounted-for system losses. In other words, water production almost always exceeds water demand, sometimes by a considerable margin. Using water production records to forecast demand may lead to serious error in computed estimates.

It is somewhat more difficult to estimate future water demand under circumstances of changing usage, especially under water conservation programs or, possibly, under an assumption of increased per-capita use. Like population forecasting, estimation of future water demand is an inexact science. The members of Canada's Inquiry on Federal Water Policy (Pearse et al. 1985) calculated that total municipal withdrawals in Canada (estimated at 4,263 million m^3 in 1981) could rise to 8.458 million m^3 or fall to 3,984 million m^3 by 2011, depending on whether conservation policies (and pricing) are or are not adopted. These figures do not include estimates of municipally drawn industrial demand, which the Inquiry estimated at an additional 811 million m^3 in 1981, nor rural residential use, which was estimated at 347 million m^3.

In the early 1970s, the United States National Academy of Sciences examined a range of technological advances with the potential to increase or decrease future water demand. Among the latter group were increasing reliance on advanced nuclear, wind, and water power, improvements in industrial cooling technology (ponds and towers), altered plant genetics to improve drought tolerance, and movement away from surface irrigation to subsurface irrigation and "xeriscaping"—landscaping for water conservation. Most of these predictions are now implicit or explicit in water management plans. For instance, recently tightened industrial discharge regulations have prompted the adoption of newer, more water-efficient technologies (one example is improved water reuse in the metal mining/refining sector), and this has had a spin-off effect in reducing overall water consumption patterns.

Efforts to promote water conservation have increased steadily over the past decade. To some degree these efforts have resulted in decreased water demand, particularly among residential customers. Industrial, commercial, and institutional water use patterns have been more difficult to change, although some reduction in demand is also apparent in these uses. Beard (1996), Heathcote (1993, 1995), and many others have suggested that the abundance of water in most of Canada and the eastern United States has made water managers complacent about water supplies. As the Ontario MISA Advisory Committee (1991) wrote:

> [In Canada, t]he traditional assumption of water is there is always enough, it's always clean and it's always free. Today's reality is that there is not enough, it is not always clean, and it will never again be free.

In many watersheds, it takes a water supply crisis to focus public and institutional attention on water demand and possible mechanisms for demand reduction; few water-rich jurisdictions currently undertake aggressive demand reduction programs voluntarily.

To address the problem of demand forecasting, water managers must therefore make a number of critical planning assumptions; often these will be specific to a particular region or even a particular watershed. These assumptions include:

1. Estimated population growth rate over the planning horizon
2. Estimated precipitation entering the area
3. Estimated volume of surface and/or groundwater available for future extraction
4. Nature of water-using industries, commercial establishments, and institutions in the basin

Table 3.5 Principal Considerations in Estimating Demand for Nonwithdrawal Water Uses

Water Use	Demand Considerations
Hydroelectric Power Generation	Minimum head required for expected power generation
Navigation	Minimum depths required for navigation Minimum flows required to maintain depths
Fish and Wildlife Habitat	Effect of water depths and flow/level fluctuations on fish and wildlife species Impact of water levels on water quality and, indirectly, on health of biota
Water-Contact Recreation	Required water depth for satisfactory swimming and bathing opportunities Impact of water levels on water quality and, indirectly, on human health
Recreational Boating	Minimum depth required for safe and aesthetic boating Minimum flows required to maintain depths
Preservation of Wetlands	Minimum surface and groundwater flows required to maintain wetland habitat Effect of water level fluctuations on wetland species

5. Attitudes toward water use versus conservation in the basin; i.e., likelihood of uncontrolled versus controlled water use

When the majority of water withdrawals in a basin are for domestic potable water supply, forecasts can be made on a uniform per-capita basis. If industrial, commercial, or institutional uses are important in the basin, however, each major water demand category—often each individual major user—must be predicted separately.

Water demand for nonwithdrawal uses such as navigation and fishing can be expressed in similar fashion using estimates of population and industrial growth. Table 3.5 shows some of the primary considerations in estimating water demand for nonwithdrawal uses.

As with population forecasts, it is often wise to make several estimates of future demand to bracket a range of possible outcomes.

3.2 IDENTIFYING CURRENT WATER USERS (STAKEHOLDERS)

The water uses present in a basin are simply the physical expression of human behavior patterns. Implicit in an understanding of water uses, therefore, is the need to understand water *users*—their attitudes, their needs and wants, and the priorities they set on different features and activities. Water managers ignore users at their peril: community support is one of the most important elements in successful implementation of management schemes, and lack of that support can be one of the most formidable obstacles.

Water users include a broad array of individuals and groups with an interest in basin water resources. The particular mixture of users differs from watershed to watershed, but in most cases includes some or all of the entities discussed in the following paragraphs.

3.2.1 Government Agencies

Some government agencies have a direct role in water use and must usually be included in decision making about watershed management. These often include municipal governments and their public utilities (because of their role in water supply and wastewater treatment and disposal), natural resources agencies (through their interest in protection of fish and wildlife species and habitat), public health agencies, and agencies involved with shipping and navigation. These agencies may have a statutory or regulatory responsibility to oversee water use within their jurisdiction and may thus be in a position to forbid or place constraints on certain management approaches.

A second category of government agencies includes those that oversee and regulate water-using activities, such as the United States Environmental Protection Agency (EPA), Environment Canada, and state and provincial pollution control agencies. Others in this category are electric power generation agencies and utilities and those engaged in overseeing agriculture and urban development.

Still other agencies have a less direct, but nevertheless important, interest in water use. These may include agencies overseeing tourism (recreational water use and aesthetic considerations) and industrial development (which could be limited by scarce or low-quality water supplies).

It is important to realize that different arms of government may have different agendas in water management, and that interagency conflict in making decisions about water resources is therefore not uncommon. For instance, government departments concerned with timber management and extraction may be much less concerned with the impact of logging on streamflows than they are with issues of safety and efficiency. Similarly, it is not uncommon for public health agencies to recommend that domestic water users flush their taps for several minutes before drawing water in order to reduce concentrations of lead and other contaminants that may have accumulated in standing water. Such a recommendation, however, is in direct conflict with local programs to reduce domestic water use and conserve scarce water supplies. These differences are rooted in agency mandates and are entirely understandable. Yet they have the potential to create major obstacles in allocation of funds—and funding responsibilities—for watershed management activities.

3.2.2 Industrial Water Users

In some ways industrial water users are among the most straightforward to understand. There are usually a limited number of major industries in a basin, and these are relatively easy to list and characterize. Although the concerns of one industry may differ from those of another, in general industries are interested in obtaining water of adequate quality and quantity for process and cooling water purposes, and in their regulatory or voluntary responsibilities in discharges of wastes to receiving water.

Industrial water users are often accessible through sector-based or other interest groups. These include sectoral associations, Chambers of Commerce, and regional industrial societies. As a result, it is usually easy to make a first contact with these users and to sustain that contact through decision-making and implementation phases.

3.2.3 Commercial Shipping and Fishing Interests

In areas with larger river or lake systems, commercial shipping and fishing interests may be important. Like land-based industries, these are often accessible through sector associations and are in any case likely to be well known in the watershed area. A critical consideration for shipping interests is water levels, because if these are not maintained at adequate depth for navigation, commercial interests can be severely impacted. In major systems such as the Great Lakes, commercial fishing interests employing larger vessels may also be concerned about water levels. More frequently, however, fishing interests relate to the preservation of fish stocks and, thus, to other water conditions, particularly water quality.

There is strong potential for commercial shipping and fishing interests to come into conflict with those of private landowners, particularly around the issue of water levels. High water levels may, for instance, be very advantageous for shipping and navigation but can increase property damage through flooding and erosion of waterfront lots. Commercial fisheries may also come into conflict with recreational fisheries or wildlife habitat preservation efforts. The most prominent example of this may be the collapse of the Atlantic groundfish fisheries. Commercial fishers blame what they believe to be a large harp seal population for preying on scarce cod (or cod prey species) and thus preventing a faster recovery. Detailed food web analysis demonstrates, however, that Atlantic cod comprise only a small part of the seals' diet, and that fish populations may be more strongly affected by physical conditions such as temperature and salinity than by predation. The scientific evidence underlying this debate is not clear-cut, allowing the conflict around fishing rights, sealing rights (and protections), and employment issues to escalate in Atlantic Canada (Hutchings and Meyers 1994).

3.2.4 Residential (Private) Water Users

By far the most numerous water users in most areas are private residents. As a group, their primary concern is usually the quality of water for consumption: whether the water they buy or draw privately is safe to drink and tastes and looks good. A second concern, almost as important, is that of water supply:

whether they will be able to use the quantity of water they believe they require, not only for drinking, washing, and cooking, but for activities like lawn and garden watering, car washing, and foundation watering. Many also have an interest in recreational water-based activities such as swimming, boating, and nature enjoyment.

Residential water users are difficult to consult, in that their interests are diverse and, as a group, they are seldom well organized or easily accessible. Yet their property interests and strong links to valued local features make residential users very sensitive to changes in water management practice. As its members become better informed and more involved in environmental issues, the public has also become less trusting of government and more willing to challenge public policy in consultation forums and the media. Increased opportunity for public participation over the past 10 years has also been instrumental in teaching private citizens that there is a difference between public information sessions and meaningful involvement in decision making. In most areas citizens now expect to have a substantive role in the development of water management policy for their region. Meaningful involvement of the public in watershed planning has tremendous advantages for the water manager as well. First, it is an opportunity to learn from the vast range of knowledge and expertise in the community. Second, it is helpful in identifying local issues and values, such as those related to valued features and social systems (see Sections 2.8 and 2.9). And it provides a forum in which water managers can clarify and resolve conflicts before the implementation stage.

3.2.5 Public Interest Groups

Virtually all communities and watersheds have some representation by organized "public interest" groups, sometimes termed nongovernment organizations (NGOs). (Occasionally the acronym ENGO is used, for environmental nongovernment organization.) It is important to note that these groups do not necessarily reflect all or even a majority of public opinion on a given issue or watershed system. Often, they have strong and clearly stated agendas of their own. The World Wildlife Fund, for example, is dedicated to the preservation of wildlife and, by extension, wildlife habitat. This laudable goal has, however, sometimes placed the organization in conflict with watershed residents where the goals of preservation conflict with the needs of human communities (see, for example, Alpert 1993; Richardson 1993).

Public interest groups thus, as the name implies, represent specific, often narrow, interests within a community. Some have as their primary goal a heightened public awareness of issues central to their cause. Although not always backed by in-depth research, these organizations have value in the watershed planning exercise simply because they highlight sensitive or controversial issues. Other groups may have access to impressive research resources and be in a position to produce comprehensive reports on key themes. The expertise in these groups is, again, a valuable adjunct to the planning process and may include perspectives and insights that have largely been ignored by conventional planning approaches.

It is essential to remember that however valuable in watershed planning, representation by public interest groups *does not* constitute representation by the public. A meaningful consultation regime, therefore, includes both public interest groups and members of the general public. Chapter 4 discusses the consultation process in more detail.

3.2.6   Aboriginal Communities

Absent in some watersheds, at least as an organized group, aboriginal communities are extremely important in others. These groups have all of the same concerns and interests as the nonaboriginal residential water user, but may have others as well. Critical to many aboriginal societies is the relationship between humans and the land—in a sense, a landscape and its resources are respected not so much for what they can yield for human use, as for what they mean in terms of human history and values. Aboriginal communities may also be subject to different laws and agreements, especially treaty agreements, than would be the case for nonaboriginal groups. Reservations, for instance, may be under the jurisdiction of the federal government or may be self-governed. By contrast, neighboring communities outside the reservation are more likely to be answerable first to state and provincial regulators. Historical nonwater issues, such as unresolved land claims, may also influence aboriginal attitudes and decision making in a watershed.

Native self-government is an issue of increasing importance in water management institutions. Some aboriginal groups have expressed interest in, and jurisdiction over, the setting of water quality standards on aboriginal lands. Although these issues are rarely fully resolved, they raise questions about jurisdiction and harmonization of standards that may become contentious and divisive in watershed management.

3.3   SETTING TARGETS FOR FUTURE USE

The next step in identifying water management issues within a basin is to develop targets for each water use. A comparison between existing conditions (as revealed by the watershed inventory; see Chapter 2) and ideal conditions (discussed in this section) can indicate areas where beneficial uses are currently impaired and restoration may be desirable.

This process should involve the setting of targets not only for current uses but also for any envisioned over the planning horizon, which may be 10, 20, or 50 years into the future. An area may, for example, hope to encourage the establishment of a particular kind of industry in the future; the likelihood of the industry's actually settling in the basin may depend very much on the quality and quantity of water available for its use.

In some cases the ideal conditions for two water uses will differ and may even be incompatible, as in the conflict between navigation authorities and private landowners over water levels. These conflicts, if not already apparent, will likely emerge in public consultation and may be resolved through conflict resolution processes, by mediation, or even by unilateral government decision (see Chapter 4). At the scoping stage (see Section 3.4), it is more important to understand where each beneficial water use is currently, or may in the future be, constrained by existing water conditions.

The following sections provide an overview of how targets are set for major water uses.

3.3.1 Water Quantity

Water quantity targets are closely linked to estimated water demand, whether for withdrawal, consumptive, or nonwithdrawal uses. This topic has already been covered in some detail in Section 3.1.2. The question of whether or not the watershed has, or will have, enough water for all intended uses is fundamental to watershed planning. Indeed, it is often a water supply crisis that prompts concerted water management planning. In setting targets for water quantity, the water manager must be able to answer several key questions:

1. What proportion of reliable annual flows is currently required to meet withdrawal demands and service nonwithdrawal uses?
2. What proportion of estimated withdrawals is consumed, that is, not returned to the hydrologic system for reuse?
3. How are these supply-to-demand ratios likely to change in the future in response to population growth, consumption rates (which may be related to various factors including per capita income and conservation policies), or climate change?

The answers to these questions will provide a preliminary assessment of whether available flows are adequate to service existing and projected needs, and the estimated excess (if any) of supply over demand.

3.3.2 Water Quality

As discussed in Chapter 2, the term *water quality* is necessarily a subjective one, interpreted differently by different agencies and user groups. Over the past quarter-century, the United States and Canada have often worked together in setting water quality targets, most commonly through the Canada-U.S. International Joint Commission (IJC), a binational body set up under the Canada-U.S. Boundary Waters Treaty of 1909. (There is also a U.S.-Mexico Border Environment Cooperation Commission (BECC), established in 1993, whose goals relate to the coordination and implementation of environmental infrastructure projects under the North American Free Trade Agreement (NAFTA). BECC is not currently involved in the development of binational water quality standards.)

The IJC is an organization whose mandate and structure make it almost unique. Charged with oversight of the boundary waters, the IJC has for decades provided a forum in which scientists and water managers from both countries can meet and discuss technical issues of joint interest. One such has been the setting of water quality objectives, most recently contained in the 1987 revision of the Canada-U.S. Great Lakes Water Quality Agreement (IJC 1987). Table 3.6 gives some of these objectives, which were formally ratified by the two countries on November 18, 1987.

Over the past decade, however, there has been less agreement between the two countries, and between these countries and others in the world, on what water quality objectives are appropriate and manageable. In the United States, for example, the Great Lakes Initiative strove to set a full range of water quality targets, some (but not all) more stringent than those in the Great Lakes Water Quality Agreement.

Possibly the greatest disagreement between jurisdictions exists in the area of targets for trace substances such as heavy metals and industrial organic compounds. Different jurisdictions have very different approaches to this task, in part because ambient levels of the substances are so low and toxicity is so difficult to define. One jurisdiction may define *toxic* as causing an observable effect such as a tumor or lesion, while another may include more subtle effects, such as reduced reproductive success and changes in perception and behavior, in the definition of toxicity.

A further complication arises in that different jurisdictions employ different bases for their target setting. Some set "maximum allowable" levels, others set targets for ideal conditions, some for drinking, some for the protection of aquatic biota, and so on. As a result, few standards or targets are directly comparable; rather, each must be interpreted through its rationale and intended application.

Finally, there is a difference between water quality *standards*, which are legally binding requirements (for instance, for drinking water supply in some systems), and water quality *targets*, *objectives*, *criteria*, or *guidelines*, which carry no weight under the law and are intended to reflect an ideal condition. Where standards are in place, they must be met or those responsible for meeting them will suffer sanctions. If only guidelines are available, water managers are expected to work toward them with the understanding that achieving targets may take some years. In some cases multitier systems exist, with a minimum water quality level specified as a standard and more protective (but harder to achieve) levels designated as objectives. Some health-related criteria specify

Table 3.6 Selected Water Quality Objectives Under the Great Lakes Water Quality Agreement (Annex 1) (IJC 1987)

Substance	Specific Objective
Dissolved oxygen	In the connecting channels and in the upper waters of the lakes, the dissolved oxygen level should not be less than 6.0 milligrams per liter at any time; in hypolimnetic (bottom) waters, it should be not less than necessary for the support of fish life, particularly cold water species.
pH	Values of pH should not be outside the range of 6.5 to 9.0, nor should a discharge change the pH at the boundary of a limited use zone by more than 0.5 units from that of ambient waters.
Phosphorus	The concentration should be limited to the extent necessary to prevent nuisance growths of algae, weeds, and slimes that are or may become injurious to any beneficial water use.
Settleable and suspended solids and light transmission	For the protection of aquatic life, waters should be free from substances attributable to municipal, industrial, or other discharges resulting from human activity that will settle to form putrescent or otherwise objectionable sludge deposits or that will alter the value of Secchi disk depth by more than 10%.
Microbiological constituents	Waters used for body-contact recreation activities should be substantially free from bacteria, fungi, or viruses that may produce enteric disorders or eye, ear, nose, throat, and skin infections or other human diseases and infections.
Lead	The concentration of total lead in an unfiltered water sample should not exceed 10 micrograms per liter in Lake Superior, 20 micrograms per liter in Lake Huron, and 25 micrograms per liter in all remaining Great Lakes to protect aquatic life.
Zinc	The concentration of total zinc in an unfiltered water sample should not exceed 30 micrograms per liter to protect aquatic life.
Mercury	The concentration of total mercury in a filtered water sample should not exceed 0.2 microgram per liter, nor should the concentration of total mercury in whole fish exceed 0.5 microgram per gram (wet weight basis) to protect aquatic life and fish-consuming birds.
Polychlorinated Biphenyls (PCBs)	The concentration of total polychlorinated biphenyls in fish tissues (whole fish, calculated on a wet weight basis) should not exceed 0.1 microgram per gram for the protection of birds and animals that consume fish.
DDT and metabolites	The sum of the concentrations of DDT and its metabolites in water should not exceed 0.003 microgram per liter. The sum of the concentrations of DDT and its metabolites in whole fish should not exceed 1.0 microgram per gram (wet weight basis) for the protection of fish-consuming aquatic birds.

one level for children and another for adults; others give a short-term exposure limit and a longer, or chronic, exposure limit.

Table 3.7 gives some examples of the variations in water quality objectives used by different jurisdictions.

The wide variation in water quality targets, their rationales, and the manner in which they are applied makes it imperative that water managers determine applicable standards and objectives for their regions of interest. This information is most easily obtained from federal, state, or provincial pollution control agencies.

3.3.3 Fisheries

Fisheries activity in a watershed may be commercial (in which case it is likely an important basin employer and economic force as well), recreational, or both. Fish stocks are also a good measure of overall aquatic ecosystem health. Generally speaking, fisheries are assessed through measurements of fish community composition, both in terms of species and in terms of age and sex distribution (see also Section 2.7). This information may then be compared against some ideal condition, including both species and numbers of individuals. Often (but not always), the ideal will be that which was believed to have existed before widespread human settlement and disturbance of the natural environment. But because fish are living organisms not simply responsive to chemical and physical change, the challenge in fisheries management and fisheries restoration is to achieve not only species presence (which can be maintained by artificial stocking), but a self-sustaining fishery.

Setting targets for watershed fisheries therefore encompasses two primary elements—desired species composition and desired stocks for each species—but also secondary considerations in terms of the physical, chemical, and biological conditions required to support a self-sustaining population of each desired species. Physical conditions include aspects such as water depth, temperature, and the availability of appropriate spawning and nursery habitat (often degraded by excessive siltation in areas with dense urbanization or intensive agriculture). Good chemical condition implies an absence of compounds that are acutely or chronically toxic to fish; biological requirements include an abundance of healthy prey species.

Federal and state/provincial natural resource agencies are excellent sources of information about the current status of fisheries in the region of interest, and previously identified issues in fisheries management. The British National Rivers Authority has published a series of "strategy" documents (NRA 1993a), including one on fisheries throughout England and Wales. Table 3.8 illustrates the major components of that strategy.

In addition to setting targets for fisheries restoration, the NRA Fisheries Strategy illustrates several features essential to watershed planning. First, it has a general goal or vision toward which to strive. (In some systems, this goal might be more explicit—for instance, itemizing ideal species composition and popula-

Table 3.7 Differences in Standards for Major Water Quality Parameters

Parameter	Jurisdiction	Standard Type	Limit
Total Dissolved Solids	World Health Organization	Drinking water— guideline level	1,000 mg/L
	U.S. Environmental Protection Agency	Drinking water	500 mg/L
	Health and Welfare Canada	Drinking water	500 mg/L
	State of New Jersey, USA	Groundwater protection (aesthetic)	100 mg/L
Total Phosphorus	European Economic Community	Drinking water— maximum admissable concentration	5,000 mg/L
	European Economic Community	Drinking water guideline	400 mg/L
	Ontario Ministry of Environment and Energy (MOEE) (Canada)	Provincial water quality guideline (ambient)	lakes—20 mg/L rivers—30 mg/L
Total Zinc	U.S. Environmental Protection Agency	Drinking water	9,000 mg/L
	Health and Welfare Canada	Drinking water	5,000 mg/L
	State of New York, USA	Ambient	300 mg/L
	European Economic Community	Drinking water guideline	100 mg/L
Nitrate	European Economic Community	Drinking water guideline	25 mg/L
	U.S. EPA	Drinking water	10 mg/L
	Health and Welfare Canada	Drinking water	10 mg/L
Total DDT	National Academy of Science (USA)	Drinking water— suggested "no adverse response" level	0.083 mg/L
	Ontario MOEE	Drinking water— maximum acceptable concentration	0.030 mg/L
	State of New York, USA	Ambient	0.010 mg/L
	U.S. Environmental Protection Agency	Ambient	0.024 ng/L
	World Health Organization	Drinking water— guideline level	0.001 mg/L

(*Source:* Ontario Ministry of the Environment and Energy Accessible Standards Information System, 1994)

Table 3.8   Key Elements of the NRA Fisheries Strategy

Basic strategy goal: to maintain, improve and develop fish stocks, in order to optimize the social and economic benefits from their sustainable exploitation.	
Number of native species of freshwater and estuarine fish	>50
Number of introduced species of freshwater fish	>10
Length of river fisheries	c. 19,000 km
Length of canal fisheries	>2,000 km
Number of still water fisheries	c.15,000
Area of coastal waters	3.6 million ha
Number of prosecutions for fisheries offences	c. 5,500/year
Current impairments	• Declining salmon and sea trout catches • Indiscriminate stocking affecting genetic integrity of native stocks • Rise in coarse fish species and related diseases • Declining elver (juvenile eel) catches
Major sources	• Poor water quality due to changing land use • Excessive abstractions of water aggravating natural low flow conditions • Land drainage works and channel straightening • Illegal netting (poaching) • Escape of nonnative species from fish farms • Obstacles to migrating fish from in-channel works like barrages
Key indicators of progress	• Number of licensed fishermen • Number of license checks made • Number of fisheries offenses detected • Number of monitoring surveys undertaken • Number of habitat improvement structures built • Number of fish reared and stocked • Salmon and sea trout catches • Coarse fish abundance • Juvenile salmonid fish abundance

(*Source:* NRA 1993a)

tion sizes.) Second, it identifies areas of current fisheries impairment (problem definition). Third, it identifies a number of causes of impairment, a step that will help in developing effective management actions. And finally, it lists "key indicators," some very specific, that it will use to monitor progress in maintaining, improving, and developing fisheries. (The strategy also includes a discussion of the legislative, financial, and institutional framework within which fisheries management is practiced; such factors are discussed in more detail in Chapters 8 and 9.)

Together, these elements provide the NRA with a management framework that is implicitly dynamic. That is, the organization is able to measure the status of fisheries, on a continuous basis if desirable, and compare progress against targets. Regular feedback also allows it to modify or refine management actions in response to changing conditions.

3.3.4 Ecology and Conservation

Virtually every watershed subject to management scrutiny has experienced a long history of alteration through human activities. Yet many, perhaps most, watersheds contain features that are rare, of particular scientific interest, sensitive to disturbance, or simply valued as relics of predisturbance times. Sometimes an area is valued simply because it is beautiful. These elements properly form part of a management strategy, but setting targets for them is a more challenging task than setting water quality objectives or desirable water depths.

One commonly expressed target for conservation uses is simply "more"—that is, to protect more watershed area from further disturbance. The difficulty with such a goal is that its vagueness may allow it to be overridden by more explicit, and thus apparently more urgent, watershed goals. Yet even this simple approach is useful in establishing a benchmark against which progress can be measured: any improvement is a step toward the goal of "more."

An alternative approach is to divide the broad concept "conservation" according to separate and distinct uses. These are likely to include some or all of the following (some may overlap):

- Rare or endangered species and/or habitats
- Valued historical or archeological remains
- Areas of great natural beauty
- Habitats especially susceptible to disturbance, such as wetlands or flood plains
- Designated park lands
- Nature reserves
- Scientific research stations

Having established which of these uses is present and valued in the basin, targets can be set by deciding, perhaps based on community consensus, whether existing resources are adequate or additional resources are needed. For example, the community may decide that an existing park land area of 50 km² is adequate for current uses but should ideally be expanded to 75 km² over the next 20 years. The target of 75 km² then becomes a measurable goal against which progress can be judged over time. Similarly, a manager could set explicit targets, for instance, to protect a given area of sensitive habitat in the future, or a certain list of historical monuments. It is important to remember that the definition of these various uses may change with time. What today appears an innocuous wood structure may tomorrow be revealed as an early example of pioneer architecture, and the definition of "wetland" may change according to prevailing scientific views.

Several major environmental management agencies, including the United States Fish and Wildlife Service, Environment Canada, and the English National Rivers Authority, have implemented management programs based on inventories (e.g., of various classes of wetland) and have set targets to protect or enhance specified additional resources in each category. An advantage of this approach is that resource-specific targets lend themselves more easily to public education and outreach programs, including the land stewardship programs currently promoted throughout North America and Europe.

3.3.5 Navigation

Shipping and boating are essential uses in many watersheds. Indeed, recreational boating may be one of the most widely practiced water sports in some areas, with demand steadily increasing. Yet the terms "shipping" and "boating" cover a vast range of activities. Setting targets for these uses requires careful consideration of the water craft using local waterways and the uses to which they will be put. Lund et al. (1995) discuss the problem of developing dredging schedules for the lower Mississippi River, because that river has "uncertain and varying channel bottom elevations (due to sedimentation and scour), river stages, and vessel traffic of various drafts." Commercial navigation may be constrained at a given depth-limited crossing by:

- Uncertain draft
- Cargo
- Destination
- Each vessel's light-loading, delay, lightering, and diversion options

Any or all of these factors may contribute to the costs of delayed or avoided navigation. For commercial shipping and navigation, the issue of certainty of draft is as important as estimated draft itself. Water managers must therefore set goals for commercial navigation that reflect the expected maximum draft of relevant vessel classes, the required duration of the navigation season, and the required certainty of draft and related considerations such as turning area.

Typical categories of recreational use include:

- Open-water boating, including motor boating, sailing, and windsurfing
- Still-water boating, including canoeing
- White-water boating, including rafting and kayaking

Each of these uses is also constrained in some way by water conditions such as channel depth, channel width, turbulence, availability and condition of locks, and similar features. Here again, the requirements of individual categories of use can form the basis for specific use-restoration targets.

Finally, some navigable waterways are administered by navigation authorities whose requirements must be met in any watershed plan. Other lakes and streams are navigable in some sense but are not overseen by a special authority.

3.3.6 Hydroelectric Power Generation

In many systems electric power generation is one of the largest water users, both in terms of volume of water withdrawn and in terms of the economic value of the products of that water use. Hydroelectric power imposes special requirements on water management; as McDonald and Kay (1988) have said, geography may, in fact, be the most serious limitation on the development of hydropower.

Until very recently the limitations of hydroelectric power generation technology required a minimum head of about 7 m for a viable installation. Over the past 15 years technologies employing heads as low as 5.2 m have been used in the Mississippi River basin (Keevin 1982), and newer, "micro-head" low-capacity (18 to 125 kW) stations in China use only about 2 m as the smallest head (McDonald and Kay 1988). According to these authors, only a few nations in the world possess the right combination of flow and topography for hydroelectric power, but for these nations this water use is of primary importance. In order of importance, based on production capacity under *average* flows (so-called G_{av}), they are China (with 13.5% of global hydroelectric power capacity), the countries of the former USSR (together 11.1%), the United States (7.2%), Zaire (6.7%), Canada (5.5%), and Brazil (5.3%). If capacity is calculated based on flows that obtain 95% of the time (so-called G_{95}), the order of importance and capacities are somewhat altered: Zaire (13.97%), China (10.84%), countries of the former USSR (together 9.02%), Brazil (8.58%), Canada (5.45%), and the United States (4.89%).[*]

Geography is not the only consideration in siting hydroelectric installations. Upstream development can affect river sedimentation and flow regimes, with

---

[*]Total capacity for the six nations is much reduced with G_{95}: 15.96 exajoules as compared with 35.3 exajoules for G_{av}. An exajoule, 10^{18} joules, is the amount of energy that would power a domestic 1 kW heater for 3 million years.

detrimental effects to river hydrology and reservoir capacity. Even without examination of the many environmental implications of dams and inundation, it is clear that there is a need to determine the desired capacity of any hydroelectric facility in the planning area, the necessary average or 95% frequency flows to maintain that capacity, and the present and future risks of changes to flows that may result from upstream activities. These factors will be important performance targets for any watershed plan incorporating hydroelectric power generation.

3.3.7 Recreation

The vast range of potential recreational activities makes it difficult to develop a generalized list of targets for ideal recreational use. Furthermore, the requirements of one use, such as bird-watching, will differ from those of other uses, such as hiking and picnicking.

Generally speaking, recreational water uses can be broken into several categories, which can be helpful in determining targets for restoration. Even though these distinctions may seem arbitrary, they provide useful insight into the ways that water-related resources are, or could be, used.

Distance-Related Activities Activities such as hiking, mountain biking, off-road vehicle driving, and bicycling usually require some sort of prepared trail or path. The costs of building and maintaining such trails is usually proportional to their length. Similarly, the extent of opportunity for these activities—whether an existing condition or some future ideal—can be measured in terms of kilometers of available trail.

Area-Related Activities Other activities, such as habitat preservation or simple nature enjoyment, require a certain area in order for them to be effective. (The term *area* may be qualified by some characteristic related, for instance, to habitat type: e.g., so many square kilometers of native dune habitat.) Land area can therefore be used as a direct measure of the "quantity" of such activities currently available or planned for the future.

User-Days The term *recreation* implies infrequent leisure or hobby use for activities such as bird-watching or recreational fishing. The number of available user-days can therefore be a good measure of the current state of the resource or of some desired future conditions.

3.4 SCOPING THE PLAN

3.4.1 The Meaning of *Scoping*

This chapter has developed a simple framework for identifying ideal water conditions in a watershed:

1. Identify the categories of water users in the basin.
2. Determine desired water uses.
3. Set short- and long-term goals (targets) for each use.

The challenge now is to determine which uses and watershed areas require management intervention and which do not. This process is often called "scoping" the watershed plan.

Scoping has two important components: boundary setting and focusing. Boundary setting refers to the conscious limitation of the plan to a specified geographic area, a particular time period, and, often, a small number of species or populations of particular interest. Focusing is the identification of issues that have overriding importance in the basin and should therefore be treated in depth in the planning exercise.

Scoping can be a more challenging process than may be apparent, partly because of differences in perception and priority among participants. It is not usually difficult, for example, to generate a comprehensive list of use impairments (areas where beneficial uses are currently constrained by inadequate quality or quantity of water). Much more difficult is setting priorities among the various use impairments. Which urgently demand attention? Which can be deferred until a later, perhaps more affluent, time? These are questions that must be resolved at the community level, through appropriate consultation processes, to avoid acrimony and reduced implementation success.

There are three keys to successful scoping:

1. A clear definition of each problem to be solved
2. Specific goals for restoration
3. Community consensus on the importance attached to each element of the plan.

Defining the "problem" may seem the simplest element of these three, but it may, in fact, be the most difficult. The simplest and perhaps best approach may be to state problems in terms of use impairments, with specific attention to the timing and spatial extent of those impairments. When the problem is clearly defined, it should be relatively easy to identify suitable indicators of impairment or restoration and thus to define clear targets for solving the problem. The following sequence of questions may be helpful in defining problems.

1. What are all the use impairments presently observed in the basin (or anticipated under future growth conditions if long-term planning is involved)? What is the geographic extent of each impairment? (Create a long list.)
2. Why is each use considered to be impaired? What parameters do people use as criteria in deciding that the use is no longer viable? What standards or objectives for the use are not currently met?
3. Which of these parameters can be measured easily? For which are existing

data available? Which can be simulated using computer models or other methods?

4. Of the long list of use impairments developed in (1), which are the most important to the community? (Aim for a list of two to three but no more than five. These will become the focus of your planning initiative.)

5. What specific numerical targets do you wish to meet in order to consider an impaired to be restored (e.g., total phosphorus <0.02 mg/L)? (Be specific—identifying "bacteria" is not good enough; "fecal coliforms" is better, but "*E. coli*" is better still.)

6. During what time periods do you want these targets to apply? All the time? During the summer only? During dry weather only?

7. Over what area do you want these targets to apply? Over the full area currently impaired or some smaller portion of that area?

Clear and specific problem definition is also helpful in community debate. For example, it is not uncommon to hear statements to the effect that "industrial pollution" or "agricultural drainage" are problems in a watershed. These terms cover a wide range of activities and a variety of potential impacts. Typically, some of these activities do cause use impairment—but others do not. Framing the "problem" in these terms invites conflict without providing a practical basis for problem resolution: it is simply not clear what needs to be fixed. On the other hand, if the problem is described as beach closures due to elevated bacteria counts in rural areas, few observers would argue that a problem exists or that its source—and its solution—can be found in human or animal waste management practices.

The targets developed in Section 3.3 provide the goals toward which the plan is directed. For each problem (impaired use), it should be possible to state a clear goal, as well as a clear statement of the existing condition as a benchmark of impairment. Impaired uses can rarely be restored with a few months' activity, so this approach also creates a framework for periodic reporting of progress toward the goal (this idea will be discussed more fully in Section 5.5).

The final component of successful scoping is community consensus about the relative importance of the various problems identified in the basin. The process by which this consensus can be achieved is discussed in more detail in Chapter 4. One approach that has been used to develop consensus in scoping is Adaptive Environmental Assessment and Management (AEAM). This technique is described in Section 3.4.2.

3.4.2 Examples of Scoping

Adaptive Environmental Assessment and Management (AEAM) in the Latrobe River Basin, Australia AEAM is both a philosophical and a methodological framework aimed at using existing knowledge of an environmental system to develop sound management approaches for it (cf. Holling

1978). Like many other management processes, it centers on the development of a computer simulation model of the system and the application of the model in assessment of alternative management strategies.

A key feature of AEAM, consistent with the idea of community consensus, is use of a series of structured workshops to collect the information needed for the simulation model and to develop the model itself. People with a range of interests attend these workshops, including those involved in policy development, management, and technical issues, and lay people.

Workshop participants undertake a series of structured activities to define the following elements:

1. Model scope, including possible management actions to be simulated
2. Indicator variables of the system, to test the effectiveness of management options
3. The required spatial scale of the model
4. The simulation time step (model time interval at which conditions will be recalculated; e.g., daily, seasonal, annual) and overall period of simulation (total time period to be simulated; e.g., months of April, May, and June 1993 or 20-year period 1967–1986)

As the process continues, subgroups are formed to examine more detailed technical and management issues and submodels. Eventually the various submodels are linked, and the integrated model is tested and validated by the group. Alternative management strategies are then evaluated, using the model in a process called "gaming" (essentially, trying different strategies and responses to determine which is most effective). The model can become a tool for ongoing management in the basin, and the planning process a semicontinuous one.

Grayson et al. (1994) applied AEAM to the management of the Latrobe catchment in Victoria, Australia. These authors note that the direct involvement of a range of stakeholders in model development demystifies computer simulation and ensures that outputs are understood by all. They also note that:

> The computer model is the tangible outcome of the process, but the modeling workshops are of primary importance. They are a highly efficient medium for the accumulation of information about the system and require participants to focus clearly on problems and achievable solutions from the outset. By choosing the temporal and spatial scales at the start, it is possible to avoid the temptation to "over-model" those aspects about which there is detailed information and ensure that the complexity of sub-models is in keeping with the available data and the questions to be asked of the final model.

Table 3.9 shows the possible management actions developed by the Latrobe workshops and the water-quality indicator variables they would predominantly affect. (It should be noted that in the AEAM system management options are developed more or less at the same time as indicator variables, through

Table 3.9 Management Actions and Water Quality Indicators for the Latrobe Catchment AEAM Study (adapted from Grayson et al. 1994)

Management Action	Water Variable Affected
1. *Management of point source pollution*, including review of existing discharges, treatment of urban wastewater, land disposal of wastes, ocean disposal of wastes, promotion of "best management practices" in industry, improved domestic sewage systems, and introduction of a "polluter pays" system or transferable discharge entitlements	Water quality variables of point source inputs and receiving waters
2. *Management of diffuse source pollution*, including land use controls, adoptions of sustainable agricultural practices, control of extractive industries, forestry controls, erosion controls for construction sites and roads, retention ponds for urban, industrial, and rural runoff, vegetation conservation and revegetation, regulation of dairy waste disposal, encouragement of recycling/reuse of water, establishment of buffer zones (revegetation of riparian zones and stream banks, controlling grazing, controlling access), and policy on leases of streamside reserves	Water quality variables of surface runoff and in-stream habitat variables
3. *Management of river flows*, including storage operations (flow regimes, environmental allocations, rates of release, hydropower operation), diversions (within catchment and between catchments), controls on farm dams, water recycling, regulation of wastewater discharges, introduction of new pricing policies, groundwater use, desnagging, and meander reinstatement	Streamflow variables and in-stream habitat variables
4. *Management of the in-stream environment*, including streambed and bank stabilization, willow control, and establishment of macrophytes	River water quality and in-stream habitat variables

the workshop process. This approach differs somewhat from that described in Chapter 5, on development of workable management options, but reflects the same societal consensus as to problems and indicators and thus achieves the same end as more structured processes. It can be argued that the AEAM process, while less systematic or even perhaps "scientific," has the definite advantage of developing strong community support and tapping a wide range of information sources in the community.)

The Latrobe AEAM process took this analysis a step further to set priorities among the various indicator variables and to reject some as not significant or as having insufficient data. Table 3.10 presents the results of these decisions.

Although the Latrobe AEAM process identified a large number of possible management actions and a wide range of "problems" (Table 3.9), ultimately the model encompassed only total suspended solids, total dissolved solids, total phosphorus, macroinvertebrate community index, and index of potential fish habitat. This scope reflects the decisions of a great many participants over a

Table 3.10 Indicator Variables Used in the Latrobe Catchment Planning Process (adapted from Grayson et al. 1994)

Category	Variable	Priority	Comments
Hydrology	Streamflow	High	
	Surface runoff	High	
	Reservoir levels	High	
	Groundwater depth	Not included	Not important to the catchment behavior
	Aquifer pressure	Not included	Not important to the catchment behavior
	Wetland water depth	Not included	Outside the catchment
	Period of wetland inundation	Not included	
Water Quality	Suspended solids	High	
	Phosphorus (soluble or attached)	High	
	Total dissolved solids	High	
	Total nitrogen	Low	
	Chlorophyll A	Low	
	Temperature	Not included	Not significant
	pH	Not included	Not significant
	BOD	Not included	Insufficient data
	Color	Not included	Not significant
	Dissolved metals	Not included	Insufficient data
	Total organic residues	Not included	Insufficient data
	Total hydrocarbon residues	Not included	Insufficient data
	Bacteria levels	Not included	Insufficient data
Ecology	Macrophytes	Not included	Insufficient data
	Diatoms	Not included	Insufficient data
	Macroinvertebrates	High	Community index used
	Native fish	Not included	Insufficient data
	Introduced fish	Not included	Insufficient data
	Platypus	Not included	Insufficient data
	Potential fish habitat	High	
Economic	Agricultural productivity	Not included	Insufficient data
	Industrial productivity	Not included	Insufficient data
	Land prices	Not included	Insufficient data
	Employment	Not included	Insufficient data
	Production costs	Not included	Insufficient data
	Gross regional product	Not included	Insufficient data

series of workshops and thus can be said to represent community consensus. Model calibration, validation, and use in scenario testing vary somewhat from traditional approaches, the AEAM process employing user judgment (where the user is a group of participants), rather than optimization of some objective function. AEAM dictates that participants in fact "never take a hard copy of the model output" (Grayson et al. 1994), reinforcing the view that AEAM is essentially a consensus-building process centered on a predictive model and its scope. As such, AEAM provides a framework within which basin residents can meet, share information and opinions, and explore the utility of different management approaches. In the context of this discussion, AEAM meets all the requirements of successful scoping: a clear definition of each problem to be solved, specific goals for restoration, and community consensus on the importance attached to each element of the plan.

The Grand River Basin Water Management Study and the Don River Water Management Strategy
The Grand River and the Don River flow through southern Ontario into Lake Erie and Lake Ontario, respectively. In the early 1980s watershed management plans were prepared for these systems, with participation in both by many of the same agencies and individuals (Grand River Conservation Authority 1982; Ontario Ministry of the Environment 1987). Yet the impact of the two studies differed greatly, in part because of differences in scoping.

The Grand River flows through southwestern Ontario, passing through large areas of intensive agriculture and five major cities before draining into Lake Erie. The river once supported a cold-water trout fishery, but in recent years only less desirable warm-water species have been present. A major impediment to the restoration of the fishery has been summertime low dissolved oxygen levels, thought to result from excessive plant growth encouraged by high nutrient concentrations. High phosphorus levels in sewage treatment plant discharges were also of concern to local regulators, who were considering limiting development in the basin until effluent quality could be improved. A separate set of concerns centered on persistent flooding and high flood damages in some communities along the river.

The proponent of the Grand River Basin Water Management Study was the Grand River Conservation Authority (GRCA), a basin management agency that operates at the provincial level with the cooperation of local municipalities. One of GRCA's first tasks in this study was to set up a two-tiered advisory committee structure. The senior advisory committee consisted of local decision makers, including local politicians and senior government bureaucrats. A second tier of technical committees provided specialized advice on topics ranging from hydrology to economics. These committees were unusual at that time in that they included representation from the general community. The advisory committees were charged with developing a scope of work for the basin management plan, beginning with problem definition. Over a period of months they considered a wide range of documents from scientists, economists, and local residents. The issues they debated were also diverse, ranging from flooding, through degraded aesthetics (largely a result of algae blooms), to the presence of toxic organic compounds.

Ultimately, the advisory committees came to several important conclusions:

1. That several of the problems were interrelated
2. That available resources would allow them to solve only three or four of the most pressing problems.

3. That the problems most urgently in need of correction were flooding, dissolved oxygen depletion, and elevated phosphorus concentrations in the stream.

4. That solving these problems would likely have some indirect benefit for other issues not directly considered in the plan

Using a purpose-built dynamic simulation model, the advisory committees developed and tested a number of management alternatives that would have direct benefit for flood control, would raise in-stream oxygen levels, or would lower in-stream phosphorus. Some of these alternatives, notably a proposal to build a major dam and inundate a large area of farmland, were fraught with controversy. Yet through a series of regional consultations and direct involvement of community members in decision making, even the most controversial plans were evaluated and screened without acrimony. In the end, the committees chose a medium-cost alternative that included sewage works upgrades at several locations and a variety of other channel improvements (but no dam). These improvements were all in place by the mid-1980s, and local municipalities continue to collaborate with GRCA as the plan and the simulation model are continually updated to reflect changing conditions and technology.

The Don River Water Quality Management Strategy was a project of the Ontario (Canada) Ministry of the Environment, in partnership with the Metropolitan Toronto and Region Conservation Authority (MTRCA), a group like GRCA charged with managing the several river basins in the Toronto area. The problem was clearly seen to be frequent closures of swimming beaches along the Toronto waterfront. These closures were ordered by the Toronto Medical Officer of Health because of high concentrations of indicator bacteria, especially fecal coliforms, and the implicit human health risk these created. There were several sources of the bacteria to the waterfront, primarily the number of storm and combined sewer overflows and bypasses from the Main Toronto Sewage Treatment Plant.

Targets for solving the problem were, however, less clear. Some people thought that the swimming beaches ought to be open continuously throughout the May-to-October swimming season. Others felt that it would be enough to have beaches open within 48 hours after a rainfall event, thus allowing time for wet-weather discharges to dissipate in the near-shore zone. Still others felt that swimming was not an appropriate use for an urban waterfront with a high level of industrial activity and storm-related discharges.

The scoping process for this plan attempted to involve a range of interested parties drawn from the six local municipalities that make up the metropolitan Toronto area. In the early 1980s, the Ontario government, like many bureaucracies, tended to discourage, or at least discount, citizen involvement in public decision making. So the participants in the Don River plan were drawn from municipal engineering and health departments, from interested federal agencies, and from Ministry of the Environment staff. It was expected that this group (which, as in the Grand River study, was arranged in a two-tier committee system) was close enough to the industrial, commercial, and residential interests in the basin to represent them effectively.

In fact, the Don River study made several important mistakes. By excluding nonagency representatives from the process, the government effectively cut off access to a range of interests and solutions that might have helped to inform the plan and clarify its scope. Although the problem was clear, the targets for remediation were not. As a result, project staff spent a vast amount of time and money developing and testing solutions that might or might not be feasible. In the end, no accurate simulation model of the basin was ever developed, and no AEAM-type qualitative analysis was possible because of the lack of consensus on key issues. The final plan consisted of a long and costly list of solutions that would undeniably improve waterfront water quality. There was, however, little consensus on who could or should pay for these, or on which could or should be done first.

The Don River study remains to this day a paper exercise that has never been implemented. Individual local municipalities continue to proceed, as they always have, with their own sewer management plans. Beaches continue to be closed throughout much of the summer. In the late 1980s, the Don River situation was revisited through the International Joint Commission's Remedial Action Plan (RAP) process, the Toronto waterfront being one of the IJC's Areas of Concern under that program. From that time until the present, the RAP process has been engaged in redoing much of the work of the Don River study (as well as plans for several neighboring basins). Section 4.3 describes the RAP process and committee structure, which involves a large and active public advisory committee and extensive public outreach activities. With the growing awareness effected by the committee and its outreach, the RAP is achieving a higher level of public consensus about the Don's problems and solutions than was achieved a decade earlier and, thereby, giving politicians public support for spending money on plan implementation.

Although the Grand and Don Rivers are separated by only a few tens of kilometers, their basin management experience has differed widely. The problem of scoping, and community buy-in to that scoping, lies at the heart of these differences and explains much of their failure, or success, in creating implementable solutions.

REFERENCES

Alpert, P. 1993. Conserving biodiversity in Cameroon. *Ambio* 22(1): 44–49.

Beard, D. P. 1996. Keynote address, Conserv96 Conference: Responsible Water Stewardship. Orlando, Florida, January 4, 1996.

Eckenfelder, W. W., Jr. 1970. *Water Quality Engineering for Practicing Engineers.* New York: Barnes and Noble.

Goodman, A. 1984. *Principles of Water Resources Planning.* Englewood Cliffs, N.J.: Prentice-Hall.

Grand River Conservation Authority. 1982. *The Grand River Basin Water Management Study.* Summary Report. Cambridge, Ont.: Grand River Conservation Authority, in conjunction with the Ontario Ministry of the Environment.

Grayson, R. B., J. M. Dooland, and T. Blake. 1994. Application of AEAM (Adaptive Environmental Assessment and Management) to water quality in the Latrobe River catchment. *J. Envir. Management* 41: 245–258.

Heathcote, I. W. 1993. An integrated water management strategy for Ontario: Conservation and protection for sustainable use. In *Environmental Pollution: Science, Policy and Engineering*, edited by B. Nath, L. Candela, L. Hens, and J. P. Robinson. London: European Centre for Pollution Research, University of London.

———. 1995. Conflict resolution in Ontario water resources policy. In *Water Quantity/Quality Management and Conflict Resolution*, edited by A. Dinar and E. T. Loehman. Westport, Conn.: Praeger.

Heathcote, I. W., H. R. Whiteley, and K. E. Morrison. 1996. *Structuring an Efficient Water Management System: Case studies from France, England, and Ontario (Canada).* Proceedings, American Water Works Association Conserv96 Congress, Orlando, Florida, January 4–8, 1996. Denver: AWWA.

Holling, C. S. 1978. *Adaptive Environment Assessment and Management.* Chichester, U.K.: John Wiley & Sons.

Hutchings, J. A., and R. A. Meyers. 1994. What can be learned from the collapse of a renewable resource? Atlantic cod, *Gadus marhua*, of Newfoundland and Labrador. *Canadian J. Fish. Aquat. Sci.* 51.

International Joint Commission. 1987. Great Lakes Water Quality Agreement of 1978 (revised, as amended by Protocol signed November 18, 1987). Windsor, Ont. International Joint Commission.

Keevin, T. M. 1982. The Corps of Engineers planning process as it relates to the assessment of the environmental impacts of low head hydro-power development. In *Energy Resources and Environment*, edited by S. W. Yuan. London: Pergamon.

Lauria, D. T., and C. H. Chiang. 1975. Models for municipal and industrial water demand forecasting in North Carolina. Raleigh: Water Resources Research Institute of the University of North Carolina and the North Carolina State University.

Lund, Jay R., Vini Vannicola, David A. Moser, Samuel J. Ratick, Atul Celly, and L. Leigh Skaggs. 1995. Risk-based decision support for dredging the lower Mississippi River. In *Integrated Water Resources Planning for the 21st Century*, edited by M. F. Domenica. Proceedings of the 22nd Annual Conference. American Society of Civil Engineers, Water Resources Planning and Management Division, May 7–11. New York: American Society of Civil Engineers.

McDonald, Adrian T., and David Kay. 1988. *Water Resources Issues and Strategies.* Harlow, U.K., and New York: Longman Scientific and Technical.

Metcalf and Eddy Inc. 1991. *Wastewater Engineering: Treatment, Disposal, and Reuse.* 3d ed. New York: McGraw-Hill.

MISA Advisory Committee. 1991. *Water Conservation in Ontario Municipalities: Implementing the User Pay System to Finance a Cleaner Environment.* Technical Report. Ontario: Ontario Ministry of the Environment.

National Rivers Authority. 1993a. *National Rivers Authority Strategy* (8-part series encompassing water quality, water resources, flood defense, fisheries, conservation, recreation, navigation, research and development). Bristol, U.K.: National Rivers Authority Corporate Planning Branch.

National Rivers Authority. 1993b. *Upper Wye Catchment Management Plan Consultation Report.* Cardiff, Wales: National Rivers Authority, Welsh Region.

Ontario Ministry of the Environment. 1987. *The Don River Water Quality Management Strategy.* Toronto: Ontario MOE.

Ontario Ministry of the Environment and Energy. 1994. Ontario Accessible Standards Information System. Toronto: Ontario MOEE.

Pearse, P. H., F. Bertrand, and J. W. MacLaren. 1985. *Currents of Change: Final Report of the Inquiry on Federal Water Policy.* Ottawa: Queen's Printer.

Richardson, M. 1993. Wrestling with the preservation of the Korup rain forest. *Our Planet* 5(4): 4–7.

van der Leeden, F., F. L. Troise, and D. K. Todd. 1990. *The Water Encyclopedia.* 2d ed. Chelsea, Mich.: Lewis Publisher.

World Resources Institute. 1993. *The 1994 Information Please Environmental Almanac.* Boston: Houghton Mifflin.

World Water 1989. Who pays what for water. *World Water* (Washington, D.C.), December.

4

The Consultation Process

4.1 THE NEED FOR PUBLIC INVOLVEMENT

A quarter of a century ago—not long in the life of regulatory institutions—public involvement was seen as unnecessary and superfluous, if not downright invasive, by many decision-making bodies. Even a decade ago, meaningful public involvement was rare in the development of public policy. Yet today the public voice is heard routinely and, indeed, is mandated by law in many jurisdictions.

What has caused this change? Seminal thinkers like Bruce Bishop suggested as early as 1970 that water resources planning is in fact a process of creating a program for social change (Bishop 1970). Implicit in this notion is the idea of consensus: social change will not occur unless the affected society agrees that change is necessary. Bishop emphasizes the importance of focus not just on the end product of planning, but on the planning process, "in order to produce a product which achieves a more widely accepted solution to the wants and needs of society." Goodman (1984) extends this idea in stating that a trend toward more open decision making is consistent with ethical behavior in democratic societies. Bishop believes that water development is also an *instrument* of social change because of its ability to alter social and economic growth patterns.

There is no question that the trend toward openness has accelerated over the past decade. In large part, this is due to heightened public concern about the environment and deteriorating public trust in government following debacles like the industrial organic contamination at Love Canal and the pollution-related fires on the Cuyahoga River. Public opposition to some projects has sometimes been fierce and prolonged, to the point of forcing abandonment or, at minimum, costly delays. Indeed, it was Congress's concern about such delays that prompted its introduction of action-forcing provisions in the National Environmental Policy Act of 1969, to the effect that:

> All agencies of the Federal Government shall ... identify and develop methods and procedures ... which shall ensure that presently unquantified environmental amenities and values may be given appropriate consideration in decision-making, along with economic and technical considerations (S. 102(2)(b)).

and a companion section, S. 102(2)(c):

> All agencies of the Federal Government shall ... include in every recommendation or report on proposals for legislation and other major Federal actions significantly affecting the quality of the human environment, a detailed statement ... on (i) the environmental impact ... (ii) adverse environmental effects which cannot be avoided ... (iii) the alternatives to the proposed action ... (iv) the relationship between local short-term uses ... and ... long-term productivity, and (v) any irreversible and irretrievable commitments of resources.

It is unlikely that Congress foresaw the impact that these few phrases would have on environmental management in the United States and Canada. Together, these brief provisions now form the basis of modern environmental assessment legislation in the United States and, indirectly, in Canada (Gibson and Savan 1987). Furthermore, their implicit message, that government must be able to anticipate the environmental impacts of its actions, forced a fundamental change in the ways that environmental information was gathered. Environmental assessment legislation generally requires consideration not only of a proposed project but of alternatives to that project, and thus forces a balanced and comprehensive data collection exercise that would have been unheard of before 1969. And the collection of such a broad database further requires consideration of a wide range of viewpoints, which can be assayed only by extending debate outside government, into the wider public. It can therefore be argued that with these few sentences Congress not only changed government's preparedness in advancing public projects, but vastly increased the transparency and accountability of the decision-making process.

With these provisions as a backdrop, later amendments to the U.S. Federal Water Pollution Control Act and similar legislation included clear mention of the need for public participation in water management, especially with respect to the setting of standards and priorities.

These changes in turn raised a need for guidance on exactly how the public was to be involved. This need was answered by a range of documents such as the U.S. Water Resources Council's *Principles and Guidelines* (WRC 1983) and the U.S. Army Corps of Engineers' Institute for Water Resources report *Public Involvement Techniques* (Creighton et al. 1982). More important, regulatory requirements and associated guidance documents steadily raised the public's awareness of opportunities for involvement in public policy-making. This trend, coupled with the increasing prominence of environmental issues in the media and on the public agenda, has continued virtually uninterrupted to the present day.

Although opportunities for public involvement come and go, public concern about environmental issues, perhaps especially water, remains higher than it has ever been, while public awareness of when and how to be heard continues to grow. Recent emphasis on government "downsizing" and deregulation has not reversed this trend: the various interested publics remain well informed, skeptical, and vocal.

It is perhaps unfortunate that public involvement through the 1970s and 1980s came to be viewed as adversarial: antigovernment and antiproject. Policymakers have also expressed concern about the lengthy delays and high costs occasioned by some public participation processes. Priscoli (1982) notes that as adversarial positions became entrenched, participants increasingly turned to litigation to resolve their conflicts, with the result that extreme positions solidified, progress slowed or halted, and costs escalated still further. In fact, Priscoli argues, this form of public involvement subverts the goals of open planning: information sharing, negotiation of reasonable trade-offs, and the creation of new approaches for the development of public policy. To regulators, public involvement may therefore appear inefficient and wasteful, and thus unnecessary. Nothing could be further from the truth. The following section describes how well-managed public involvement is an essential tool for determining community consensus on community issues and, thus, for meeting community needs.

4.2 PRINCIPLES OF CONSULTATION

4.2.1 Watershed Planning in a Pluralistic Society

In a democratic society, public decisions should reflect broad social values and changing policy should equally reflect changing values. Implicit in this is the simple notion of pluralism—decision making by the many, not the few. At the watershed level, the concept is equally valid: watershed planning, as a process to develop a program for social change, should reflect the values of a majority of the community. In this sense, the watershed plan should not be considered a "product" but rather a process of achieving social change—a way of moving from the present, presumably unacceptable, state to some future, more desirable, condition. As societal values and priorities change, so must the plan, and the process, adapt to accommodate them. To accomplish this end, all major viewpoints—publics—must be heard, consensus must be built, and the views of the majority reflected in the change that takes place. Yet for a variety of reasons relating to time and resource demands, unwillingness to relinquish authority, or simple ignorance of opportunities, this does not always happen.

Table 4.1   Typology of Change Processes (from Bishop 1970; adapted from Bennis 1961)

	Intentional Goal-Setting	No Intentional Goal Setting
Mutual goal setting	Planned change	Interactional change
Bureaucratic (unilateral) goal setting	Technocratic change	Change without goals

Drawing from social and political science in his analysis of planning processes, Bishop (1970) concluded that the ways that goals are set, and the clarity of goals, have a direct impact on the nature of change that occurs (Table 4.1).

Clearly, successful achievement of change will be more difficult if the change appears to be forced or aimless, so processes involving mutual goal setting are to be preferred over unilateral or "bureaucratic" processes.

4.2.2   The Meaning of *Involvement*

Watershed planning is essentially a sequence of activities that occur over time, each leading to the next. It is generally accepted that planning initiatives begin with problem awareness and progress through various information-gathering stages to a point of decision or action. Problem awareness may come from agency representatives, from interested professionals such as consulting engineers or planners, or from regulatory directives, but may most frequently come from the community, whose members are, after all, most directly affected by water resources quality and quantity in the basin. Figure 4.1 illustrates a typical planning sequence, showing opportunities for public involvement.

There is, arguably, a role for the public in each stage of this sequence, yet few public involvement processes allow participation over the full range of opportunities. Many, perhaps most, processes in fact limit public involvement to the last two stages: stabilization of change (implementation) and (possibly) maintenance and monitoring.

Arnstein (1969) has described a wide range of public involvement styles, from "tokenism" to full citizen control over decision making. Although Arnstein's work was less concerned with the temporal sequence of events (and opportunities) than with the quality of opportunity at any one stage, her hierarchy of involvement styles (Figure 4.2) illustrates the wide range of interpretation placed on the term *involvement*.

What, then, is the ideal form of public involvement? The answer may differ according to the perceived problem, the community and its values, and the willingness of decision makers to delegate authority.

A variety of factors are important in designing a public participation program. In essence, these factors relate to the concepts of democratic decision

Problem Awareness

Desire for Change

Establishment of Change Process

Diagnosis of the System

Community Goal Setting

Development of Alternatives for Change

Selection of Preferred Alternative

Stabilization of Change

Maintenance and Monitoring

Figure 4.1   *Typical sequence of public involvement in water resources planning.*

making discussed earlier, including sharing of information, building of trust and credibility, relationships between the public and the decision makers, and conformance with preexisting requirements such as laws and funding availability. Probably the most critical factors in effective public participation are the following:

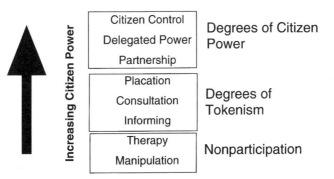

Figure 4.2   *Styles of public involvement (after Arnstein 1969).*

Before the Process Begins

1. Mutual respect
2. Clearly stated expectations about:
   > Proposed project scope and key issues
   > The nature and timing of public involvement
   > Consultation and communication mechanisms
   > The level of power citizens will have in the process
   > Explicit proposals for selecting citizen representatives for the planning process
3. Inclusion of all interested publics, including:
   > Staff of public agencies at the federal, state/provincial, regional, and local levels of government (as appropriate for the planning exercise)
   > Elected officials at all levels of government
   > Private corporations and other organizations with an economic interest in the plan
   > Public interest groups (that is, groups formed to represent specific interests in the general public), including both high-profile *leaders* and the more general membership of those groups
   > Other groups and individuals in the community, including private citizens, legal and medical professionals, and others with a general but not necessarily economic interest in the plan

Plan Administration

1. A single program manager and clear reporting relationships
2. Program staff who are well informed about the project and the community, who are skilled in public involvement techniques, and who are receptive to the ideas of community representatives
3. Specialized expertise where necessary, for instance, in conflict resolution and facilitation
4. Adequate funding to achieve stated program goals for public participation

Data Collection and Analysis

> Joint collection and analysis of data on:
> > Community values, systems, and interested publics
> > The proposed project and any likely impacts on community values and/or life-style
> > The experiences of similar projects
> > The costs of proposed measures and possible funding sources for those measures
> > The environmental and economic impacts of proposed measures

Communication

1. Clearly written documents and legible graphics
2. Text written in semitechnical or lay language so as to be easily understood by all participants, *and/or* opportunities for technical education during the consultation period
3. Full and unrestricted access to all data, documents, and other materials for all participants, whether electronic or in text form
4. Well-structured advisory and consultation groups with:

    A variety of skills represented (e.g., innovators, detail checkers, encouragers, moderators, etc.)

    A balance of activity between plan-related tasks and team-building activities. (The latter, although essential for long-term group function, should comprise no more than 40% of total activity or the group may be unproductive.[*])

5. Prompt, sensitive, and respectful review of citizen submissions and thoughtful and timely responses from bureaucrats to public representatives. (The goal is to create an ongoing dialogue whereby ideas and suggestions from a variety of sources are welcomed and thoughtfully considered, not to create an adversarial sense of "us" and "them.")

Priscoli (1982) cites the example of a woman who attended a large public meeting and "pour[ed] out her heart in tears over a proposed project" in front of several hundred participants. After hearing her submission, the chair of the meeting responded, "Thank you, ma'am. Now, do we have any factual comments?" Priscoli goes on to say that, in fact, the woman had just given the meeting the most important "facts" necessary for designing successful implementation alternatives. But, he wryly adds, "For us bureaucrats, armed with advanced engineering and scientific degrees, this is a hard pill. To us, emotions are irrational; facts can be separated from values."

To place this example in a different context, the woman had given the meeting a clear and strongly expressed message about her objection to the plan. No amount of scientific data or persuasion would be likely to change her position on the project. Furthermore, if she felt this strongly about the issue, it is unlikely that she was alone in her views. The watershed planners involved would have been well advised to consider the substance of her objections and develop alternatives that were more palatable to her and her fellow objectors. This approach is the heart of conflict resolution, discussed in more detail in Section 4.4.7: the respectful acknowledgment of differing perspectives, an attempt to understand each party's true motives, and a creative search for alternatives that accommodate all parties' concerns.

---

[*]Several authors discuss the need for a mixture of talents and skills on any team. The interested reader is referred to Deutsch (1960); Lumsdaine and Lumsdaine (1995); and Herrmann (1995).

4.3 IDENTIFYING INTERESTED PUBLICS

4.3.1 Finding Those Who Are Interested

One of the most daunting prospects in designing a public participation scheme is to decide who should be involved. The risk of engaging too narrow a range of participants is that some interested parties will be excluded and will become adversaries rather than collaborators, thus creating new obstacles in the planning process. Yet too large a group is equally ill suited for committee work. It may also be that a large number of possible participants will remain silent and distant from the planning process, even though they have useful contributions to make. Creighton (1983) notes that this silence (of potential participants) usually stems from one of three factors:

1. They feel adequately represented by some other group, such as a neighborhood association or environmental public interest group.
2. They are unaware that they have a stake in the decision or somehow view the decision as of minor importance in their lives.
3. They do not believe that they can influence the outcome of the process.

These factors have different roots and thus different solutions. In the first case, we can assume that such an individual is sufficiently well-informed to make the decision not to participate; no further action may be necessary. In the second case, the decision under discussion may or may not have major importance to the individual; a comprehensive public information program will ensure that people making this decision will do so in command of all the facts. The third case, which Creighton terms apathy, is not surprising in view of declining confidence in government. Here again, a good public information program, focusing not only on technical issues but on the proposed public involvement process, may allay concerns and encourage wider participation.

Willeke (1976) and Creighton (1983) suggest three ways to identify interested publics:

1. *Self-identification* (voluntary or to express opposition)
2. *Third-party identification* (for instance, existing committees are used to generate suggestions as to individuals and organizations who would be appropriate participants.)
3. *Staff identification*; that is, identification of potential participants by project staff based on surveys, consultation with other agencies, or analysis of data on community composition, associations, user groups, geographic and demographic data, tax records, newspaper and magazine archives, and similar sources

All of these methods are in use to some degree. Self-identification may be perceived as more in keeping with a spirit of openness and transparency than identification by some (possibly biased) external group. Self-identification may be more complex than Willeke suggests, however, because of the difficulty in providing adequate notice to the public that opportunities for involvement exist. This difficulty has been overcome in various ways, each of which has advantages and disadvantages.

Open Invitation   One method of soliciting participants in a public involvement program is to issue an open invitation to the community, for instance, to attend a public meeting. Such invitations must be widespread to be effective and would likely include advertising in local newspapers, television, and radio. Printed invitations can be sent to the membership of local interest groups, Chambers of Commerce, neighborhood associations, and other groups with potential interest in the plan. It is important to ensure that the invitation is seen as truly open, for instance, by asking recipients to pass along the notice to others who may be interested. There must be no sense that only a select few have received the invitation while others have been intentionally ignored.

Once attendees have gathered at the public meeting, the selection process can be handled in several ways. One approach is to ask participants to group themselves according to various constituencies—regulators, public institutions, private corporations, nongovernment organizations (public interest groups), and private citizens. Then ask each group to elect a stated number of representatives. Another approach is to hold a series of public meetings and simply allow attendance to drop off, on the assumption that those who continue to come constitute the truly interested public.

The advantage of the open invitation approach is that, properly conducted, it is perceived as open, equitable, and fair. The disadvantages are that the process of selecting individual representatives can be chaotic and even acrimonious, and that there is no guarantee that the desired number of representatives from each constituency will actually participate. This method may have greatest utility in small-scale plans and close-knit communities where the individuals and issues are well known. It becomes more unwieldy as the potential number of participants increases and the issues, and therefore conflicts, become more complex.

Selection by Application   Another method of soliciting representatives has been used with success in the Remedial Action Plan (RAP) process of the Canada-U.S. International Joint Commission. The RAP process is intended to generate cleanup plans for each of 43 "Areas of Concern" (pollution hot spots) in the Great Lakes system through a prescribed process of problem identification, source identification, development of cleanup alternatives, and selection of a preferred plan. The process, which is to be community based, makes extensive use of public input through Public Advisory Committees. (Technical and Scientific Advisory Committees are also part of the process, providing detailed technical advice on issues like fisheries biology, hydrology, and water chemistry to the agency-based RAP team.)

Representatives for the Public Advisory Committees were chosen from a pool of applicants solicited through newspaper, television, and radio advertising. The selection was made by the RAP team (essentially a group of regulatory agency representatives) in each RAP area and was based on applicants' understanding of the issues, background preparation, availability, and other considerations. A similar approach was used with success in the Grand River Basin Water Management Study, described in Section 3.4.

The advantage of this approach is that it allows much better control over the total number of participants and the number of representatives from each interest sector. The disadvantages are minor: some administrative effort in receiving and screening applications and the possibility that the selection process will be viewed as biased or preferential. The latter can be overcome if a multistakeholder committee—perhaps even a "blue ribbon" committee of respected local citizens—conducts the screening.

Limited Invitation to Potential Opinion Leaders   Any planning process benefits from continuity of leadership and clarity of vision. Depending on the issues under study and the community involved, it may be appropriate to invite participation from key public and private organizations, with the suggestion that leaders in those organizations participate in the process. It is then up to the organization in question as to which individual is chosen to participate, the planners having ensured the involvement of groups whom they believe to be central in decision making.

Creighton (1983) makes the valid point that different publics may be required throughout the planning process, some stages requiring broad review by the widest audience possible and others having a greater need for technical focus and continuity. As discussed earlier, this difference has been accommodated through multitier processes. This approach is discussed more fully in Section 4.4.

4.3.2   Finding Leaders

Ultimately, a watershed plan is intended to build public consensus about actions that will result in an improved quality of life in the community, for instance, through restoration of impaired water uses. As discussed in previous sections, change can occur through mutual goal setting or can be imposed by public agencies; in either case, it can be planned (following a predetermined pattern or sequence) or unplanned.

Although consensus can develop gradually in the absence of discernible leadership, there is no question that the active participation of opinion leaders can vastly accelerate the building of a clear community agenda. There is considerable debate in the social science literature as to what constitutes "leadership," because people exert leadership in several ways. This is apparent to anyone working in community activities or, indeed, within any given organization. Those who ostensibly hold positions of power are not always those viewed as leaders. Similarly, individuals remote from the formal power structure may in fact be extremely important leaders in community opinion. Some of the main categories of leadership include the following:

Positional leaders hold elected or appointed positions at a rank that implies influence over their constituencies. One irreverent analyst has referred to these as "signboard" positions because the participation of positional leaders in a process, or even a single event, is often valued more because it implies support by the organization than because of the individual's personal attributes or skills. Positional leaders may include elected officials, presidents of major corporations, directors of major public agencies, and high-profile leaders of major nongovernment organizations.

Reputational leaders are those who are generally thought to be the key decision makers in the community. These could be influential elected officials in a local government, visible and respected representatives of public agencies or private corporations (but not necessarily the most senior or powerful members of those organizations), or any others who are considered to have clout in the local community, for instance, through personal wealth or family connections. Reputational leaders can include particularly active or specialized members of nongovernment organizations who are respected for their knowledge of key issues.

Decisional leaders have demonstrated their leadership in earlier community decisions. While positional and reputational leaders have potential to wield power, they may not have chosen to do so in previous watershed planning initiatives. Decisional leaders are those who have both the ability and—equally important—the interest to influence public opinion on water issues, based on their involvement in similar processes in the past.

Most processes will benefit from the involvement of leaders of all three types. Although continuity of representation is often important, some processes will extend over many years and, therefore, beyond the term of office of many positional leaders and perhaps beyond the tenure of any single reputational or decisional leader. When a change of representation must take place, there is sometimes a formal transfer of responsibility from one representative to another; for instance, the president of a university may announce his or her retirement from a process, while simultaneously introducing the incoming president as the new representative. Where possible, this formal "handoff" is to be encouraged as a public mark of continuity of interest and support, not just a matter of administrative convenience.

4.3.3 Limiting the Number of Participants

Section 4.4 discusses a variety of public participation techniques and processes, some of which can involve a large number of people. Ideally, however, working committees should never be larger than 30 people, with 20 or fewer a much more desirable size. The reasons for this will be obvious to anyone who has ever been involved in committee work: large committees are difficult to administer (for instance, in arranging meetings) and, because of the space they must occupy and the difficulty of communicating across that space, often tend to splinter into small discussion groups rather than to function as a single unit.

Smaller committees are also better able to build strong working relationships and produce useful outcomes. This is particularly important in lengthy planning processes. Deutsch (1960) and others have described how strong teams employ a range of skills and personality types to build and maintain the team and perform project-related tasks. It takes time for a team to learn how to work together—how to make best use of the array of talents available in the group and find the most appropriate style of discussion and conflict resolution for their needs. This phenomenon may indeed be an argument for standing committees that can provide a degree of continuity in public decision making, rather than short-term ad hoc committees that must be built and rebuilt each time a new issue arises.

4.3.4   The Policy Profile: Forecasting Outcomes

Coplin et al. (1983) describe a technique called policy profiling that is useful in assessing the likely outcome of planning processes and the probable impact of individual representatives and organizations on policy decisions. Their technique includes the following steps:

1. Identify the issue to be decided (should be a clear and explicit definition using terms such as "restrict," "permit," or "build," rather than "improve" or "protect").

2. Identify the individuals, groups, and organizations that should be included in making the decision.

3. Group together participants with the same economic interests, such as developers. Do not group participants with similar veto power, especially government participants. Do not group participants whose position on the issue differs or whose power in the decision-making process will vary significantly.

4. Structure the grouping so that the actual power distribution is reasonably well reflected. Do not include an unreasonable proportion of participants from one sector, thus skewing the power structure unnaturally. If one group of participants has an immense amount of power, divide that group into several smaller groups so that the total power structure is accurately reflected.

5. For each actor, estimate:

    Issue position: score support using numerical values such as +1, +2, and +3 to indicate support; 0 to indicate neutrality; and −1, −2, and −3 to indicate opposition. Larger numbers indicate more extreme positions.

Power, expressed as a number from 0 to 3, where 0 indicates no power or influence and 3 reflects substantial influence or veto power.

Salience, expressed as a number from 0 to 3, where 0 indicates no interest or concern for the issue and 3 indicates participants who consider the issue to be of the very highest priority.

6. Calculate the weights for each participant group and the whole system by multiplying issue position times power times salience for each, then calculating total positive scores and total negative scores.

7. Calculate the policy profile ratio, the net weight between those supporting and those opposing the decision under discussion. Coplin et al. (1983) note that this ratio in a sense is a measure of the political benefits and costs of the decision. A ratio greater than 1.00 indicates net benefit (net community agreement) from a political and social point of view, while a value less than 1.00 indicates a net cost. A value of 1.00 reflects an equal balance of benefits and costs.

An example of these authors' policy profile analysis is shown in Figure 4.3. This fairly simple analysis of participants' positions and power reveals 1.71 times as much support for the decision as opposition, and therefore community acceptance of the proposal is predicted. The analysis also reveals other information of interest, such as the fact that the only serious opposition comes from the federal government and that the support of state agencies, local environmental groups, and influential citizens is only moderate. This gives the analyst clues that, although the decision to issue the permit should probably be made, continued support for the measure will depend on retention of support from these participants. The analyst may also wish to explore the reasons for the federal government's firm opposition to the issue, and determine whether con-

Issue: Whether or not to issue a general permit concerning residential landfill operations

Actors	Position		Power		Salience		Positive	Zero	Negative
1 City Government	3	×	2	×	3	=	+18		
2 County Government	3	×	1	×	1	=	+3		
3 State Government	1	×	2	×	1	=	+2		
4 US Fish and Wildlife	−2	×	3	×	2	=			−12
5 US EPA	−2	×	3	×	2	=			−12
6 Land Developers	3	×	1	×	2	=	+6		
7 Environmental Groups	1	×	1	×	3	=	+3		
8 Influential Citizens	1	×	3	×	3	=	+9		
Total Scores							+41		−24

Policy Profile Ratio = Positive Scores/Negative Scores = 41/24 = 1.71

Figure 4.3 *Example of policy profile analysis (after Coplin et al. 1983).*

flict resolution (described in Section 4.4.7) may be of benefit in removing such objections.

Like many other factors in a planning process, participant interest and position can wax and wane with time, especially if issue position is weak (−1, 0, or +1). The policy profile analysis should therefore be repeated from time to time as the process continues. Clues as to when reassessment is needed may include altered behavior in meetings, increased conflict among participants who were formerly in agreement, and changes in external forces, such as state or national politics or major community events.

4.4 PUBLIC INVOLVEMENT TECHNIQUES AND PROCESSES

4.4.1 Defining the Purpose of the Involvement

There are many reasons that public involvement may be desirable in a planning process, and many times in the process when that involvement will be useful. Yet, as discussed in Section 4.2.2, the nature of that involvement, and thus the techniques appropriate for it, can differ markedly from process to process. In part, this is because the planners designing the public participation scheme have different end points in mind. In some processes, the purpose of public involvement is simply to give information to the public—what is often termed "public information." No public response is sought or even desired; the agency has the facts and has made the decision and seeks only to communicate its decision to the public.

In other cases, the planning process may be well advanced, but a decision point has been reached and the agency believes that input from the public would be helpful in making that decision. The purpose of public involvement in such an instance is for the agency to receive information, a process that could be conducted as easily (and perhaps more thoroughly) by mail or telephone than in a face-to-face meeting.

A third type of process intends two-way communication between the decision makers and the public—for instance, so that each side can ask and answer questions. An extension of this interactive process is negotiation or conflict resolution (Section 4.4.7).

Although, generally speaking, interaction is a beneficial part of planning, there may be a place in any planning process for simple information giving or receiving. Whatever approach is chosen, the purpose of the public participation may limit the techniques available. Even the arrangement of furniture in a room may differ, depending on the goals of the process. For example, a simple public information meeting would likely be designed to attract as many citizens as possible and thus would not be well suited to an informal seminar setup or even a boardroom setting. The purpose of the event therefore dictates several important factors:

- The number of people who should be involved
- The type of people who should be involved
- The room layout that will best accomplish the goals of the event
- The best public involvement technique(s) for the purpose of the event

The following sections discuss a range of public involvement techniques currently in use in water management planning.

4.4.2 Techniques for Information Dissemination

As discussed previously, one form of public involvement can be termed information giving—mechanisms by which public agencies communicate information to the interested public. The literature abounds with information on such techniques, probably because for many years they were the only ways in which government formally communicated with the public. Some of the most widely used techniques for disseminating information to the public are given in Table 4.2.

The techniques presented in Table 4.2 are of two types: those that allow the agency considerable control over the information the public receives, and those that are beyond the agency's control. The former group, including public information meetings, open houses, and various printed materials, are useful methods of transferring technical information to the public but can be high in cost. They are not, however, usually as effective as the media in raising the profile of a planning exercise. For this reason, most information dissemination programs employ a mixture of methods and rely on the media for widespread advertising of events, key issues, and decision points.

4.4.3 Techniques for Receiving Information

Sometimes an agency requires detailed information—feedback—from the community, for instance, to supplement or update census data or municipal records. Often this will require contact with as many people as possible, so that a detailed and comprehensive database may be built. Like information dissemination techniques, techniques for receiving information are one-way and do not permit dialogue or negotiation. Unlike information dissemination techniques, they tend to be private rather than public and employ the written word rather than speech and images. Some classic techniques for receiving information are given in Table 4.3.

4.4.4 Two-Way Communication Techniques

By far the most common public involvement techniques currently in use provide for two-way communication. This may be informal, in the nature of a conversation, or formal, like a citizen lawsuit against a public utility. Two-way

Table 4.2 Techniques for Information Dissemination

Technique	Description	Room Setup	Advantages	Disadvantages
Public Information Meeting	Large open meeting to inform public about proposed agency action.	Auditorium set up with seating for large number of participants for presentations.	Efficient, usually inexpensive method of transferring information to a large number of people.	Impersonal and distant. Attracts only those with the time and interest to come. Can be difficult to control if audience becomes restless or hostile to proposals.
Open House	An informal, all-day opportunity for public to tour displays and chat with agency representatives.	Open setting with poster displays, tables with books and pamphlets; project staff circulate informally.	Informal and friendly. Attendees can obtain a wide range of information while focusing on topics of interest to them.	Somewhat more expensive than a public meeting due to preparation of displays and printed materials. Requires more staff time than a public meeting.
Permanent Information Center	A permanent office where interested citizens can chat with staff and pick up project information.	A permanent office, not necessarily large, equipped with bookshelves and tables for the display of information.	Informal and friendly. Visitors can obtain a wide range of information while focusing on topics of interest to them. Visitors can come at any time convenient for them.	Considerably more expensive than meetings or open houses. Requires a long-term fiscal and staff commitment, which may be justifiable only for large or long-term projects.
Newsletters	Regular information releases from the project team, usually mailed to a specific distribution list.	Not applicable.	Allows dissemination of more detailed information than is usually possible in meetings. Provides continuity and builds profile.	Costly to edit and produce. Readership may be much less (or more) than distribution. High costs may limit distribution and, thus, impact.

Information Brochures	Printed booklets with brief descriptions of project activities and progress, distributed in schools, libraries, and by mail on request.	Not applicable.	An easy-to-understand format for the lay reader. Provides a permanent record of project activities for later reference.	May oversimplify complex issues. Time-consuming and costly to produce. Costs may limit distribution and, thus, impact.
Newspaper Articles	Easily understood articles about the project in local newspapers.	Not applicable.	Reaches a wide audience at no cost to the project. A very useful method of announcing future meetings and other events.	Reporter may not present issues as the project team would like to see them presented. Project credibility and trust can be damaged by "bad press."
Television and Radio Coverage	Very brief (usually less than one minute) overviews of project goals and activities.	Not applicable.	Reaches a wide audience at no cost to the project. Excellent way to raise project profile, especially useful for key events.	Coverage is very brief and may trivialize complex issues. Reporter may not present issues or events as the project team would like to see them presented.

Table 4.3 Techniques for Receiving Information

Technique	Description	Room Setup	Advantages	Disadvantages
Public Hearing	Formal public meeting held in courtroom-like setting. Sometimes required by law.	Usually courtroom setting (chair or tribunal presides over meeting, with many participants seated in rows).	Provides a formal mechanism for the public to ask questions and voice concerns before agency representatives. Many participants may make presentations at a hearing. Relatively inexpensive.	Usually a one-time event that does not allow participants to see direct results of their presentations. Can be excessively formal and adversarial and thus may discourage participation. Costs may rise if legal representation becomes necessary or the hearing is prolonged.
Surveys	Formal assessment of public views on key issues using mailed, telephone, or in-person questionnaires.	Not applicable.	Properly designed, a good survey produces quantitative data on public attitudes and values. Can be a cost-effective way of determining the opinions of a large number of people.	Costly to prepare and distribute. Requires an expert for questionnaire design (improperly worded questions may result in answers that are difficult to interpret or that do not address the issue of interest.)
Key Informant Interviews	Agency staff conduct detailed interviews with people who are considered central to the project or the decision-making process.	Interviews can be conducted in any small private location such as a private office, home, or restaurant.	Provides a detailed and immediate assessment of how various interests want to be involved in the process, and about their interests and concerns. Allows the building of personal relationships with those who may be key later in the process.	Clumsy interviewing style can alienate key informants, who may also be key participants. Key informants may not necessarily represent wider public opinion. Expensive and time-consuming.

communication is very useful in a planning process because it allows for the gradual modification of positions and thus may encourage satisfactory resolution of controversial issues.

Not all two-way techniques are alike in terms of the access they provide to decision makers or the tools required to make them work effectively. Some, like phone-in programs on television or radio, may simply be useful as a sounding board for public opinion; feedback from the agency may be limited to terse answers, and some questions may not be answered at all. Other mechanisms, like joint field trips, may have special value beyond their technical content, in building trust and consensus between the parties to a decision.

Two-way techniques are of three main types: small group meetings, large group meetings, and what might be termed detached methods, whereby the various parties do not actually meet face to face. Table 4.4 shows a variety of two-way communication techniques suitable for use in watershed planning.

4.4.5 Small-Group Discussion Processes

Within a small-group meeting, there are many ways to structure discussion and achieve progress. While the most familiar of these may be so-called parliamentary procedure, with a chair, a fixed agenda, and formal motions that must be approved by a majority, this is by no means the only choice available. Formal meeting structures of this type may in fact inhibit participation by those who are shy or uncertain of appropriate procedure. Many other methods have come into use simply because they offer more flexibility, and are less daunting to the uninitiated, than traditional meeting formats. Most of these are intended to open up discussion and allow participants to be creative in identifying problems and suggesting solutions. The challenge to the meeting organizer is, therefore, to maintain a balance between creativity and chaos. Yet, as many observers have pointed out, most small groups are quite effective in controlling their own behavior, even without a formal chair. The following are some of the more commonly used small-group discussion processes.

Brainstorming   *Brainstorming*, a common term in everyday language, identifies what may be the most familiar of the group processes currently in use. Brainstorming seeks to generate a large number of creative, innovative ideas in a short space of time. The idea is to go beyond conventional approaches and thought patterns. To do this, the group must decide to suspend evaluation completely, even to the point of disallowing laughter at really farfetched proposals. The group simply suggests any solution that comes to mind, with a facilitator recording each suggestion on a flip chart or blackboard. The facilitator can encourage creativity by the use of:

- Images—"Here's a photograph of the earth from space. What does it make you think of?"

Table 4.4 Two-way Communication Techniques

Technique	Description	Room Setup	Advantages	Disadvantages
		Small-Group Meetings		
Community Presentations	Formal or informal presentations to local groups by agency representatives or specialists, with opportunity for open discussion provided.	Depends on size of group. Up to 30 people may be accommodated around a large table. Larger groups should be set up in auditorium style.	Efficient, usually inexpensive method of transferring information. Smaller groups allow good opportunity for discussion of issues and concerns, and for answering specific technical or process questions.	May have to be repeated many times with different groups to capture the desired audience; therefore potentially costly and demanding of staff time. Groups may vary in their response, but format does not allow resolution of differences or building of consensus.
Joint Field Trips	Visits to sites where action is proposed or problems exist. Participation by a range of interested publics with agency representatives.	Usually requires some form of transportation, depending on size of group (e.g., minivan, small bus, car pooling) but no other facilities needed.	Excellent opportunity for joint evaluation of site conditions. Small group size can facilitate sense of team building and trust. May be required under some legislation.	Can be costly, depending on location of site and nature of transportation. Benefit usually limited to a small group of potential participants. Visit may not reveal all major issues or concerns.
Working Committee	Citizen representatives are included in the committee structure of the project and work on project tasks alongside agency representatives.	Usually boardroom setup, with chairs organized around a table large enough to accommodate the expected number of participants.	Reduces need for costly public meetings and open houses. Builds expertise in the public. Face-to-face contact on a regular basis can build trust and respect on both sides. Can open up and resolve other community issues.	The need to keep committee membership small limits number of public who can participate; not all viewpoints may be represented. Agenda may drift to unrelated topics. Different levels of technical preparation may demand a long learning curve for the committee and slow progress.

	Description	Setting	Benefits	Drawbacks
Workshops	Small-group meetings with a mixture of presentation and hands-on activities, designed to address specific issues or learn about new concepts.	Seminar setting, with desks or tables arranged informally before a podium. Format may change as workshop progresses from formal presentations to working sessions.	Intimate, informal atmosphere encourages learning. Allows building of trust among participants. A good approach for joint evaluation of data or new institutional options. Creates a sense of equality and teamwork among participants.	Must be well focused, usually on a single issue or task. Benefits from leadership by a trained facilitator. Without these elements, the workshop may drift and fail to produce useful results.

Large-Group Meetings

	Description	Setting	Benefits	Drawbacks
Public Meeting	Large open meeting to present information about proposed activity and provide a forum for public commentary.	Auditorium set up, with seating for large number of participants and front podium for presentations.	Efficient, usually inexpensive method of meeting with a large number of people. Allows many viewpoints to be heard in a large public forum.	Impersonal and distant. Attracts only those with the time and interest to come. Can be difficult to control if audience becomes restless or hostile to proposals.
Litigation	Legal action (lawsuit) between two parties, conducted before a judge in a courtroom setting.	Courtroom setting.	Allows for systematic presentation of information by both sides in a disagreement. Impartial arbiter (judge) decides case on the basis of evidence presented.	Time-consuming, costly, and strongly adversarial. Destroys trust and confidence in opponent. May not resolve all key issues in the dispute, yet may create significant barriers to future progress.

Table 4.4 *(Continued)*

Technique	Description	Room Setup	Advantages	Disadvantages
		Detached Techniques		
Exchange of Views by Correspondence	An exchange of letters, e-mail, videotape, or audiotape in which questions are asked by one side and responded to by the other side.	Not applicable.	Allows systematic evaluation of a range of issues. No time pressure, so format allows research to be conducted before response must be prepared. Inexpensive. Yields accurate information on technical issues and concerns about process.	Remote and distant, so does not build trust among the parties. No joint analysis of data, so misconceptions or misinterpretations. Format limits responses to questions asked, so related community issues may not be raised or resolved.
Television or Radio Phone-in Programs	A TV or radio host and guest(s) invite and respond to questions and comments phoned in by members of the public.	Appropriate broadcast facilities.	Open to any interested participants. Anonymity may encourage callers to ask difficult or controversial questions. Reaches a very wide audience at no cost to the project. Good for raising project profile.	Brief questions and answers may oversimplify complex issues. Format is very difficult to control—agency representative may end up in the "hot seat," answering questions that were not anticipated or are unrelated to the advertised topic. Poor responses can create an impression of incompetence or indifference.
Environmental Ombudsman or Advocate	A respected individual is appointed as liaison between the public and the agency on environmental/planning matters.	Office and meeting space for the ombudsman and any associated staff.	If public mistrust of government is high, an ombudsman can create a bridge between the public and the agency. One-to-one contact allows detailed discussion of concerns in an informal setting.	Costly (because of staff and resource demands). Ombudsman may not be seen as truly impartial or trustworthy by public or agency and therefore may not have all requisite information. Ombudsman's power to resolve disputes may be limited.

	Description	Requirements	Advantages	Disadvantages
Public Cleanup Events	The public is invited to spend a weekend cleaning up some valued local site (beach, river) under the auspices of the project.	None. Refreshments may be provided for participants.	Very attractive to the media and to the public. Creates a sense of personal contribution to a community goal. Generally low in cost, although some staff supervision is advisable. Raises public profile of the planning exercise.	Event achieves primarily cosmetic effects and does not usually resolve fundamental issues of resource management or impaired environmental quality. Turnout may be disappointing in poor weather, and the media may interpret this as lack of public interest in the project.
Telephone Hot Line	Dedicated telephone line staffed by knowledgeable agency personnel. Interested members of the public may call at any time and have questions answered.	Requires office space and one or more dedicated telephone lines.	Convenient and easy to use. A good way for the public to obtain information about forthcoming events and meetings.	Inadequate knowledge on the part of staff can lead to an impression of incompetence or indifference. If callers must be directed to another individual to have a question answered, there may be a sense of "getting the runaround." Relatively costly as compared with other methods.

- Fantasies—"In the best of all possible worlds, what solution would you like to see here?" or "If you had all the money in the world, what would you do about this problem?"

- Analogies—"Think of the river as a human body. What is its heart? What about its hands?"

When a long list of suggestions has been compiled, the group discusses and rates each. A skilled facilitator can assist the group in linking ideas from different suggestions to create new options, and can help participants work through apparent obstacles to otherwise good suggestions. Impractical suggestions can be discarded at this point, unless someone cares enough about them to champion them in the discussion. Ideas that receive no support can be quickly eliminated.

Delphi Process   The Delphi process is applied, in various forms, to develop consensus among a group of participants. It offers a quick and low-cost way to assess group preferences using simple, intuitive techniques.

The basic approach is one of iterative questionnaires, each summarizing and building on the results of the previous one. In one version, participants are offered pairs of choices until all possible pairs are exhausted (pairwise comparison). Each participant chooses one option from each pair. At the end, all "votes" in favor of each option are totaled and the option with the most support is adopted. A simple example of selection by pairwise comparison is shown in Figure 4.4.

In another version, questionnaires are distributed by mail or in a meeting setting to a group of participants, seeking their estimates, forecasts, or preferences on key issues. Responses are then summarized and sent to each participant for review. With all participants' responses known, individuals are asked to estimate the probable occurrence of the various estimates or forecasts that have been submitted, or in the case of preferences, to rank the various possibilities.

	Steak	Fish	Salad
Steak	---	Steak	Steak
Fish	Steak	---	Salad
Salad	Steak	Salad	---

Figure 4.4 *Example of pairwise comparison. In this example, the participant is asked to indicate a preference from three menu choices: steak, fish, or salad. The matrix summarizes all possible pairs, and the participant simply fills in the preferred option in the open cells. The total number of "hits" is then summed for each choice. In this case, fish is never chosen, salad is chosen twice, and steak is chosen four times. Steak is therefore the clear preference.*

This round of answers is again summarized and returned to the participants. Those whose answers do not conform to the majority may be asked to supply explanations for their positions. A final round of questionnaires may be circulated, asking for each participant's final estimate, forecast, or preference.

The advantages of the Delphi process are its low cost and its ability to circumvent obstacles created by individual personalities or strong issue positions. It has several disadvantages, among which is potential complexity of application, especially in the second version, discussed previously. Another disadvantage is that in applying this process participants learn from, and may unconsciously try to accommodate to, the positions of others, with the result that consensus may appear to have been achieved when substantive support for the agreed position may in fact be lacking. To be effective, Delphi processes require skilled facilitators and careful questionnaire design, both of which may increase costs. Finally, the group must be committed to reaching consensus and must have the power to make commitments on behalf of their constituencies, or the process will be simply a paper exercise.

Breakout Groups   The "breakout" process, a variation on brainstorming, uses a workshop setting in which participants are grouped into clusters of six to eight (sometimes called breakout groups), each of which is led by a facilitator or discussion leader. The leader poses a question (or several questions in turn, allowing for responses after each), and participants respond either by voicing suggestions, which are then recorded by the leader, or by silently noting ideas on a pad in front of them. In the latter case, each participant is then asked for a single idea (taken from his or her notes), which is recorded on a flip chart. Several rounds of responses may be offered before the leader moves on to the next question. The pooled ideas are then considered one at a time by the breakout group and modified or deleted as the group wishes. The results of the discussion are then brought back to the larger workshop in plenary session at the end of the allotted time.

This process works best if the questions posed are very clear and explicit, and if the group is strongly motivated to find solutions. Skilled facilitation can help elicit consensus from the group, for instance, around common elements of different possible solutions.

Breakout groups are used in many regulatory agency activities, as well as in nonagency symposia on emerging issues or long-term planning. The value of this process lies in the sense of teamwork created by a very small group and the creativity that is often unleashed in such a setting. When several breakout groups are used (the preferred format), results often differ dramatically from group to group. Where similar positions emerge, there is clear evidence for consensus. Where positions or solutions differ markedly, the larger group may make use of the differences to generate new, hybrid solutions that draw from each breakout group.

Like the Delphi process, however, breakout groups benefit from trained facilitation, which can increase costs. When different groups develop widely different solutions, it may be difficult to find common elements and generate useful products from the workshop; the result may instead be confusion about which direction is best.

Values Clarification Exercise Values often underlie conflict in water management issues. Values clarification processes attempt to make explicit the values that underlie behavior and, thus, expectations in a planning process. They may be most useful when opinions differ but the reasons for these differences are not clear. Planning processes involving aboriginal groups have found values clarification useful in resolving disputes over land claims and resource management (e.g., fisheries) rights. The actual process can be any of the techniques listed in this section, and may include structured written exercises, role play, or mediated dialogue. Priscoli (1989) gives an example of a values clarification exercise in which the various parties are asked to mark their value positions on a figure like that in Figure 4.5. The participants have a choice of position along a continuum from economic development to environmental quality (along the horizontal axis) and from low to high levels of government control (along the vertical axis). For instance, an environmental group might take a position close to the "environmental quality" end of the continuum and close to the "high government control" end, reflecting its members' interest in having strict rules to protect the environment. Business interests may choose very different positions.

When all groups have marked their positions on the grid, value clusters usually become apparent. These clusters can be used, as Priscoli suggests, as a

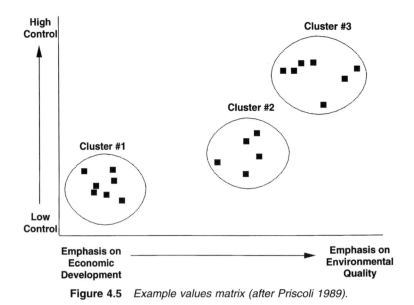

Figure 4.5 *Example values matrix (after Priscoli 1989).*

basis for understanding competing values in the watershed and for generating creative alternatives to accommodate those values.

Values clarification can be a real eye-opener in some planning processes: participants may have thought that everyone shared common (obvious!) values, but learn that values in fact differ widely. The technique is therefore a useful learning tool and provides a good foundation for later conflict resolution or negotiation. Like several of the other techniques described in this section, values clarification usually requires trained mediators. Also, as a relatively recent addition to the decision-making tool kit, it may not be as well understood or as widely practiced as some other techniques. Therefore, there is the potential for clumsy or insensitive application, with resulting alienation of one or more parties to the process. A more complete discussion of values in decision making is contained in Lumsdaine and Lumsdaine (1995).

Circle Processes   Circle processes are gaining favor in many difficult negotiation situations, some of which involve aboriginal groups and strong economic forces. The idea behind circle processes is to reduce the sense of "us" and "them" by eliminating the role of chair or moderator. This is its special value in strongly adversarial meetings, where any discussion leader may be seen immediately as partisan by some members of the group.

Circle processes are intended to allow everyone who wishes to present information or opinions, but not necessarily to resolve conflicts or arrive at consensus. As a result, they may become an important prelude to later processes. Circle processes are usually used with smaller groups (fewer than 30 people) but have also been used in groups of several hundred individuals. The larger group size is possible because the heart of the process is a single table with five or six chairs—the rest of the room is set up as "audience," usually with concentric rings of chairs sufficient to accommodate the expected turnout. Aisles arranged as spokes in the "wheel" of chairs are helpful in allowing people to move easily to and from the central table.

The process begins with someone (for instance, a representative of the agency who is sponsoring the meeting) calling the group to order and announcing that the purpose of the meeting is simply for all to learn from one another. This person should then review the rules of the process, but have no further role in the meeting other than to call it to a close when the allotted time has elapsed.

Typical rules for circle processes are as follows (adapted from Aggens 1983):

1. Only people seated at the central table may speak.
2. Anyone present may take an open seat at the table at any time in the meeting.
3. Speakers may ask or answer questions, state opinions, change the current topic of discussion, or alter the current course of discussion.

4. Speakers may interrupt other speakers or may wait for a lull in the discussion to introduce their points.

5. Speakers may stay at the table only until they have made their points. Once finished, a speaker is expected to leave the table so that another person can have a chance to speak. Speakers may return to the table as often as they like during the meeting.

6. If there are no vacant seats at the table, would-be speakers should stand close to the table, thus signaling those at the table that a seat is wanted.

7. Those not seated at the central table are expected to listen in silence. They should not laugh, cheer, groan, or boo another until they are seated at the table. Only those at the table may comment in this way.

Aggens (1983) points out that it is helpful to have one or two people experienced with the technique to start off the discussion. Ending the meeting is usually a matter of allowing discussion to run its course, with the audience gradually slipping away until only the group at the table remains.

This type of process can feel uncomfortable to those used to formal meeting structures, but is not unusual for discussion formats in tribal groups. It is certainly advantageous to have agency representatives present at such a meeting, but they should be encouraged to suspend their impatience with long-winded speakers and allow the meeting to run itself. An implicit advantage of this technique is exactly this: that it allows the respectful hearing-out of those who may have felt marginalized by more formal processes.

If there is a time limit on the discussion, someone who is responsible for meeting that limit can take a seat at the central table and remind the other participants of the dwindling time available. When the time comes to end the meeting, the person who opened it should approach the table and wait for a seat. As each seat is vacated, the person should tip the chair forward, signaling that it is no longer available for use. The audience—and the remaining speakers—thus receive a gentle and nonthreatening reminder of the need for closure. This process should not be rushed; each speaker should be allowed to finish and leave the table at his or her own speed.

Circle processes, while seen as unconventional and even risky by some, can be extremely useful in controversial processes or where there is a great deal of pent-up emotion and thus a need for "venting." They are inexpensive, easy to run, and usually leave participants with a good feeling about the process and the intentions of those who sponsored the meeting. As a result, they may be quite valuable in establishing a foundation for later, more traditional, discussion processes.

Role Play   Role play, one of the most effective techniques for teaching conflict resolution and related topics, has real value in consultation processes outside the classroom. In a role-play process, the participants adopt the roles of different stakeholders in the planning situation, usually roles that differ from their own "real life" positions. Each individual role is usually described in a brief written handout that is seen only by the person playing the role. In this way, the player can be given information that is not known to others in the process. Each player may also be given props suited to the role, and perhaps a badge stating his or her "name" and "affiliation." The discussion leader begins the process by describing a scenario in which the various players find themselves. The leader then sets a time limit and describes any products or milestones expected of the group.

Role-play exercises are often most productive when they are structured around a mock hearing or other formal process—in other words, when the players are expected to act as they would in a real situation. This approach provides a useful structure for group dynamics (especially a temporal sequence of events that may be reassuring to participants) and some idea of the type of product that is expected (for instance, a decision on a proposed action).

Role-play exercises are usually a lot of fun (everyone likes to play dress-up) and a good icebreaker. They may not, for those same reasons, be seen as appropriate in highly controversial situations or those in which formal processes have been the norm. If role play is seen as beneficial in such processes, it may be wise to use an external facilitator to introduce and oversee the exercise. Otherwise, there is a risk that jaded agency representatives and cynical citizens will reject the exercise as frivolous or a waste of time.

In fact, role play is an excellent way for people to learn what it is like to be in someone else's shoes. Citizens can learn that decision making is perhaps more complex than they had realized, while agency representatives can come to understand the reasons for citizens' apparently emotional positions. Role-play exercises are, however, surprisingly time-consuming to develop, especially if each role is well researched and equipped. (An exercise employing 12 roles requires 12 separate documents, one for each player, and thus 12 separate research exercises.) Use of a professional facilitator will certainly raise costs, but will also increase the potential of a productive outcome.

Simulation Small groups may benefit from the use of simulation games or joint evaluation of computer simulations. Simulation games combine role play with random forces to create a realistic sequence of events mimicking real life. For instance, a simulation may take the form of a board game in which various players act out roles (e.g., state government, local government, citizens' group, industry) while various random factors (e.g., major storm, stock market crash, change of government) force them to take actions affecting the community in which they live and work. Simulation games may be proprietary—that is, available on the market—or developed by the project team.

Many computer simulation models are also in use in watershed planning (see Chapter 7) and can form the basis of a group process if sufficient computer equipment is available for participants to work alone or in pairs. In this case, participants are given some basic constraints on a system, such as areal size and current land use, and then given free reign over future scenarios to see the impact of alternative management approaches. This is sometimes termed "what if?" analysis, because it allows participants to explore the response of the system to actions within their control: altering the characteristics of discharges, shifting the pattern of future development, changing target water levels and water quality criteria, and so on.

These simulations have value for several reasons. First, the very small group size (usually fewer than five participants) allows people to get to know each other as individuals, a good way to build trust and team feeling. The technical content of these exercises is also important, because it permits joint analysis of complex technical issues and the building of mutual understanding about system response to stress and management intervention. Simulation games are inexpensive but may not adequately represent conditions in the real situation. Computer simulation is more costly because it requires several computers and associated software, but can provide more realistic insight into management issues and responses.

4.4.6   Organizational Considerations

As discussed in Section 4.2.2, public involvement can and, as many believe, should occur throughout the planning process. Although it is possible to obtain that involvement through a series of unrelated events, media reports, public meetings, and similar techniques, it is often considered advisable and efficient to establish citizen representation as a formal part of the decision-making process.

Typically, water management decisions are made by groups of people, often primarily bureaucrats, through a series of meetings held regularly over the course of a planning exercise. The beginning of the planning process may involve more data collection and analysis (as well as establishment of working relationships on committees), while later stages may be more concerned with the development of consensus and the evaluation of implementation mechanisms. One commonly used approach for public involvement is simply to appoint public representatives on advisory committees.

Two types of committee organization are possible, single-tier and multitier. Simpler and shorter-term projects may be adequately served by a single-tier committee structure, in which a single committee oversees the entire planning exercise, including problem definition, goal setting, an inventory of the existing system, development and testing of alternative management strategies, and selection of a preferred approach. (The technical work for this committee may in fact be done by staff engineers, scientists, and economists or by external consultants. The point here is that decision making is limited to a single committee.)

There are numerous advantages to single-tier processes, primarily relating to administrative simplicity and the development of strong team dynamics within a long-standing committee. In single-tier processes, public representatives can be appointed as full members when the process is begun and continue to serve until the planning exercise is complete.

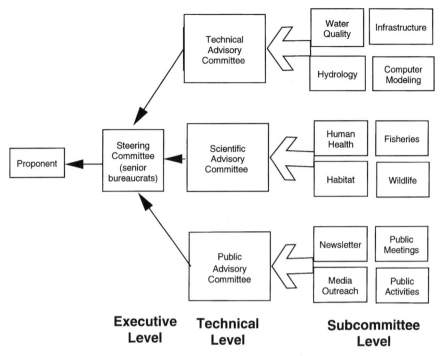

Figure 4.6 *A typical two-tier advisory committee structure.*

For more complex planning problems, larger basins, or longer planning initiatives, however, it may be advisable to organize staff and consultative resources into several tiers. This may be done in several ways.

The most common approach to multitier structures is to set up a senior committee made up of positional leaders, or their delegates, and one or more technical working committees composed of senior technical experts from inside and outside the agency. Both committees can include representatives from the public. A typical two-tier committee framework is shown in Figure 4.6.

Figure 4.7 shows an alternative structure modeled on that used in the IJC's Remedial Action Plan process. These illustrations are only intended to show possible approaches to committee configuration. Many such configurations are possible and will work well (indeed, in many processes a very simple one-committee structure will serve the purposes of the planning exercise). The advisory framework adopted by any given planning group will depend on the goals of the planning activity, the interests and resources of the proponent and any affiliated regulatory agencies, and the extent to which groups outside the agencies are allowed to participate in the process.

As discussed in Section 4.3, it is desirable to include community leaders in the decision-making process, because these individuals are in a position to explain and influence community opinion and thus build (or destroy) support for

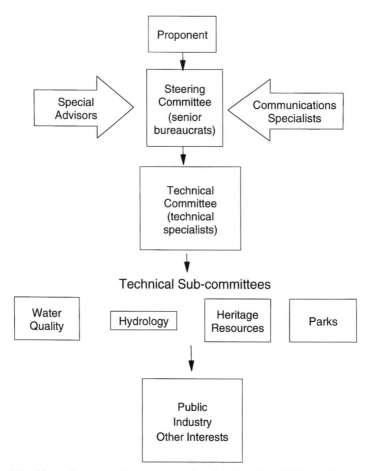

Figure 4.7   *Alternative committee structure (modeled on the IJC Remedial Action Plan process).*

change. Both committee structures illustrated in Figures 4.6 and 4.7 reflect an awareness of this point. The structure shown in Figure 4.7 emphasizes inclusion of senior bureaucrats and public officials, while the technical committee is made up of senior technical specialists, engineers, and scientists. In this structure, public liaison is conducted through a designated committee, but there is no public membership on the committees themselves.

The advisory framework shown in Figure 4.7 includes a significant level of public involvement through a public advisory committee that is parallel to technical and scientific advisory committees. Although this approach gives the public a substantive voice in the process, it essentially excludes its members from technical and scientific deliberations. In the IJC's Remedial Action Plan sites, a typical committee structure includes a project team (primarily, if not entirely, agency representatives) that receives advice from a Technical Advisory Committee (sometimes also a Scientific Advisory Committee) and a Public Advisory Committee. Generally speaking, the Public Advisory Committee is excluded from detailed discussions of technical matters (the province of the Technical and Scientific Advisory Committees) and is more concerned with the review of documents and proposals and the design of an implementation strategy. The Technical and Scientific Advisory Committees may have nonagency representatives, but are unlikely to include members of the general public. In this structure the exclusion of the public, and public interest groups, from technical decision making may limit the building of expertise in the public and impair the building of solid public-agency relationships.

4.4.7 Conflict Resolution (Alternative Dispute Resolution)

Unlike negotiation, which implies distinct and entrenched bargaining positions and the gradual yielding of concessions on each side, conflict resolution, sometimes known as alternative dispute resolution (ADR), seeks the common ground in a dispute. This technique has been in use for many years but has recently gained prominence through its application in high-profile cases such as native land claims, acrimonious labor disputes, and timber management issues. Increasingly, conflict resolution and mediation are being offered as formal alternatives to tribunal processes, for instance, under the 1990 Administrative Dispute Resolution Act (U.S.) and the 1994 Canadian Environmental Assessment Act. Conflict resolution is based on several key assumptions:

1. That there is agreement on the vast majority (90% or more) of issues in any dispute. Properly facilitated conflict resolution exercises will quickly reveal this fact.
2. That each party has a true interest, or stake, in the issue. However, this interest is not necessarily, or even usually, stated outright.
3. That each party in a dispute has a "fallback" position—the position that will be taken if consensus on the issue is not reached. Again, this position, which can be interpreted as the party's "threat" in the process, is seldom stated outright.
4. That each party's true interest is valid and important in the process and should be accommodated by whatever solution is developed by the group.

The conflict resolution process is focused on learning each party's true interest and likely fallback position and developing innovative solutions that accommodate the true interests of the various parties. It is important to realize that the positions stated at the outset are likely to change as the process continues. This differs markedly from traditional negotiation, in which positions are instead very likely to become more entrenched and intractable with time. A typical conflict resolution process works like this:

1. Representatives from a range of constituencies—agencies, public interest groups, industries, or whatever groups are involved in the dispute—are invited to attend a "retreat" over one or more days. Those invited should have the power to make decisions on behalf of their constituencies.

2. The meeting begins with introductory remarks from the sponsor, including a statement to the effect that consensus is the desired outcome of the process, that the process attempts to value and accommodate all major viewpoints, and that everyone will have an equal opportunity to speak.

3. Each of the participants in turn is invited to speak about the reasons for their participation, their views on the issue, and the reasons for those views. (Time limits may be set on this "venting" process if the group is comfortable with that.) This step serves to clear the air and ensure that everyone's views are known to the group.

4. The facilitator then emphasizes the importance of full understanding in the process and invites the participants to talk to one another, learning more about them as individuals and about their views and values. The proviso in this step is that discussions may be held only on a one-to-one basis. Each participant is encouraged to learn from each other participant. The facilitator may set a time limit, probably several hours, on this activity.

5. After this initial icebreaking and information-gathering step, the facilitator recalls the group into a plenary session. The mood of this session is usually much more relaxed and friendly than that of the initial meeting, reflecting the participants' new understanding of their colleagues and a building sense of team. The facilitator invites the group to speak, again one at a time, on any insights they have had or any questions that remain in their minds. The participants should, however, be discouraged from talking about any "deals" that have been struck (as will inevitably happen) between stakeholder groups. (The reason for this is to avoid antagonizing those who have not yet formed firm opinions as to possible outcomes.)

6. The facilitator then allows the group plenty of time for free discussion, this time allowing larger groups to converse. In this step, what typically happens is that small clusters of people form, and as the discussion—and consensus—begins to develop, the clusters gradually coalesce into larger and larger groups until the whole group is together.

7. Even at this encouraging stage, consensus may be fragile. The group should choose a quiet and diplomatic individual as recorder. (A loud or forceful personality may provoke argument among the group or may be seen to be biased in representing the views of the group.) The group then calls out the points of the consensus for the recorder to write on a flip chart or blackboard. This list is systematically reviewed and edited by the group until the wording is acceptable to everyone.

8. As a final step, each participant formally signs the agreed-upon position to indicate full support.

Conflict resolution often requires more than a single day (although surprising progress can be made in a day or less, given that the group is in agreement on the goal of consensus) and may in fact continue over a period of weeks or months. This time may be necessary to allow the various parties to check back with their constituencies, to obtain necessary funding, or simply to think. Where positions have become entrenched through past processes, a skilled facilitator is needed to encourage the parties to see beyond their stated positions to other possibilities still consistent with their true interests.

Conflict resolution is the basis of the Harvard Negotiation Project's techniques, described by Fisher and Ury (1981), and is the focus of a growing number of publications, including the newsletters *Consensus*, published by the Harvard Law School's Public Disputes Network, and *Resolve*, published by the World Wildlife Fund's RESOLVE Center for Environmental Dispute Resolution. (It should be noted, however, that the term *conflict resolution* is also used in the sense of international, including armed, conflict. The term *alternative dispute resolution* or simply *dispute resolution* may be helpful in separating one group of sources from another.)

REFERENCES

Aggens, Lorenz. 1983. The Samoan circle: a small group process for discussing controversial subjects. In: *Public Involvement Techniques: A Reader of Ten Years' Experience at the Institute for Water Resources*, edited by J. L. Creighton, J. Delli Priscoli, and C. Mark Dunning. IWR Research Report 82-R1. Fort Belvoir, Va.: Institute for Water Resources, U.S. Army Corps of Engineers.

Arnstein, Sherry. 1969. A ladder of citizen participation. *J. American Institute of Planners* 35: 216–224.

Bennis, Warren G. 1961. A typology of change process. In *The Planning of Change*, edited by W. G. Bennis, K. D. Benne, and R. Chin. New York: Holt, Rhinehart, and Winston.

Bishop, Bruce. 1970. *Public Participation in Planning: A Multi-Media Course.* IWR Report 70-7. Fort Belvoir, Va.: U.S. Army Corps of Engineers Institute for Water Resources.

Coplin, William D., Donald J. McMaster, and Michael K. O'Leary. 1983. Creating a policy profile. In *Public Involvement Techniques: A Reader of Ten Years Experience at the Institute for Water Resources*, edited by J. L. Creighton, J. Delli Priscoli, and C. Mark Dunning. IWR Research Report 82-R1. Fort Belvoir, Va.: Institute for Water Resources, U.S. Army Corps of Engineers.

Creighton, James L. 1983. Identifying publics/staff identification techniques. In *Public Involvement Techniques: A Reader of Ten Years Experience at the Institute for Water Resources*, edited by J. L. Creighton, J. Delli Priscoli, and C. Mark Dunning. IWR Research Report 82-R1. Fort Belvoir, Va.: Institute for Water Resources, U.S. Army Corps of Engineers.

Creighton, James, Jerry Delli Priscoli, and C. Mark Dunning. 1982. *Public Involvement Techniques: A Reader of Ten Years Experience at the Institute for Water Resources.*

U.S. Army Corps of Engineers Institute for Water Resources Research Report 82-R1. Fort Belvoir, Va.: US Army Corps of Engineers.

Deutsch, A. 1960. The effects of cooperation upon group process. In *Group Dynamics—Research and Theory*, edited by D. Cartwright and A. Zander. Evanston, Ill.: Row Peterson.

Fisher, Roger, and William Ury. 1981. *Getting to Yes: Negotiating Agreement Without Giving In.* New York: Houghton Mifflin.

Gibson, R., and B. Savan. 1987. *Environmental Assessment in Ontario.* Toronto: Canadian Institute for Environmental Law and Policy.

Goodman, Alvin S. 1984. *Principles of Water Resources Planning.* Englewood Cliffs, N.J.: Prentice-Hall.

Herrmann, Ned. 1995. *The Creative Brain.* Lake Lure, N.C.: Brain Books, The Ned Herrmann Group.

Lumsdaine, Edward, and Monika Lumsdaine. 1995. *Creative Problem Solving: Thinking Skills for a Changing World.* New York: McGraw-Hill.

Priscoli, Jerry Delli. 1982. The enduring myths of public involvement. *Citizen Participation* 3(4): 5–7.

———. 1989. Public involvement, conflict management: Means to EQ and social objectives. *J. Water Resources Planning and Management* 115(1): 31–42.

U.S. Water Resources Council. 1983. *Economic and Environmental Principles and Guidelines for Water and Related Land Resources Implementation Studies.* Washington, D.C.: U.S. Water Resources Council.

Willeke, Gene E. 1976. Identifying the public in water resources planning. *J. Water Resources Planning and Management Division* (American Society of Civil Engineers) 102(WR1) (April).

5

Developing Workable Management Options

Chapter 2 described the process of developing a watershed inventory: learning how the watershed systems operate and interact, and developing a sense of the components and processes that are most influential and highly valued. Chapter 3 carried that analysis through to decisions about problem definition and the "scoping" of a watershed management plan. These two chapters therefore form the foundation of a watershed management strategy. Along with Chapter 4, on public involvement, they provide the basis for focused problem solving in the management area.

This chapter describes the next stage in developing a watershed management strategy: developing workable management options—actions that cause intervention in the existing system and result in movement toward a desired target. Management options may include measures that use technology or structures to change existing conditions (sometimes called "structural" measures) and those that rely on changes in human behavior or management practices (sometimes called "nonstructural" measures). Generally speaking, structural measures are easier to implement, although considerably more costly, than nonstructural measures. Nonstructural measures, for instance, encouragement of crop rotation or contour plowing, tend to be inexpensive but difficult to implement, because they ask people to change entrenched behavior patterns. Most watershed management schemes include a mixture of structural and nonstructural measures geared to the control of different problems and sources.

5.1 IDENTIFYING THE SOURCES

The sources of flow or pollutants to a watercourse are of four types:

1. Dry-weather point sources—discharges that occur more or less contin-

uously and that enter the watercourse from a single, distinct point, usually a pipe or diffuser. Examples of dry-weather point sources are sewage treatment plant effluents and treated industrial wastewaters. The name stems from the fact that these are usually the major sources operating during dry weather conditions.

2. Wet-weather point sources—urban discharges of stormwater runoff and combined sewer overflows from a collection/distribution system outflow point. Although urban stormwater runoff is diffuse in origin, most cities now have efficient runoff collection systems and discharge collected stormwater from pipe outlets into receiving waters. These sources operate only during rainfall events and for a short period following rainfall and are generally negligible during dry weather. They are distinguished from wet-weather diffuse sources, such as overland runoff, because of their suitability for end-of-pipe treatment and control technologies.

3. Wet-weather diffuse (nonpoint) sources—diffuse drainage of rainwater from urban or agricultural lands. These sources require control over a large area, not just a single point, and may have important implications for bank and channel stability as well as flow and quality.

4. Internal (in-stream) sources—chemical, physical, or biological sources internal to the watercourse, such as sediment resuspension (which can elevate suspended solids concentrations) or decaying biomass (which can affect the levels of nutrients, solids, oxygen-demanding materials, and other water quality constituents). These sources can be difficult to find, quantify, and control and are therefore often omitted from watershed planning initiatives.

Chapter 3 gave some general guidance as to the scope of the management plan—for instance, deciding, as in the case of the Grand River study (Section 3.4), to focus on reducing flood frequency and magnitude and on improving instream dissolved oxygen levels. The next logical step is to identify all sources within the basin that contribute to the problem of interest. To avoid unnecessary effort and expenditure of resources, it is important to focus only on sources directly related to the problem of interest and not allow the discussion to drift into other areas of possible interest. In the Grand River study, these might have included control of persistent toxic substances or protection of wildlife habitat.

The scoping exercise described in Section 3.4 must provide clear guidance, and a restricted focus, for subsequent planning activities, or limited resources will quickly be dissipated and little progress will be made. There must be clear agreement among the parties to the plan as to the scope of the plan and the problems that are to be solved. If several problems are defined, then the relative importance of each must also be described.

If there is not clear consensus on the scope and focus of the plan—if the problems to be solved are not clearly and explicitly defined—then the decision-making group should take time to reexamine issues and priorities and attempt to develop a clearer focus. The techniques described in Sections 4.4.5 and 4.4.7 may be helpful in developing this consensus. Indeed, disagreement about project goals at this stage may suggest that exposition of values and interests has not been complete or effective and that there are hidden interests and agendas remaining to be discussed.

Once agreement on the focus of the plan has been achieved, the description of sources becomes a fairly routine task. The following steps may be used to develop a reasonably comprehensive list.

1. List the land uses in the watershed, ranking them in order of land area.
2. For each land use, estimate the proportion of land surface that is permeable (e.g., cropland, park land, forests, and woodlots) versus impermeable (e.g., roads, parking lots, roofing).
3. From available sewer maps, determine which areas of the basin are drained by storm and sanitary sewers, and the location of discharge points for each system.
4. Identify the known dry-weather (continuous) point sources to the system. These should include known discharges from any industrial or commercial operations and from any municipal sewage treatment facilities.
5. Identify the known wet-weather (intermittent) point sources to the system, including stormwater discharge points and combined sewer overflows. Sewage treatment plant bypass may also occur in wet weather and should be included in this category. Estimate the volume and frequency of discharges from these sources.
6. Estimate the pollutant-generation potential of land uses that have a high proportion of pervious surfaces (see, for example, Table 2.6). For the most part, these will be diffuse sources active only, or mainly, in wet weather.
7. Estimate the potential for internal sources or processes to generate or remove flow or pollutants to/from the system.

The identification of these factors will often provide clear signals as to dominant sources and even appropriate correction measures. Nevertheless, it is useful to develop consensus about the importance of each source as guidance for later priority setting; in other words, it is helpful to determine which sources are likely to contribute a large portion of the total load, and which a relatively small portion. This is best done with the advice of agency personnel and community representatives through joint analysis of available data.

Pollutant Loadings Versus Concentration   This is perhaps an appropriate point to introduce the concept of pollutant "load," as opposed to concentration. Both concepts are of interest in water management, but the terms differ in meaning and implications.

The term *loading* refers to the total mass of a substance, expressed either as mass alone (e.g., kilograms or tonnes) or as mass per unit time (e.g., tonnes per year). *Concentration* refers to the mass of a substance contained in a given volume of water and is expressed as mass per unit of water volume (e.g., milligrams per liter).

Historically, most water quality management programs focused on controlling pollutant concentration. There was good reason for this, in that concentration is probably the most critical consideration in assessing potential impact on an aquatic system. A species of fish may, for instance, tolerate a certain concentration of ammonia in water, but die at high concentrations. A regulatory agency may specify both an average concentration allowed in a discharge and a maximum allowable concentration, to avoid toxic effects on aquatic biota. Concentration therefore gives an intuitive measure of potential impact on aquatic species.

In recent years the notion of limiting pollutant loadings has become more important in water management. This probably reflects a growing understanding that some pollutants, especially persistent toxic substances like DDT and dioxin, are not readily assimilated and degraded by aquatic systems, but accumulate over time in sediments and biological tissue. It also reflects an increasing interest in controlling diffuse sources of pollution through land-based activities, controls whose in-stream impact cannot readily be predicted. For instance, conservation tillage is believed to reduce the loss of solids and nutrients from agricultural land. But although it is possible to estimate the total pollutant loading reduction afforded by such a measure, it is much more difficult to predict changes in in-stream water quality that result from activities undertaken over a large area. By contrast, industrial effluent controls usually result in easily measured changes in pollutant concentration at the "end of the pipe," so both concentration and loading reductions can be estimated with accuracy.

5.2 CREATING A LONG LIST OF MANAGEMENT OPTIONS

Once the sources have been identified and their relative importance is known, it is possible to create a "long list"—an unedited and comprehensive list—of all possible solutions to the problem of interest. Some of the small-group techniques discussed in Section 4.4.5 may be helpful in this regard. Brainstorming, for example, is frequently used to generate a list of possible solutions. Lumsdaine and Lumsdaine (1995) make the point that creative thinking is often limited by habits, emotions, and false assumptions. They describe a wide range of techniques for opening up what they refer to as the "artist's mind-set," as compared with the "detective's mind-set," which is used in defining the problem to be solved, and the "explorer's mind-set," which is used to place the problem in context.

Lumsdaine and Lumsdaine's emphasis on creativity is important at this idea-generation stage of problem solving. The challenge for the decision-making team is to overcome preconceptions about workable options and create, in the literal sense, a broad and imaginative range of solutions for further investigation. Creativity is central at this stage, simply because subsequent steps emphasize screening and reduction of options rather than creation of new approaches. (The term *long list* is used to differentiate this group of options from a "short list" of options that will be subjected to detailed analysis and scrutiny.)

Various sources may be used to generate options for the long list. These include experts drawn from agencies and the local community and from more distant places that have undergone similar planning exercises. The scholarly (peer-reviewed) and "gray" literature (e.g., government reports) can be a rich source of inspiration and practical experience. Novel or even apparently ridiculous proposals should not be discouraged at this stage; indeed, it is perhaps the most appropriate time for "what if?" and "in the best of all possible worlds ..." speculation. This is also an important time to clarify the values and priorities of the decision-making group and to ensure that all values are encompassed by the evolving list of options.

Lumsdaine and Lumsdaine (1995) provide an excellent annotated bibliography of texts on idea generation and creative problem solving. Key among these are LeBoeuf (1980), Osborn (1963), Prince (1970), and Van Gundy (1984). The reader seeking detailed guidance on idea generation is referred to these and other references in Lumsdaine and Lumsdaine's (1995) bibliography.

5.3 TYPES OF OPTIONS

The long list of potential management actions can, in theory, contain four types of practice: the "do nothing" or status quo option; structures and other built technologies (structural measures); vegetative approaches; and so-called best management practices—nonstructural options. Each of these categories is described in more detail in the following sections. Most long lists include at least three of the four categories. Individual measures are not described in this section and are, indeed, beyond the scope of this work. The sources listed in Tables 5.1 through 5.4 provide additional detail on a range of individual measures.

5.3.1 The Option of Doing Nothing

In virtually every management planning exercise, one management strategy is to keep on doing what you are currently doing; in other words, to maintain the status quo. This option has much to recommend it. Often, although not always, it is cheap: no structures need be built, no education programs funded, no additional fees or expenses to pay. It is easy for decision makers and lay people to understand. And it provides a useful basis of comparison with other intervention-based options. In a sense, the "do nothing" option simply furthers the existing state of the watershed. If selected as the desired approach in the planning exercise, it speaks to societal consensus that change is not necessary or desired at the present time.

In some systems, for instance, those with contaminated aquatic sediments from historical discharges, just leaving the system alone will result in some improvement through natural processes. In most systems, it will be useful to model the "do nothing" option as a base case, or foundation, against which other management scenarios can be compared.

5.3.2 Structural Measures

In most systems it is possible to build something, such as a treatment plant, a stormwater detention pond, or a grassed waterway, that will result in some water quality or flow improvement.

Structural solutions are well described in the literature and, indeed, in every basic environmental engineering text. They may include both end-of-pipe solutions, which seek to treat or remove pollution that has already occurred, and preventive options, which are implemented to prevent or reduce the creation of waste within a process. Table 5.1 shows some typical structural solutions for point sources of water pollution.

Table 5.2 gives some examples of structural measures for the control of nonpoint-source (diffuse) pollution.

5.3.3 Vegetative Practices

Sometimes included among nonstructural ("management") options, practices that alter the vegetative cover of the land are defined by Novotny and Olem (1994) as a separate class of options and are treated as such here. These include measures that change the extent, nature, and/or timing of vegetative cover and therefore change the rate and quality of water flowing over the land surface. Vegetative measures are almost exclusively used in the control of non-point sources of pollution. They are particularly important in agricultural applications because they are readily controlled by the farm operator, are often low in cost, and can provide secondary benefits in terms of crop production. Vegetative measures are increasingly used in urban settings as well, particularly with the recent trend toward "bioengineering"—the use of grading and vegetation in place of older structural solutions such as riprap and gabion baskets in stream bank stabilization. Table 5.3 gives some examples of common vegetative measures.

5.3.4 Nonstructural Options (Best Management Practices)

In some systems, nonstructural measures—sometimes called "best management practices" to differentiate them from best management technologies—may be as or more effective than structural measures. They are, however, often harder to implement because they require people to change the way they behave. A classic example of a low-cost, highly effective, but nevertheless often unsuccessful management measure is a "stoop and scoop" ordinance or bylaw for

Table 5.1 Examples of Structural Controls for Point-Source Pollution

Wastewater Pretreatment	Primary Treatment	Secondary Treatment	Tertiary Treatment	Sludge Treatment	Waste Disposal
Screening	Oil separation	Activated sludge	Granular activated carbon absorption	Gravity thickening	Evaporation lagoons
Grit removal	Sedimentation	Trickling filters	Chemical oxidation	Flotation thickening	Incineration
Flow equalization	Flotation	Aerobic lagoons	Air stripping	Centrifugal thickening	Starved air combustion
Neutralization	Filtration	Anaerobic lagoons	Nitrification	Aerobic digestion	Landfilling
	Ultrafiltration	Facultative lagoons	Denitrification	Anaerobic digestion	Land application
		Effluent polishing lagoons	Ion exchange	Byproduct recovery	Composting
		Rotating biological contactors	Polymeric adsorption	Thermal (heat) conditioning	Deep well injection
		Steam stripping	Reverse osmosis	Disinfection	
		Solvent extraction	Electrodialysis	Vacuum filtration	
			Distillation	Filter press dewatering	
			Disinfection	Belt filter dewatering	
			Dechlorination	Centrifugal dewatering	
			Ozonation	Thermal drying	
			Chemical reduction	Drying beds	
				Lagoons	

(*Sources:* Davis and Cornwell 1991; Metcalf and Eddy 1991; Ontario Ministry of Environment 1987)

Table 5.2 Examples of Structural Controls for Nonpoint-Source Pollution

Urban Stormwater Quality	Urban Stormwater Quantity	Agricultural Runoff Quality	Agricultural Runoff Quantity
Street sweeping	Grassed channels and waterways	Grassed channels and waterways	Grassed channels and waterways
Sediment retention basins	High-side spill weirs	Terraces and diversions	Terraces and diversions
Grassed channels and waterways	Sewer separation	Runoff retention ponds	Runoff retention ponds
Porous pavement	Stilling pond regulators	Sedimentation basins	Grade stabilization
Bermed and covered materials storage areas	Computerized collection system controls to optimize storage and treatment	Stream bank stabilization and channel integrity repair	Wetland rehabilitation
Barracks	In-line storage using dams and gates	Composting facility	Irrigation water management with seepage control (ditch lining) and tailwater pits
Swirl concentrators/regulators	Off-line storage using ponds, lagoons, deep tunnels, tanks, bags, etc.	Manure storage	Drop inlet structures
Fine screens	Roof leader disconnection	Activated carbon filtration	Weirs and spillways
Sedimentation	On-site detection in swales, roof storage, ponds, or detention basins	Milkhouse was water treatment systems	Subsurface (tile) drainage
Dissolved air flotation	On-site retention (total containment) to allow refiltration to groundwater	Waste treatment lagoons	
High-rate filters	Porous pavement	Livestock exclusion fencing	
Biological treatment	Inflow and infiltration controls in sewers		
High-rate disinfection	Sewer deseparation in overloaded sanitary systems		
	Diversion channels and berms		

(*Sources:* Novotny and Olem 1994; Switzer-Howse 1982; Waterhouse 1982)

Table 5.3 Examples of Vegetative Measures for the Control of Nonpoint-Source Pollution

Agricultural Systems	Urban Systems
Filter strips and buffer zones	Filter strips and buffer zones
Critical area planting	Critical area planting
Constructed wetlands or reed beds	Constructed wetlands or reed beds
Restoration of aquatic habitat	Restoration of aquatic habitat
Crop/plant selection for maximum nutrient uptake and/or minimum pesticide use	Crop/plant selection for maximum nutrient uptake and/or minimum pesticide use, or for physical stability
Strip-cropping and intercropping	Xeriscaping
Cover crops	
Range management	
Crop rotations	

(*Sources:* Novotny and Olem 1994; Switzer-Howse 1982; Waterhouse 1982)

the control of pet excrement in urban systems. Despite the obvious efficacy of this measure, it has been difficult to implement in many cities because of public resistance to change and because municipalities simply cannot enforce such ordinances effectively. Successful implementation has occurred where the ordinance has been combined with a comprehensive public education program.

In agricultural systems, best management practices may include measures such as conservation tillage and contour plowing. Here again, the water manager may meet with resistance from the farm operator, even though the measure is low in cost and likely to be effective. Family traditions, for example, may dictate that a particular tract of land is plowed in a given way. More commonly, farmers express an understandable reluctance to risk lower productivity by changing tried-and-true farming methods. Sometimes financial incentives such as grants, loans, or tax relief can be helpful in encouraging a shift toward more protective practices. Table 5.4 gives some examples of nonstructural, or "management," practices commonly used in agricultural and urban systems.

5.4 DEVELOPING MUTUALLY EXCLUSIVE MANAGEMENT ALTERNATIVES

In most cases the management goals of a watershed plan can be accomplished in many different ways. In one situation a "high-tech" solution, employing a range of built technologies, may work best; in another, simply changing operating practices will be the preferred approach. But how are these strategies developed in the first place? Although management strategies are, in practice, developed in many ways, often ad hoc, careful and systematic formulation of alternatives can vastly simplify subsequent analysis and improve decision clarity.

The principles of developing alternatives are straightforward:

1. All possible management options must be listed separately (these can be

Table 5.4 Examples of Best Management Practices for the Control of Nonpoint-Source Pollution

Agricultural Systems	Urban Systems
Contour plowing	"Stoop and scoop" ordinances
Range management	Storm drain marking and public education to reduce waste disposal into sewers
Appropriate pesticide and herbicide application practices	Appropriate pesticide and herbicide application practices
Drip irrigation	Drip irrigation
Recycle/reuse of irrigation return flow and runoff water	Mulching and surface protection in new construction areas
Timely lagoon pump-out	Catch basin cleaning
Proper application rate and timing of manure application	Proper site grading
Proper site selection for animal feeding facility	Control of deicing chemicals and abrasive materials
Integrated pest control	Antilitter ordinances
Appropriate stocking rates	Sewer flushing

(*Sources:* Novotny and Olem 1994; Switzer-Howse 1982; Waterhouse 1982)

considered separate and independent "projects" in that each would have a unique budget, schedule, and associated considerations).

2. All possible *combinations* of options are listed (for the purposes of this discussion, we can consider that these will be the feasible combinations of options—the "short list" to be developed through screening procedures discussed in Chapter 6).

3. All options (and combinations of options) are evaluated on the basis of a common planning period, discount rate, and basis of comparison (evaluation criteria).

This framework ensures that all options are enumerated and that all management strategies, whether comprising one or several individual actions, are mutually exclusive. This is important, so that acceptance of one management strategy clearly precludes acceptance of another.

An example may illustrate this important point. A hypothetical river basin currently experiences heavy loadings of sediment and nutrients, primarily from agricultural activities. Nutrient enrichment has, in turn, led to excessive aquatic plant growth (eutrophication), with associated nighttime oxygen depletion and fish kills. The following (simplified) individual management actions are possible:

1. Do nothing.

2. Construct riparian buffer strips to capture sediment and attached nutrients before they enter the stream.

3. Construct livestock exclusion fencing to prevent cattle from entering the stream and trampling stream banks.
4. Encourage conservation tillage in the basin.

These individual options may then be combined in various ways to create several new management strategies:

5. Construct riparian buffer strips *and* livestock exclusion fencing.
6. Construct riparian buffer strips *and* encourage conservation tillage.
7. Construct livestock exclusion fencing *and* encourage conservation tillage.
8. Construct riparian buffer strips *and* livestock exclusion fencing *and* encourage conservation tillage.

This list of options is exhaustive (of all possible combinations of options), and the alternatives listed are mutually exclusive—that is, acceptance of one automatically precludes acceptance of another. The fact of this complete list thus allows systematic and objective comparison of each possible management strategy to determine which is the preferred alternative. Table 5.5 illustrates the process graphically.

Table 5.5 Mutually Exclusive Combinations of Four Independent Projects

Management Alternative (Strategy)	Project				Explanation
	Do Nothing	Buffer Strips	Fencing	Conservation Tillage	
1	Yes	No	No	No	Accept "do nothing" option
2	No	Yes	No	No	Construct buffer strips
3	No	No	Yes	No	Construct fencing
4	No	No	No	Yes	Encourage conservation tillage
5	No	Yes	Yes	No	Construct buffer strips and fencing
6	No	No	Yes	Yes	Construct fencing and encourage conservation tillage
7	No	Yes	No	Yes	Construct buffer strips and encourage conservation tillage
8	No	Yes	Yes	Yes	Construct buffer strips, fencing, and encourage conservation tillage

This example does not include the case of alternatives that are not independent—that are in some fashion interdependent. An example of a dependent alternative is provided by a sewage treatment plant upgrading scheme. One alternative is, of course, to do nothing—to decide not to upgrade. Another option is to upgrade from primary treatment to secondary treatment. A third option is to upgrade from primary treatment to tertiary (advanced) treatment. Although the upgrade-to-secondary option is clearly independent of the other two, the upgrade-to-tertiary option implies installation of secondary treatment as well: the upgrade-to-tertiary option is therefore dependent on the upgrade-to-secondary option.

Dependencies between management alternatives are not common but do arise in the development of remedial strategies. They must be treated carefully, so as to avoid double counting or, as is more likely to be the case, omission of linked options in evaluation.

Further discussion of this topic is available in most introductory engineering economics texts (which are generally concerned with the economic analysis and comparison of engineering projects) and in major works such as Bussey (1978).

5.5 EVALUATION CONSTRAINTS AND CRITERIA

The problem of determining which of a long list of management options is "best" occupies much of Chapters 6, 7, and 11 in this book. Fundamental to that screening is a need for evaluation constraints—limits to the solution—and criteria—measures of the effectiveness or suitability of a possible management action. In some decision processes, these constraints and criteria are not made explicit, with the result that participants disagree about the acceptability of an option without a clear understanding of the reasons for their dissatisfaction. Explicit discussion of design constraints and evaluation criteria encourages better citizen understanding and more focused decision making and can strengthen (or at least clarify) agency support.

There is an extensive literature available on multiobjective planning techniques. Some introductory engineering texts (e.g., DeGarmo et al. 1997) offer a simplified overview of this literature, and there are also review papers such as that by Cohon and Marks (1975) that summarize and evaluate possible approaches for use in situations where there may be multiple decision makers and disparate planning goals.

5.5.1 Development and Application of Planning Constraints

Constraints are the practical limitations to a solution, the limits beyond which the manager may not go. Constraints are typically expressed as "musts," rather than "shoulds," and are usually stated in terms of a specific numerical or spatial boundary. The most obvious example of a constraint is financial: there are only so many dollars available to achieve the goals of a project. Other constraints may include time (the project must be completed within a specified time period), space (must fit a specific site), or performance (must meet a minimum standard—for instance, of effluent quality).

In many countries, including the United States, Canada, and the United Kingdom, federal and state/provincial regulations limit the use of certain kinds of land or the development of certain resources. Typical among these are protections on wetlands and estuaries, wilderness preserves, scientific reserves of various kinds, and similar lands. As discussed in Chapter 2, valued historical or cultural resources in a watershed may be protected in ways that preclude certain water management options. Rare or endangered species may similarly be protected, often under law, from the impacts of near or encroaching development. These kinds of regulatory limits therefore place clear constraints on a plan.

Application of constraints is the first stage of option screening. It is the step in which infeasible options are excluded from further analysis and the long list of possibilities is reduced to a short list of feasible options that will be subjected to detailed scrutiny.

A simple example illustrates the application of constraints. Using the hypothetical river situation presented in Section 5.4, we can further assume that solutions must meet the following constraints:

- Annualized capital and operating costs less than $1,000
- Fully implementable within four months of plan approval

Table 5.6 presents some hypothetical data for the several mutually exclusive management alternatives under consideration.

Table 5.6 illustrates that *if* costs must be kept under $1,000, *and* any solution must be fully implemented within four months of plan approval, only three alternatives are really feasible: do nothing, construct buffer strips, or construct livestock exclusion fencing. The rest are either too costly or take too long to implement. This analysis thus allows us to reject the remaining five alternatives and proceed with detailed analysis of the three feasible options. (This simple example also illustrates the need for constraints to be chosen carefully. Is it really essential that costs be kept under $1,000? Or that all implementation be completed within four months? If so, these are appropriate constraints. If not, the analyst would be well advised to reconsider the limits to the solution and determine which are truly necessary for the planning situation.)

5.5.2 Choosing Appropriate Evaluation Criteria

Having eliminated options that are clearly not feasible, the analyst can proceed to evaluate the worth of the remaining options using a set of evaluation criteria. The term *criteria* seems intuitive: we know, for instance, what criteria we might use to choose a new car. But careful thought reveals that

Table 5.6   Application of Planning Constraints in the Screening of Management Alternatives

Management Alternative (Strategy)	Description	Annualized Cost	Estimated Time to Implement	Meets All Constraints?
1	Accept "do nothing" option	$0	Immediate	Yes
2	Construct buffer strips	$750	1 month	Yes
3	Construct fencing	$500	4 months	Yes
4	Encourage conservation tillage	$250	1 year	No
5	Construct buffer strips and fencing	$1,250	4 months	No
6	Construct fencing and encourage conservation tillage	$750	1 year	No
7	Construct buffer strips and encourage conservation tillage	$1,000	1 year	No
8	Construct buffer strips, fencing, and encourage conservation tillage	$1,500	1 year	No

effective evaluation criteria must—even in the case of a new car—have two important characteristics. First, they must be measurable by some agreed-upon method. In this sense, "capital cost" is readily seen as a good criterion; we can easily imagine the components of capital cost and how they might be measured. By contrast, "social impact" is a more nebulous criterion: do we mean employment rate? average income? or some other measure such as "quality of life"? "Environmental impact" is just as problematic—it can be taken to mean anything from habitat disruption to the presence of persistent toxic chemicals. A criterion is most useful when it expresses a single, quantifiable attribute.

The second important characteristic of an effective criterion is that it does, in fact, separate alternatives. Again, "capital cost" is effective in this regard, because no two alternatives will have exactly the same capital costs. But a criterion such as "phosphorus removal capability" is less useful if all options under consideration have similar removal capability.

This second characteristic is often harder to achieve. In some cases, it will not immediately be apparent until the decision process is well advanced that the chosen criteria are not effective in separating available options. Even late in the game, however, it is possible (and indeed advisable) to refine, replace, or simply drop problematic criteria, as long as the change has the support of the full decision-making group.

Table 5.7 Typical Evaluation Criteria

Economic Criteria	Environmental Criteria	Social Criteria
Capital cost (construction and equipment)	Pollutant removal capability (best expressed separately for each pollutant)	Average level of education in the community
Operating and maintenance costs	Concentration or loading of a key water quality constituent	Population density
Administrative costs; administrative complexity	"Footprint" as a measure of extent of habitat disruption	Land area available for recreation
Probable life of structures and equipment	Species diversity	Number of days available for recreation/leisure
Estimated revenues	Available bather days (for swimming beaches) as measured by bacterial counts	Number of available fishing permits
Estimated cost savings (including avoidance of fines and litigation)	Available fishing days	Aesthetic potential (as measured by survey or key informant interview)
Number of commercial navigation days available	Kilometers of habitat for desired fish species	User satisfaction (as measured by survey of key informant interview)

Lumsdaine and Lumsdaine (1995) suggest that criteria should attempt to answer the following concerns:

- *Motivation:* Why would people want to accept the option?
- *People:* How will people be affected by the option?
- *Cost:* What will the costs be to you and others?
- *Support:* What support is available for implementation?
- *Values:* What social values are involved? What will be the benefits to people?
- *Time:* Will the option take a long time to implement?
- *Effects:* What will be the consequences of the option?

Table 5.7 gives some examples of typical evaluation criteria.

If these criteria seem familiar, it is because they, and others like them, are drawn directly from the targets set in Chapter 3. This is as it should be: if the scoping process described in Chapter 3 is intended to reflect a growing societal consensus about the current state of the watershed, and about some desired future state, then the targets for improvement should be used as direct measures of progress toward the societal goal.

It is important to understand that there is no single evaluation criterion that is universally appropriate (although it must be admitted that cost is almost ubiq-

Table 5.8 Application of Evaluation Criteria in Screening Management Alternatives

Management Alternative (Strategy)	Description	Annualized Cost	Suspended Sediment Reduction Efficiency	Phosphorus Reduction Efficiency
1	Accept "do nothing" option	$0	0%	0%
2	Construct buffer strips	$750	55%	35%
3	Construct fencing	$500	35%	45%

uitous in evaluation). The correct mix of criteria should reflect not only the problem under consideration, but also the attitudes and values of the community undertaking the evaluation.

Table 5.8 shows an extension of the simple river example used in Sections 5.4 and 5.5.1, illustrating the analysis of alternatives using a common set of evaluation criteria.

Table 5.8 illustrates a common problem in multicriteria analysis: that a single option rarely is "best" across all evaluation criteria. (It is, however, clear from this analysis that Alternative 1, the "do nothing" option, is less effective than the other two and probably should be excluded from further consideration.) Sections 5.5.3 and 5.5.4 describe several ways to deal with the problem of multicriteria decision making.

5.5.3 Weighting Evaluation Criteria

In many decision processes, a single criterion will emerge as most important. Sometimes, although not as often as might be expected, that criterion is cost. Equally often, however, considerations such as control of a particular pollutant, or reduction of flood frequency or severity, eclipse others in reaching a final decision.

Assigning weights to each evaluation criterion serves at least two purposes. First, it allows the analyst to make full use of available information about community interests and priorities. And second, it has the result of "spreading" the field—accentuating differences between alternatives so as to make a final choice more clear-cut.

Weights can and should be chosen by consensus, not arbitrarily. It is helpful if all weights in a set of criteria sum to unity, not just because the idea of percentage weight is intuitive, but also because it avoids excessive overweighting of high-priority criteria and thus skewing of results. Using the data presented in Table 5.8, we can see that three criteria are under consideration: cost, suspended sediment reduction efficiency, and phosphorus reduction efficiency. Although we may believe cost to be more important than the other two criteria, it would

Table 5.9   Application of Weighted Evaluation Criteria in Screening Management Alternatives

Management Alternative (Strategy)	Description	Annualized Cost	Suspended Sediment Reduction Efficiency	Phosphorus Reduction Efficiency
		Weight: 40%	*Weight: 30%*	*Weight: 30%*
1	Accept "do nothing" option	$0	0%	0%
2	Construct buffer strips	$750	55%	35%
3	Construct fencing	$500	35%	45%

not be reasonable to assign that criterion 10 times the weight of the others. A more realistic approach might be to assign a weight of 40% to cost, and 30% to each of the other criteria. This would yield a modified determination of criteria like that shown in Table 5.9. But it introduces a new concern: although we can readily imagine how to multiply each score by the weight of a particular criterion, how do we compile a composite score that mixes costs with pollutant removal efficiency? Section 5.5.4 discusses this problem in more detail.

5.5.4   Multiattribute Decision Making

Inevitably, watershed management requires decision making that encompasses a variety of considerations measured in different ways. The following brief discussion of multiattribute decision making provides an introduction to the more detailed techniques presented in Chapters 6, 7, and 11.

The fundamental problem of multiattribute decision making is that of comparing apples with oranges. Some researchers have tried to address this difficulty by reducing all considerations, including not only costs but environmental and social criteria, to dollar values. Although this approach, termed "compensatory analysis," is certainly viable—the literature is full of examples—it is not the only possible approach, nor is it necessarily the best.

The clear advantage of compensatory models is that they are intuitive (everyone understands the meaning of a unit of money) and allow trade-offs among alternatives. For instance, if alternative A costs more but creates more dollar benefits (say, in terms of crop production) than alternative B, it is easy to see that net benefits—benefits minus costs—are higher for A, making it the preferred approach.

Trade-offs of this kind are not possible when we are comparing costs and, say, unemployment rate, or the incidence of swimmer's itch. When the attributes of interest cannot, or should not, be reduced to a single measure such as cost, other methods of comparison must be found.

How does one make a decision when "value" cannot be reduced to dollar terms? Several "full-dimensioned" or "noncompensatory" models suggest possible approaches.

Dominance   Where one alternative is clearly better than all others for all attributes, that alternative is said to dominate the process and is obviously to be preferred. Regrettably, this rarely occurs. More often an alternative performs well on some criteria (for example, cost) but less well on others (for example, phosphorus removal capability). The decision maker is thus faced with an unhappy choice between cost and treatment effectiveness. Some of the following methods are helpful in this more common situation.

Feasible Ranges   In this method, an acceptable range of performance is established for each attribute (for example, cost between $1 and $2 million; phosphorus removal capability between 65% and 100%). Alternatives whose performance falls outside the acceptable range on any individual attribute are rejected from further consideration. Alternatives that remain after the screening process therefore meet or exceed the specified standard of performance on every decision criterion.

There are several problems with this method. One is that it can be difficult to decide what constitutes an acceptable versus unacceptable range for any single attribute. Perhaps its greatest failing, however, is that it yields satisfactory, but not optimal, management strategies.

Lexicography   *Lexicography* is a term sometimes used for a stepwise process of option screening. It requires that attributes be weighted. The most important attribute is then used to choose the "best" option. If there are several options tied for performance on this attribute, the second and third most important attributes are used to break the tie(s).

This approach places heavy emphasis on the primary attribute, even to the exclusion of other considerations, and so must be used with care to ensure that all possible information is used in the analysis.

Standardization (Nondimensional Scaling)   It may be possible to reduce different kinds of measurements to a standardized scale, say from 0 to 10, or from -10 to $+10$. (Simple ranking is one approach to this problem, but it sacrifices information that is possible to retain with ordinal scaling.) Then the standardized variables can be manipulated as if they were indeed a common measure. If different attributes are weighted differently, each nondimensional value can be multiplied by the appropriate weight and the weighted sum of scores on all attributes calculated. (The weighted-sum approach is, of course, also applicable to compensatory models, where all attributes are measured in the same way.)

Standardization of the data presented in Table 5.9 allows the completion of alternative screening and selection of a preferred alternative, as shown in Table 5.10.

Table 5.10 Application of Weighted Evaluation Criteria and Data Standardization in Screening Management Alternatives (best = 1, worst = 3)

Management Alternative (Strategy)	Description	Annualized Cost	Suspended Sediment Reduction Efficiency	Phosphorus Reduction Efficiency	Total Score (lowest is best)
		Weight: 40%	*Weight: 30%*	*Weight: 30%*	*100%*
1	Accept "do nothing" option	rank = 1 × 40% = 0.4	rank = 3 × 30% = 0.90	rank = 3 × 30% = 0.90	2.2
2	Construct buffer strips	rank = 3 × 40% = 1.2	rank = 1 × 30% = 0.30	rank = 2 × 30% = 0.60	2.1
3	Construct fencing	rank = 2 × 40% = 0.80	rank = 2 × 30% = 0.60	rank = 1 × 30% = 0.30	**1.7** (preferred alternative)

This analysis demonstrates that, within the specified constraints and using the given evaluation criteria and their weights, the long list of alternatives presented in Table 5.5 can be reduced to a single, preferred alternative: 3, construction of livestock exclusion fencing. This analysis also reveals several other interesting aspects. Among these is the fact that there is not a lot of difference between the alternatives in terms of their weighted scores; in other words, although Alternative 3 proved to be preferable, it did not "win" by a large margin. This result is a useful clue to the analyst that some test of robustness of the solution may be desirable. For instance, if cost is weighted more heavily, say at 50%, with only 25% each for the other two criteria, the scores would be:

$$\Sigma_{alt1} = (1 * 0.5) + (3 * 0.25) + (3 * 0.25) = 0.5 + 0.75 + 0.75 = 2.0$$
$$\Sigma_{alt2} = (3 * 0.5) + (1 * 0.25) + (2 * 0.25) = 1.5 + 0.25 + 0.50 = 2.25$$
$$\Sigma_{alt3} = (2 * 0.5) + (2 * 0.25) + (1 * 0.25) = 1.0 + 0.50 + 0.25 = 1.75$$

Under these conditions, Alternative 3 remains the preferred approach, but Alternative 1 ("do nothing") appears preferred over Alternative 2.

On the other hand, if we reversed the weights of the criteria, weighting cost at only 20% and the other two at 40% each, we find that Alternative 3 still has the lowest score and remains the most attractive:

$$\Sigma_{alt1} = (1 * 0.2) + (3 * 0.4) + (3 * 0.4) = 0.2 + 1.2 + 1.2 = 2.6$$
$$\Sigma_{alt2} = (3 * 0.2) + (1 * 0.4) + (2 * 0.4) = 0.6 + 0.4 + 0.8 = 1.80$$
$$\Sigma_{alt3} = (2 * 0.2) + (2 * 0.4) + (1 * 0.4) = 0.2 + 0.8 + 0.4 = 1.40$$

These results suggest (not surprisingly, given the simplicity of the example) that the analysis would be strengthened by the addition of other evaluation criteria to "spread" the final results and make selection of a preferred alternative more clear-cut. Or the results of the analysis can be used "as is" to instruct decision makers: if cost is clearly the most important criterion, then Alternative 2 is the preferred approach. If pollutant reduction efficiency is more important than cost, then the decision makers should select Alternative 3. Alternative 1 is attractive only if cost is the sole consideration and pollutant removal efficiency is excluded from the analysis.

This brief discussion is intended to introduce the concepts of constraints, criteria, weighting, and multiattribute decision making. Specific analytical techniques for the evaluation of management alternatives and the problem of selecting a "best" management plan for the watershed are described in more detail in Chapters 6, 7, and 11 of this book.

REFERENCES

Bussey, L. E. 1978. *The Economic Analysis of Industrial Projects.* Englewood Cliffs, N.J.: Prentice-Hall.

Cohon, J. L., and D. H. Marks. 1975. A review and evaluation of multiobjective planning techniques. *Water Resources Research* 11(2): 208–220.

Davis, Mackenzie L. and David A. Cornwell. 1991. *Introduction to Environmental Engineering.* 2d ed. New York: McGraw-Hill.

DeGarmo, E. Paul, William G. Sullivan, James A. Bontadelli, and Elin Wicks. 1997. *Engineering Economy.* 10th ed. Upper Saddle River, N.J.: Prentice-Hall.

LeBoeuf, Michael. 1980. *Imagineering: How to Profit from Your Creative Powers.* New York: Berkley Books.

Lumsdaine, Edward, and Monika Lumsdaine. 1995. *Creative Problem Solving: Thinking Skills for a Changing World.* New York: McGraw-Hill.

Metcalf and Eddy Inc. 1991. *Wastewater Engineering: Treatment, Disposal, and Reuse.* 3d ed. New York: McGraw-Hill.

Novotny, V., and H. Olem. 1994. *Water Quality: Prevention, Identification, and Management of Diffuse Pollution.* New York: Van Nostrand Reinhold.

Ontario Ministry of the Environment. 1987. *Technical Guidelines for Preparing a Pollution Control Plan.* Report from Urban Drainage Policy Implementation Committee, Technical Sub-Committee No. 2. Toronto, Ont.: Ontario Ministry of the Environment.

Osborn, Alex F. 1963. *Applied Imagination—The Principles and Problems of Creative Problem-Solving.* 3d rev. ed. New York: Scribner's.

Prince, George M. 1970. *Practice of Creativity.* New York: Macmillan.

Schwab Glenn O., Delmar D. Fangmeier, William J. Elliot, and Richard K. Frevert. 1993. *Soil and Water Conservation Engineering.* 4th ed. New York: John Wiley & Sons.

Switzer-Howse, Karen. 1982. *Agricultural Management Practices for Improved Water*

Quality in the Canadian Great Lakes Basin. LRRI Contribution #82-10. Ottawa: Agriculture Canada Research Branch.

Van Gundy, Arthur B. 1984. *Managing Group Creativity: A Modular Approach to Problem Solving.* New York: American Management Associations.

Waterhouse, James. 1982. *Water Engineering for Agriculture.* London: Batsford Academic and Educational Ltd.

6

Simple Assessment Methods

Chapter 5 described the development of alternative approaches for water management problems. As discussed in that chapter, any given option will usually be considered very promising in some respects, but disappointing in others. Weighing diverse criteria in the evaluation of management alternatives is therefore one of the most challenging aspects of developing a water management plan. The amount of effort and resources put toward this task will depend on the goals of the process, the availability of resources and information, and the expertise of the analysts.

Several approaches are possible. A cursory, or preliminary, screening can inexpensively provide rapid insight into the probable effectiveness of different management strategies. It may not, however, give the kind of detailed, quantitative results necessary to justify fiscal commitment or definitively sway public, or political, opinion. Detailed assessment techniques, on the other hand, are time-consuming and costly and may not be feasible within the window of opportunity afforded by a particular funding program or political administration.

Any screening program should have the following goals:

1. To determine which of the available alternatives is feasible—that is, meets the constraints imposed by cost, space, time, or technology

2. To determine which of the remaining alternatives performs best in terms of specified evaluation criteria (see Section 5.5).

In fulfilling these objectives the analyst will, by default, accomplish two other goals:

1. To ensure that complete and accurate data are available for all alternatives under consideration.
2. To rank individual alternatives (and component projects) in terms of their overall performance.

In many, perhaps most, cases watershed managers will have a limited budget and time frame within which to develop management recommendations. Resource limitations will therefore restrict the amount of new data that can be collected in the planning process, and the development of new analytical tools such as computer simulation models for evaluating management alternatives. In other cases a preliminary or "feasibility" level analysis may be desired in advance of more detailed work, and perhaps to justify the resource allocation that would be necessary for more detailed data collection and analysis.

In these circumstances the analyst can employ a range of simple analytical techniques, with little or no collection of new data, and produce a general picture of problems, mechanisms, and solutions. The effectiveness of this analysis will depend very much on the quality of available data and the extent to which watershed processes are already understood. At the very least, a preliminary analysis should reveal:

- Significant data gaps
- The need for additional technical analysis
- The need for clarification of study goals and objectives
- Recommendations for immediate pollution abatement actions

Figure 6.1 illustrates the main components of a preliminary watershed analysis, some of which have already been discussed in Chapters 2 to 5. It will be seen from Figure 6.1 that a preliminary (simple) analysis includes the same steps of inventory, scoping, development and testing of alternatives, and outputs as does a detailed analysis. It differs from a detailed assessment primarily in the level of detail used, the quality (and often age) of data, and thus in the precision and accuracy of outputs. The following discussion illustrates this point for each component of the analysis.

The following discussion illustrates this point for each component of the process.

6.1   THE WATERSHED INVENTORY

The introduction to Chapter 2 makes the point that water movement in a system is affected by many physical, chemical, and biological features and processes.

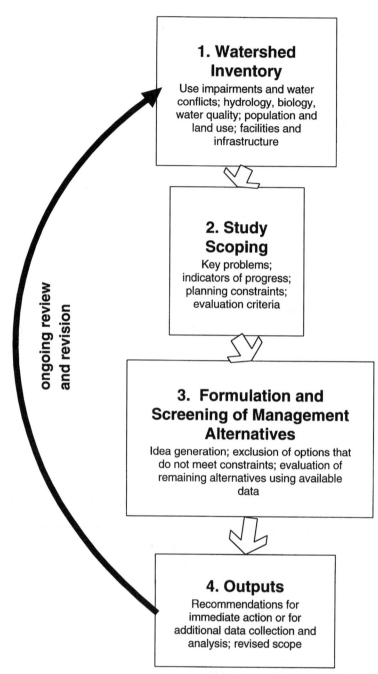

Figure 6.1 *Components of a preliminary watershed analysis.*

Preparing an inventory of a watershed is therefore directed at developing an understanding of these features and processes. At the level of a simple assessment, most information can be gleaned from existing data, from the literature, or from simple observations.

6.1.1 Using Existing Data

A simple inventory of watershed resources can be compiled with the following tools, most of which are usually readily available in major towns:

1. A foundation map of the watershed, showing

   Topography

   Location of major features with potential to influence water management, such as dams, locks, protected areas, areas of cultural or spiritual value, habitats of rare or endangered species

   Overview of land use (e.g., residential, industrial, open space, forested, agricultural, etc.)

   Location of major point source discharges (e.g., municipal sewage treatment plant(s), industrial effluent discharges)

   Spatial relationships between potential sources and receiving waters

   Political boundaries

2. Existing data on specific physical, chemical, and biological characteristics of the system (for instance, from state agencies)

3. Existing data on specific physical, chemical, and biological characteristics of pollutant sources (from public or industrial sources)

4. Media reports and public commentary about the key water use impairments and use conflicts currently present in the basin

Table 6.1 lists the types of data that are usually necessary for a basic watershed inventory.

One simple but very useful inventory technique is to use a topographic map of the watershed as a foundation and overlay that map with transparencies showing other features such as important recreational areas, key sources, and so on. (Geographic Information Systems are of great value in this undertaking and may already be in use in some jurisdictions. They are not, however, necessary to achieve the required overlays.) The overlays should reveal areas that are under particular stress, such as a major point source located near fragile wildlife habitat, or residential development encroaching on sensitive wetlands.

A composite map can be compiled for the present condition and, using land use and population growth projections, for various planning horizons in the future.

Table 6.1 Typical Data Required for a Basic Watershed Inventory

Type of Data	Potential Sources
Geology	National and state/provincial agencies
Landforms	National and state/provincial agencies
Climate	National and state/provincial agencies
Soils	National and state/provincial agencies; university researchers
Mean monthly streamflow	National and state/provincial agencies; university researchers
Minimum streamflow	National and state/provincial agencies; university researchers
Critical low streamflows and recurrence intervals	National and state/provincial agencies; university researchers
Peak monthly streamflow	National and state/provincial agencies; university researchers
Base flows	National and state/provincial agencies; university researchers
Quality and yield of groundwater aquifers	National and state/provincial agencies; university researchers
Zones of groundwater contamination	National and state/provincial agencies; university researchers
Lake morphology (depth, shoreline, area, residence time)	National and state/provincial agencies; university researchers
Quality and quantity of urban runoff	Municipal agencies; literature sources
Quality and quantity of agricultural runoff	National and state/provincial agencies; farm organizations; literature sources
Quality and quantity of industrial and municipal point sources	National and state/provincial agencies; possibly municipal agencies; industrial sources; sewer maps
Receiving water quality (average and maximum concentrations and/or loadings of key parameters)	National and state/provincial agencies; municipal agencies; university researchers
Impaired uses and water use conflicts	Municipal agencies; public consultation (see Chapter 4)
Plant and animal communities; ecology	National and state/provincial agencies; nongovernment organizations; university reseachers
Land use	National and state/provincial agencies; nongovernment organizations; university researchers; aerial photographs
Social and economic systems; urban centers	Municipal agencies; public consultation (see Chapter 4)
Valued watershed features	Municipal agencies; public consultation (see Chapter 4)
Beneficial uses	Municipal agencies; public consultation (see Chapter 4)
Existing water demand	Municipal records (e.g., water sales); national and regional statistics

6.1.2 Filling Data Gaps

Data gaps are unavoidable in any watershed analysis, but they are perhaps less critical in a simple assessment than in more detailed analysis. Because a simple analysis itself is less demanding, alternate data sources can be used to good advantage. Table 6.2 suggests some other ways to collect basic watershed information if the data suggested in Table 6.1 are unavailable.

6.2 SCOPING

6.2.1 Defining the Problem

For the purposes of a simple assessment, the most urgent problems in the watershed can be determined in several simple ways:

1. Through a listing of existing (or perceived) use impairments and their causes
2. By comparing existing receiving water or sediment quality with published national or state/provincial targets
3. Through key informant interviews and media reports

Processes like Adaptive Environmental Assessment and Management (AEAM) (Section 3.4.2) can be useful in clarifying community consensus on watershed problems without detailed quantitative analysis. AEAM draws from community knowledge of the basin and its resources and can be a very powerful tool in problem definition. The various techniques described in Chapter 4 may also be useful, but some may require resources beyond those available for a simple assessment.

6.2.2 Identifying the Sources

Identification of key problems often simplifies the identification of sources, especially if such a problem is limited to a particular area within the basin. Reference to a composite map should quickly reveal the presence of possible point sources or of land uses that may generate diffuse pollution during wet weather.

In some cases there will be several sources of a single pollutant. It makes sense to determine which of these is the most important contributor to the total load and which would repay the greatest benefit per dollar spent on remediation. The two are not necessarily the same: it may be very costly to reduce pollutant loads from a large contributor such as a sewage treatment plant. On the other hand, inexpensive measures like livestock-exclusion fencing may achieve greater pollutant reductions for every dollar spent. Depending on the situation, it may therefore be better to install many kilometers of fencing rather than upgrade

Table 6.2 Alternate Sources for Watershed Inventory Information

Type of Data	Potential Sources
Geology	Amateur observations, or omit
Landforms	Amateur observations
Climate	Typical values from published atlases
Soils	Amateur observations; farm operator observations
Mean monthly streamflow	Pro-rated data from similar neighboring stream based on watershed area; amateur observations
Minimum streamflow	Pro-rated data from similar neighboring stream based on watershed area; amateur observations
Critical low streamflows and recurrence intervals	Pro-rated data from similar neighboring stream based on watershed area; amateur observations
Peak monthly streamflow	Pro-rated data from similar neighboring stream based on watershed area; amateur observations
Base flows	Pro-rated data from similar neighboring stream based on watershed area; amateur observations
Quality and yield of groundwater aquifers	Private pumping/extraction observations
Zones of groundwater contamination	Private extraction observations
Lake morphology (depth, shoreline, area, residence time)	Fishermen's surveys; aerial photographs; rowboat survey
Quality and quantity of urban runoff	Amateur observations; field survey to identify outfall locations, size, and characteristics; literature sources
Quality and quantity of agricultural runoff	Amateur observations; farm operator observations; literature sources
Quality and quantity of industrial point sources	Literature sources for specific industrial sector
Receiving water quality (average and maximum concentrations and/or loadings of parameters)	Appearance/inference (e.g., algae blooms point to eutrophication; turbidity to high sediment loads); infer probable sources
Impaired uses and water use conflicts	Media reports; key informant interviews
Plant and animal communities; ecology	Published field guides
Land use	Drive-by surveys; school bus surveys (children note uses along bus route); self-reporting
Social and economic systems; urban centers	Road maps; municipal directories
Valued watershed features	Media reports; key informant interviews
Beneficial uses	Media reports; key informant interviews
Existing water demand	Literature values; pro-rated values from neighboring town(s)

162

a sewage treatment plant to advanced treatment levels. The problem of evaluating the relative importance of sources is discussed in the next section.

6.2.3 Evaluating the Relative Importance of Sources

Pollutant (or flow) source analysis at a coarse level is designed to set priorities on sources for further evaluation or immediate action. Source analysis can be of two types: total mass and critical levels (which, in the case of pollutants, would be critical concentrations). Section 5.1 discusses the difference between mass loading and concentration. Either or both types of analysis may be appropriate in any given situation; the decision will depend on the nature of the problem(s) under investigation, applicable laws, and the interests of the investigating agencies.

The screening techniques discussed in Section 6.3 have value in terms of estimating loads and concentrations under status quo conditions. Tables 6.1 and 6.2 gave a number of suggested sources of data on dry-weather (continuous) point sources, wet-weather (intermittent) point sources, and nonpoint (diffuse) sources. In general, for a simple evaluation, the literature provides a rich supply of information on all categories of source and on potential control approaches. Data drawn from the literature are rarely directly applicable to the area under study, however. Some judgment is required in applying literature values to the problem of interest. Local experts from government agencies, consulting firms, or universities can be helpful in making these decisions.

Ideally, the analyst should attempt to develop source analyses like that shown in Figure 6.2 for every pollutant or material of concern. (The graphical analysis shown in Figure 6.2 could, of course, be replaced with tables containing the same or similar data.)

6.3 DEVELOPING AND SCREENING MANAGEMENT ALTERNATIVES

6.3.1 Developing Alternatives

Chapter 5 has described the general process of developing alternatives. At the simple assessment level, the analyst should strive to:

- Identify the sources that contribute most to the problem under study
- For each nontrivial source (those contributing, say, more than 5% of the total load), identify at least one management option in each of the following categories:

    "Do nothing"—retaining status quo

    Structural measures—built technologies or structures

    Vegetative approaches

    "Best management practices" (nonstructural measures)

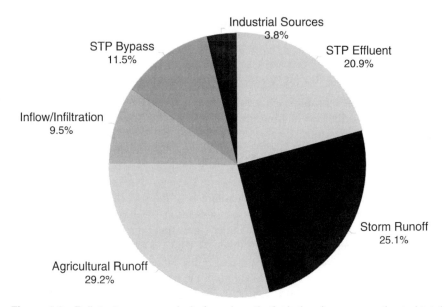

Figure 6.2 *Pollutant source analysis for a hypothetical planning area: estimated total phosphorus loads to receiving water.*

The techniques described in Chapters 4 and 5 can be used to generate these ideas. This suite of four options provides a minimum set for coarse screening. The following is an example of such a set of measures for the control of solids in urban stormwater runoff:

- "Do nothing"–Retain status quo.
- Structural measure—Construct stormwater detention pond at the intersection of Oak Street and Willow Road.
- Vegetative approach—Construct artificial wetland for stormwater treatment at intersection of Oak Street and Willow Road.
- "Best management practice"—Institute semiannual cleaning for all existing catch basins throughout the city.

This brief list is clearly a bare minimum; the tables of options given in Section 5.3 provide many more, as would the extensive literature on each source type. The point is to develop a set of possible solutions for each source of each pollutant or material of concern.

The next step is then to apply any planning constraints (Section 5.5) and eliminate any options that are clearly inappropriate or impractical for the planning situation. The remaining "short list" of options can then be subjected to screening, using a range of simple screening procedures, as discussed in the following section.

6.3.2 Simple Procedures for Estimating Pollutant Loadings from Pervious Surfaces

Section 2.3.3 described several methods for estimating stormwater runoff. Such estimates are an important first step in estimating the pollutant loadings associated with runoff. As discussed in Section 2.3.3, runoff rates, volumes, and quality are closely related to the perviousness of the soil surface. In urban areas most rainwater falls on paved or roofed surfaces and is diverted through sewer systems and to treatment or discharge points. In rural areas more of the land surface is pervious, so more of the rainwater falling over a rural area infiltrates into soils, is taken up by growing vegetation, or is returned to subsurface flows. Runoff prediction models take these differences into account.

The quality of runoff is also different in urban and rural areas. Table 2.6 illustrated some differences in pollutant loadings from different land uses. Table 6.3 presents certain values for pollutant concentrations associated with different runoff sources (note that these values can be used as values for the variable "C" values in Schueler's "Simple Method," shown in Table 6.5).

The much higher pollutant concentrations associated with older residential areas are attributable to poor "housekeeping," including accumulation of refuse and debris, decaying building fabric, high traffic volumes, and poor maintenance of open spaces and lawns.

It is certainly possible to generate detailed estimates of pollutant loadings using quantitative methods. These methods are, however, usually far too complex and time-consuming to be useful in a simple assessment. A number of authors have suggested simple techniques for estimating pollutant loadings, especially for wet-weather, intermittent sources. (These are usually the most difficult to estimate, continuous dry-weather sources being far more predictable in terms of flow and quality.)

Pollutant loads from areas with predominantly pervious surfaces (most rural areas and urban open space) are easily estimated with the Universal Soil Loss Equation (USLE) (Wischmeier and Smith 1978):

$$A = RK(LS)CP \tag{6.1}$$

where: A = average annual soil loss in mg/ha
R = rainfall and runoff erosivity index for location of interest
K = soil erodibility factor
L = slope length factor
S = slope steepness factor
C = cover management factor
P = conservation practice factor

Potential values for the USLE variables vary greatly. Graphs, tables, and nomographs showing recommended values for different conditions are widely available in the literature. Novotny and Olem (1994) provide an excellent dis-

Table 6.3 Typical Pollutant Concentrations (mg/L) (from Schueler 1987)

Pollutant	New Suburban NURP Sites* (Washington, D.C.) Source: US EPA 1983	Older Urban Areas (Baltimore) Source: BRPC 1986	Central Business District (Washington, D.C.) Source: MWCOG 1983	National NURP* Study Average Source: US EPA 1983	Hardwood Forest (N. Virginia) Source: OWML 1983	National Urban Highway Runoff Source: Shelley and Gaboury 1986
Total Phosphorus	0.26	1.08	—	0.46	0.15	—
Orthophosphate	0.12	0.26	1.01	—	0.02	—
Soluble Reactive Phosphorus	0.16	—	—	0.16	0.04	0.59
Organic Phosphorus	0.10	0.82	—	0.13	0.11	—
Total Nitrogen	2.00	13.6	2.17	3.31	0.78	—
Nitrate	0.48	8.9	0.84	0.96	0.17	—
Ammonia	0.26	1.1	—	—	0.07	—
Organic Nitrogen	1.25	—	—	—	0.54	—
Total Kjeldhal Nitrogen	1.51	7.2	1.49	2.35	0.61	2.72
COD	35.6	163.0	—	90.8	>40.0	124.0
BOD (5 day)	5.1	—	36.0	11.9	—	—
Zinc	0.037	0.397	0.250	0.176	—	0.380
Lead	0.018	0.389	0.370	0.180	—	0.550
Copper	—	0.105	—	0.047	—	—

*NURP refers to the Nationwide Urban Runoff Program, a study conducted by the U.S. federal government through the early 1980s to compile information about the quality and quantity of urban runoff.

Table 6.4 Estimated Sediment Delivery Ratios from Pervious Areas for Various Land Uses in the Menomonee River Basin, Wisconsin (after Novotny and Olem 1994)

Land Use Type	% Impervious Area	% Storm Sewering	Sediment Delivery Ratio (% delivered)
Agricultural	<5	0	1–30
Developing; under construction	<5	20–50	20–50
Low-density residential, unsewered	<20	0	<10
Parks	<10	0	<3
Medium-density residential, partially sewered	30–50	<50	30–70
Medium-density residential, sewered	30–50	>50	70–100
Commercial, high-density residential, sewered	>50	80–100	100

cussion of the USLE and its modifications, including a full range of values for the USLE variables.

The USLE cannot be used to predict loads of solids directly. To do so, the analyst must calculate potential soil loss using the USLE and then multiply that value by the delivery ratio, *DR*. Table 6.4 provides an overview of *DR* values for a range of land use types.

Table 6.4 clearly shows that as land becomes impervious and natural drainage (and infiltration) patterns are overridden by sewer systems, a higher proportion of eroded sediment is delivered to the receiving water. This may seem counterintuitive: after all, surely a higher degree of imperviousness and controlled drainage would *reduce* erosive forces. But upon understanding that sediment delivery ratio refers to the movement of sediment that has already been eroded, it becomes easy to see that paved and roofed surfaces are smooth and would not trap and retain sediment as well as pervious rural lands. In a city, therefore, sediment eroded or washed off (for instance, from construction sites) is rapidly transported to and through storm sewer systems and into a lake or stream. Unless treatment facilities have been installed to trap sediment, most (perhaps all, as Novotny and Olem suggest) of the eroded sediment will in fact reach the receiving water.

6.3.3 Simple Procedures for Estimating Pollutant Loadings from Impervious Surfaces

The preceding discussion illustrates the point that urbanization, with its paving, roofing, and sewerage, changes and overrides the natural hydrologic processes of a watershed. Typically, in systems with a high proportion of impervious surface, pollutants accumulate on surfaces during dry weather and are washed off those surfaces into stormwater runoff during wet weather. Stormwater is then conveyed through a system of pipes and traps and ultimately is discharged into a lake or stream. It is therefore possible to estimate the unit area pollutant loads from urban areas by combining estimates of pollutant accumulation on the land surface ("buildup") with estimates of runoff potential ("wash-off"). Many authors have suggested simple methods for the estimation of pollutant loads from urban areas. Most of these incorporate buildup/wash-off relationships.

Most streets are constructed with a slightly higher profile in the center of the roadway, sloping down to curbs or medians on either side. The purpose of this construction is, of course, to encourage drainage of stormwater away from traffic and toward storm drains. A variety of studies have shown that as a result of this construction, most solids tend to accumulate within a 1-meter-deep zone along the curb. This phenomenon is sometimes termed "curb storage," with solids accumulating in this way expressed as grams of pollutant per unit length of curb. The accumulation of solids on a road surface has been demonstrated to be nonlinear in most cases, with solids volume rising quickly to approach a pseudo-equilibrium level, where accumulation apparently levels off, probably as a result of displacement or movement of solids to neighboring areas. The "steady state" curb storage—the "buildup" of the buildup/wash-off relationship—ranges from about 20 to 100 g/m of curb (Novotny and Olem 1994).

Wash-off occurs when rainwater falls on an impervious surface, dislodging particles and conveying them into the stormflows. The quantity of pollutant in stormflows can therefore be estimated as a fraction of curb storage:

$$W = \left(\frac{CD}{A} \right) Sr \tag{6.2}$$

where: W = washed-off pollutant load in the sewer or at the basin outlet (kg per storm event)
   S = stored curb load of the pollutant during the preceding dry period (g/m curb length)
   CD = curb density (km/ha)
   A = subwatershed area (ha)
   r = wash-off factor (related to runoff energy)

Sartor et al. (1974) represented solids wash-off using a simple first-order relationship:

$$\frac{dP}{dt} = -k_{\mathrm{u}}rP \tag{6.3}$$

which integrates to:

$$P_t = P_0(1 - \exp(-k_u rt)) \tag{6.4}$$

where: r = rainfall intensity

k_u = urban washoff coefficient (constant, derived from street surface characteristics)

P = amount of solids remaining on the surface

P_0 = initial mass of solids in storage

P_t = mass of solids removed by rain with intensity r and duration t

This equation was further modified by staff of the Hydrologic Engineering Center (1975) in their development of the storm runoff model STORM, by adding an availability factor, A, to the equation by Sartor et al., on the assumption that not all particles are available for removal:

$$P_t = AP_0(1 - \exp(-k_u rt)) \tag{6.5}$$

where A, the availability factor, has a value between 0 and 1.0 and is determined as:

$$A = 0.057 + 0.04(r^{1.1}) \tag{6.6}$$

where r is rainfall intensity.

These equations can be used to estimate the total mass of solids that is removed by rainfall, but not the mass of pollutants removed. The most common approach to estimating pollutant loading is to multiply the mass of solids by some factor expressing the pollutant concentration of those solids. Novotny and Olem (1994) term these multipliers "potency factors." For example, if 57 grams of solids per meter of curb are estimated to have washed off during a rainstorm, and if the analyst believes that total phosphorus loads are roughly one-third of total solids loads, then the mass of total phosphorus (TP) removed by rain (and presumably available to receiving waters) would be:

$$\text{Mass}_{TP} = \text{Mass}_{\text{solids}} * 0.33 = 57.0 * 0.33 = 18.8 g/m$$

This method, although widely used in watershed planning, cannot be used to predict pollutant loadings with any accuracy, simply because the relationship between solids loading (or concentration) and the loading (or concentration) of any given pollutant is never constant nor even linear. Nevertheless, it offers a "quick and dirty" way of estimating pollutant loads from intermittent sources and is likely to be useful in comparing the relative impacts of different management alternatives. Zison (1980) provides a discussion of the potency factors for different pollutants in urban and rural areas.

Several authors have developed simple screening methods, based largely on the Universal Soil Loss Equation (for rural, pervious areas) and buildup/wash-off relationships (for urban areas). Most of these take the form:

$$\text{Load} = \frac{\text{rainfall}}{\text{area}} * \%\text{runoff} * \text{concentration}$$

Table 6.5 presents some of these methods, which may be most helpful for the estimation of flows and pollutant loadings from urban stormwater and combined sewer overflows and agricultural drainage.

The following example demonstrates the application of Schueler's Simple Method in the evaluation of development alternatives for a suburban site.

Example: Prediction of Pollutant Loading from Development Alternatives:

1. Assume a site 23 ha in area, currently forested, almost fully pervious (assume 2% imperviousness).
2. Assume that there is interest in developing this site, either as a town-house community (45% impervious surface) or as a low-density housing development (30% impervious surface).
3. Estimate the current and projected loadings of total phosphorus and nitrogen under each of these planning scenarios using Schueler's Simple Method.

Step 1: Estimate parameters for pre- and postdevelopment conditions (this example uses the values drawn from Table 6.3:

Parameter	Predevelopment	Townhouses	Low-Density Residential
P	102 cm	102 cm	102 cm
P_j	0.9 (see Table 6.5)	0.9 (see Table 6.5)	0.9 (see Table 6.5)
R_v	0.05 + 0.009(2) = 0.07 (see Table 6.5)	0.05 + 0.009(45) = 0.46 (see Table 6.5)	0.05 + 0.009(30) = 0.32 (see Table 6.5)
$C_{Total\,P}$	0.15 mg/L (see Table 6.3)	0.26 mg/L (see Table 6.3)	0.26 mg/L (see Table 6.3)
$C_{Total\,N}$	0.78 mg/L (see Table 6.3)	2.00 mg/L (see Table 6.3)	2.00 mg/L (see Table 6.3)
Site Area	23	23	23

Step 2: Estimate annual pollutant loads from stormwater runoff. (Note that Schueler's Simple Method equation works equally well whether using acres and inches, or hectares and centimeters. In either case, the product will be expressed in pounds and must therefore be multiplied by 2.2 to obtain an estimate in kilograms.)

Table 6.5 Simple Methods of Estimating Pollutant Loading from Wet-Weather Sources

Method	Description	Source
EPA Screening Procedures	$W_o = cC_vI_TA$ where: W_o = loading for period c = average pollutant concentration C_v = average runoff coefficient I_T = average rainfall divided by period of averaging A = drainage area	US EPA 1976; Mills et al. 1982, 1985
APWA unit load	$L = aP(b + cD^d)$ where: L = annual unit load (kg/ha) P = annual precipitation (m) D = population density a, b, and c are experimentally derived parameters	Sullivan et al. 1978
Schueler's Simple Method	$L = ((PP_jR_v)/12)CA(2.72)$ where: L = pollutant export (loading, in pounds) P = rainfall depth (inches) over desired time interval P_j = factor that corrects P for storms that produce no runoff; usually set to 0.9 for annual and seasonal calculations C = flow-weighted mean concentration of pollutant in urban runoff (mg/L) (see Table 6.3) A = area of the site R_v = runoff coefficient (the fraction of rainfall that is converted into runoff) = runoff volume divided by total storm rainfall, or typically $R_v = 0.05 + 0.009I$ where I = percentage of site imperviousness (12 and 2.72 are conversion factors)	Schueler 1987

Table 6.5 (Continued)

Method	Description	Source
Cornell University Procedures	$LD_{kt} = 0.1CD_{kt}Q_{kt}TD_k$ $LS_{kt} = 0.001CS_{kt}X_{kt}TS_k$ where: LD_{kt} = unit load of dissolved pollutant from land-use area k during storm event on day t LS_{kt} = unit load of suspended sedimentment from land-use area k during storm event on day t CD_{kt} = dissolved pollutant concentration in runoff (mg/L) CS_{kt} = adsorbed pollutant concentration in runoff (μg/g sediment) Q_{kt} = runoff for day t from land-use area k X_{kt} = soil loss for day t from land-use area k Td_k, Ts_k = transport factors (equivalent to sediment delivery ratio) reflecting fractions of dissolved and adsorbed solids that move from the edge of the source area to the watershed outlet	Haith and Shoemaker 1987

(Refer to Table 6.5 for explanation of equation)

$$L = ((PP_J R_v)/12)CA(2.72)$$

For total nitrogen:

$$L_{\text{predevelopment}} = [(101.6)(0.9)(0.07)/12](0.78)(23)(2.72)$$
$$= 26\text{pounds/year} = 12\text{kg/year}$$
$$L_{\text{townhouses}} = [(101.6)(0.9)(0.46)/12](2.00)(23)(2.72)$$
$$= 439\text{pounds/year} = 199\text{kg/year}$$
$$L_{\text{low-density residential}} = [(101.6)(0.9)(0.32)/12](2.00)(23)(2.72)$$
$$= 305\text{pounds/year} = 139\text{kg/year}$$

For total phosphorus:

$$L_{\text{predevelopment}} = [(101.6)(0.9)(0.07)/12](0.15)(23)(2.72)$$
$$= 5\text{ pounds/year} = 2\text{ kg/year}$$
$$L_{\text{townhouses}} = [(101.6)(0.9)(0.46)/12](0.26)(23)(2.72)$$
$$= 57\text{ pounds/year} = 26\text{ kg/year}$$
$$L_{\text{low-density residential}} = [(101.6)(0.9)(0.32)/12](0.26)(23)(2.72)$$
$$= 40\text{ pounds/year} = 18\text{ kg/year}$$

Although crude, this analysis shows that planners can expect significant increases in nutrient export from the site with either townhouse or low-density residential development. This information is useful in several ways. It can guide the analyst as to which alternative will generate the lowest nutrient loads (or create the smallest change from existing conditions). It can point to the need for mitigative measures to reduce nutrient export under development conditions—in other words, to reduce runoff from the site by encouraging retention on-site, by improving the quality of runoff by on-site treatment, or by similar measures. And it may reveal differences so dramatic (as in this simple example) that additional data collection (for instance, on runoff quality and volume) may be warranted before development goes ahead.

Simple screening methods of the type shown in Table 6.5 and in this example are not, however, useful for making accurate estimates of existing or future conditions. So the analyst could not, for example, draw the conclusion that exactly 18 kg/year of phosphorus will be exported from this site if low-density residential development is constructed. Rather, these estimates should be considered "order of magnitude" values: predevelopment phosphorus loads are probably in the range of 0 to 10 kg/year, while townhouse development would likely increase those loads by a factor of 10.

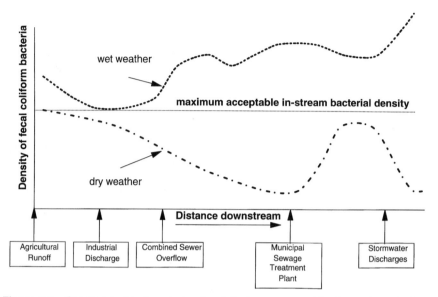

Figure 6.3 *Simple graphical analysis of pollutant sources and in stream impacts for a hypothetical planning area.*

6.3.4   Simple Procedures for Estimating Receiving Water Impacts

The simplest method of determining cause-and-effect relationships in a stream is to plot the average concentration of one or more key parameters against potential sources. Figure 6.3 illustrates this method for a hypothetical planning area. Even this type of simple analysis often reveals important relationships between pollutant sources in in-stream impacts. It will also yield clues as to the relative importance of sources for the parameter of interest.

Figure 6.3 provides a clear and visual demonstration that industrial discharge to the stream is much less important than agricultural runoff or the municipal sewage treatment plant, even in dry weather. In wet weather, the four most important sources are agricultural runoff, combined sewer overflows, the municipal sewage treatment plant effluent, and stormwater discharges. Of these, agricultural runoff and the municipal sewage treatment plant may be somewhat larger contributors than the two wet-weather sources, especially since they operate more or less continuously.

Simple quantitative methods like those described earlier for estimation of pollutants can also be used to predict receiving water impacts, such as exceedances of water quality criteria. Again, several approaches are available, mostly using simple equations for dilution with or without in-stream degradation. The following discussion, drawn from MOE (1987) summarizes these methods for conservative substances like chloride or heavy metals, which do not break down or react in the water column; reactive substances such as bacterial densities or BOD, whose concentration may change depending on physical, chemical, or biological conditions; and "coupled" substances such as ammonia, whose concentration depends directly on the concentration of another parameter (in the case of ammonia, pH).

(i) Conservative Substances (e.g., Chloride) Discharged From a Point Source

$$C = C_0 + \frac{W}{Q} \tag{6.7}$$

where: C = concentration of substance
C_0 = concentration at distance $x = 0$
W = point source loading rate (mass/time)
Q = flow (volume/time)

(ii) Conservative Substances Discharged from Nonpoint Sources

$$C = C_0 + w \, \frac{x}{Q} \tag{6.8}$$

where: C = concentration of substance
C_0 = concentration at distance $x = 0$
w = non-point source loading rate (mass/time)
x = distance
Q = flow (volume/time)

(iii) Reactive substances (e.g., BOD) discharged from a point source

$$L = L_0 e^{-K_r x/U} + (W/Q)e^{-K_r x/U} \tag{6.9}$$

where: L = concentration of reactive substance
L_0 = concentration at $x = 0$
x = distance
K_r = rate of pollutant removal
U = velocity
W = point source waste loading rate
Q = flow

(iv) Reactive Substances Discharged from Nonpoint Sources

$$L = L_0 e^{-K_r X/U} + \frac{w}{A K_r} (1 - e^{-K_r X/U}) \tag{6.10}$$

where: L = concentration of reactive substance
L_0 = concentration at $x = 0$
x = distance
K_r = rate of pollutant removal (e.g. BOD removal)
U = velocity
w = non-point source waste loading rate
A = cross-sectional area of stream

(v) Coupled Substance Discharged From a Point Source

$$D = D_0 e^{-K_a X/U} + L_0 \frac{K_d}{K_a - K_r} [e^{-K_r X/U} - e^{-K_a X/U}]$$

$$+ \frac{W}{Q} \frac{K_d}{K_a - K_r} [e^{-K_r X/U} - e^{-K_a X/U}] \tag{6.11}$$

where: D = coupled substance concentration (e.g., dissolved oxygen deficit)
D_0 = coupled substance concentration at $X = 0$
X = distance
U = velocity
L_0 = concentration of reactive substance at $X = 0$
K_d = BOD oxidation coefficient
K_r = rate of pollutant removal (e.g., BOD removal)
K_a = reaeration coefficient
W = point source waste loading rate
Q = flow

(vi) Coupled Substance Discharged from Non-Point Sources

$$D = D_0 e^{-K_a X/U} + L_0 \frac{K_d}{K_a - K_r} [e^{-K_r X/U} - e^{-K_a X/U}] + \frac{w}{A K_r} \frac{K_d}{K_a - K_r}$$

$$\cdot \left[\frac{K_r}{K_a} e^{-K_a X/U} - e^{-K_r X/U} + \frac{K_a - K_r}{K_a} \right] \tag{6.12}$$

where: D = coupled substance concentration (e.g., dissolved oxygen deficit)
D_0 = coupled substance concentration at $X = 0$
K_a = dissolved oxygen reaeration rate (i.e., rate of removal of deficit)
X = distance
U = velocity
L_0 = concentration of reactive substance at $X = 0$
K_d = BOD oxidation coefficient
K_r = rate of pollutant removal (e.g., BOD removal)
K_a = reaeration coefficient

w = non-point source waste loading rate
A = cross-sectional area of stream

Although some of these equations are lengthy, they do not require advanced mathematical techniques nor detailed information about discharges or receiving water.

It is important to note that the various reaction rates in these equations are important in influencing predictions. Basic engineering texts such as Metcalf and Eddy (1991) provide some information about reaction rates, and more detailed information is widely available in the scientific literature. Examination of the literature, however, usually reveals a substantial range in reported values. The analyst is therefore advised to calculate expected in-stream impacts using both low and (in a separate calculation) high rates, thus arriving at a likely range of outcomes rather than a single uncertain value.

(vii) Prediction of Exceedances of In-Stream Water Quality Criteria

Schueler (1987) gives an example of how his Simple Method can be adapted for use in predicting exceedances of in-stream water quality criteria. The following is adapted from that example:

Example: Prediction of Water Quality Impacts of Development Alternatives

1. Assume that a developer wishes to build a partially sewered residential development 30 ha in area within a larger tract of 130 ha of forested land.
2. Assume that the percentage of imperviousness of the developed area is 27%, while that of the forested area is 2%.
3. Assume that the receiving water into which storm runoff will drain has a hardness of 100 mg/L $CaCO_3$.
4. Assume that the concentrations of trace metals in runoff from the forest site are negligible.
5. Determine whether the projected concentrations of lead, zinc, and copper in this development situation would exceed established water quality criteria.

Step 1: Calculate the expected volume of runoff from the developed and forested portions of the watershed for a 2.5 cm storm using the first term of Schueler's Simple Method equation (previous example), as follows:

$$\text{Volume}_{\text{runoff}} = PRvA/12$$

(Note that this calculation does not contain the term P_j. As discussed in Table 6.5, P_j is not used for a single storm. Assume that R_v has been calculated to be 0.07 for the forested area and 0.29 for the development site.) Schueler's equation was designed for use with acres and yields volume in acre-feet. If

Table 6.6   Percentage of Storms in Which Given Pollutant Concentrations are Likely to be Exceeded (from Schueler 1987, based on NURP data: US EPA 1983)

Pollutant Concentration (mg/L)	50%	25%	10%	5%	1%
Sediment	31	71	151	235	545
Total Phosphorus	0.27	0.43	0.65	0.82	1.31
Total Nitrogen	2.2	3.2	4.5	5.6	8.2
COD	42	61	84	103	149
Lead	0.021	0.042	0.076	0.109	0.149
Copper	0.010	0.020	0.037	0.055	0.144
Zinc	0.06	0.101	0.1611	0.216	0.355

hectares are used for area, the product must be multiplied by 488.18 to obtain cubic meters:

$$\text{Volume}_{\text{runoff}} = (PRvA/12)(499.18)$$

Runoff volume from the two land uses for a 2.5-cm storm (i.e., 2.5 cm of rainfall) can then be calculated as:

$$\text{Development site} = \{[(1.0)(0.29)(30)]/12\} * 499.18 = 362 \text{ cubic meters}$$
$$\text{Forested area} = \{[(1.0)(0.07)(100)]/12\} * 499.18 = 291 \text{ cubic meters}$$

The total runoff to be expected from the mixed-use watershed would therefore be $362 + 261 = 623$ cubic meters.

   Step 2:   Calculate the dilution ratio of developed site runoff to watershed runoff:

$$\text{Dilution} = \text{runoff}_{\text{development}}/\text{runoff}_{\text{watershed}} = 362/623 = 58\%$$

   Step 3:   Determine the trace metal concentration in runoff that is typically exceeded in 5% of storms (about three storms a year); see Table 6.6:

   The "5%" column shows that, based on NURP data (US EPA 1983), a lead concentration of 0.109 mg/L was exceeded in the runoff from 5% of storms. The equivalent concentrations for copper and zinc were 0.055 and 0.216 mg/L, respectively.

   Step 4:   Calculate expected in-stream concentration of trace metals.

   The runoff concentrations shown in Table 6.6 are rapidly diluted by cleaner flows. In Step 2, we calculated the dilution ratio for this site to be 58% (in other words, the flow from the development site constitutes 58% of the total watershed runoff). We can therefore assume that contaminated runoff from the development site will be diluted in this proportion when it reaches the receiving stream:

$$\text{Concentration}_{\text{instream}} = \text{concentration}_{\text{runoff}} * \text{dilution}$$

or, for lead:

$$\text{Concentration}_{\text{instream}} = 109 * 0.58 = 63 \mu g/L$$

for zinc:

$$\text{Concentration}_{\text{instream}} = 216 * 0.58 = 125 \mu g/L$$

and for copper:

$$\text{Concentration}_{\text{instream}} = 55 * 0.58 = 32 \mu g/L$$

These values (again, rough estimates) can be compared against available criteria for ambient water quality to determine whether in-stream conditions will approach or exceed recommended levels following development.

Although techniques like Schueler's Simple Method were developed for single-site analyses, they can also be applied over larger areas. The larger variability—in rainfall, in runoff, in pollutant concentration—over a larger area means that estimates prepared this way may be even less accurate than those prepared for a single site. They may, nevertheless, be useful for comparing broad categories of management alternatives—for example: increase residential development in the southern part of the watershed by 15% over the next 10 years; or encourage light industrial development in that area; or retain as a mixture of open space and rural residential development.

(viii) **The Special Case of Lakes** The foregoing techniques illustrate the range of simple methods available for estimation of receiving water impacts in streams. Generally speaking, estimation of impacts in lakes is more complex, especially when a lake is large, deep, or has complicated circulation patterns.

Simple models such as those developed by Vollenweider (1976) and Dillon and Rigler (1974) use estimates of gross pollutant loads and lake residence time to determine the probability of eutrophication. Chapra (1979) developed a modified version of this approach for use in the embayment areas of large lakes. Thomann and Mueller (1987) provide a more detailed discussion of this approach and formulate Vollenweider's model as follows:

$$V \frac{dp}{dt} = W - v_s A_s p - Qp \tag{6.13}$$

where: V = volume of the lake
  p = total phosphorus in the lake, expressed in mass per unit volume (e.g., $\mu g/L$)

Q = outflow (volume per unit time)
A_s = lake surface area
W = allochthonous source of phosphorus expressed in mass per unit time
(e.g., g/sec)
v_s = settling rate of phosphorus (distance per unit time, e.g., cm/sec)

At steady state, $dp/dt = 0$ and division by the surface area of the lake yields:

$$p = \frac{W'}{q_s + v_s} \tag{6.14}$$

where: p = concentration of total phosphorus in the lake
W' = allocthonous phosphorus leading
q_s = hydraulic overflow rate $Q/A_s = h\rho$
v_s = settling rate of phosphorus
Q = outflow
ρ = $Q/V = 1/\tau_w$ = flushing rate
A_s = lake surface area
H = water depth
τ_w = detention time in the lake
V = volume of the lake

Novotny and Olem (1994) point out that the fundamental weakness in this approach is a lack of knowledge about the settling rate of phosphorus. Vollenweider (1976) used the following to approximate v_s:

$$v_s = H\sqrt{\rho} \tag{6.15}$$

which in turn yields:

$$p = \frac{W'}{H\rho \left(1 + \sqrt{\dfrac{1}{\rho}} \right)} \tag{6.16}$$

Most of the information needed for these calculations is readily available from pollutant loading estimates and simple water balance calculations. This type of approach is most useful for long-term planning purposes (it has sometimes been applied in the determination of allowable development density along a lake shoreline). It is not intended for detailed, short-term, or site-specific predictions.

Figure 6.4 illustrates the practical application of the Vollenweider model.

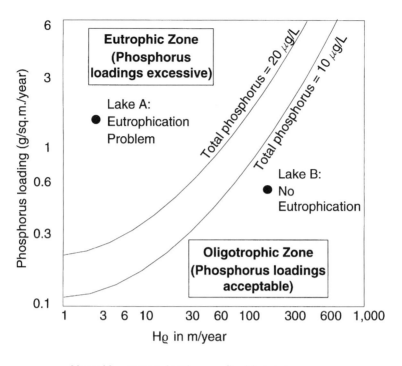

Note: H = water depth; ϱ = flushing rate

Figure 6.4 *Example of critical phosphorus loading plot for lakes (after Vollenweider 1976, MOE 1987).*

Some Cautionary Notes   It will be seen from this discussion that available simple screening models are easy to use and require little in the way of input data. Their simplicity means that they have ignored or minimized the importance of factors such as baseflow runoff, which can be significant over larger areas, and any pollutant loadings attached to that runoff. In some systems, the baseflow can account for more than half the total runoff, so use of these simple screening methods could significantly underestimate the pollutant load that would be associated with specific management actions.

A second caution relates to the generalized data that are commonly used with screening models. The NURP data (US EPA 1983), for example, represent the combined results of studies on more than 300 runoff events. They do not, however, include every condition that might be encountered in watershed planning, nor do they apply to every land use configuration or runoff situation. These methods are best used to compare management alternatives with fundamentally different influences on the land surface and water flow over it (such as in the preceding examples). It would not be appropriate to compare two very similar alternatives—for instance, a 250-townhouse development versus a 275-unit development—using these methods. The techniques simply are not sensitive enough, nor accurate enough, for accurate prediction of the differences between closely related alternatives.

Third, most of these models have been developed for a particular geographic location or set of data (for instance, Schueler's method was based on urban runoff data from recently stabilized suburban watersheds). They may be less accurate when applied in different situations, such as areas undergoing construction, industrial areas, nonfarming rural development, or agriculture.

In summary, simple screening methods are intended to be only that: simple techniques to assist the analyst with limited time and budget in assessing management alternatives. More complex basins, and/or more complex problems, will require detailed watershed and receiving water simulation, with the extensive input data required by those tools. These methods are described in more detail in Chapter 7. The reader should, however, be cautioned that even the most complex simulation models are sometimes based on the very simple screening models discussed earlier, not on precise mechanistic representations of hydrologic or chemical phenomena. The reader is advised to examine carefully the algorithms on which such models are based, so as to understand the assumptions and mechanisms used to make model predictions.

6.3.5 Simple Screening Procedures

Armed with an overview of the watershed system and some basic information about pollution sources and their relative magnitude, the analyst can proceed to screen management alternatives.

The first step in screening is application of constraints. Graphical methods are helpful here. Table 5.6 illustrates a tabular method, but it is also possible to represent the information given in Table 5.6 in graphical form. The advantage of such an approach is that it can allow the presentation of more complex information than is possible in a table. Figure 6.5 shows how this might be done for the example given in Table 5.6. This example illustrates only two variables, cost and time. With a larger number of variables, graphical methods allow visual patterns to emerge across a single management option, or a single constraint.

Other graphical approaches are possible. Figures 6.6 to 6.8 show methods that are helpful in displaying information with a higher degree of uncertainty—in other words, variables in which performance may be feasible under some conditions, clearly not feasible under others, but marginally feasible in some intermediate zone. Graphical methods allow this gray area to be included in the analysis.

6.4 OUTPUTS OF A SIMPLE ASSESSMENT PROCESS

The outputs of a simple assessment process may be of several types. Egregious problems may have been identified in the data collection and plan screening process, leading to recommendations for immediate abatement action. Frequently,

Acceptable
Marginal
Unacceptable

Option	Cost	Time to implement
Do nothing.	$0	immediate
Construct buffer strips.	$750	1 month
Construct livestock exclusion fencing.	$500	4 months
Encourage conservation tillage.	$250	1 year
Buffer strips and fencing.	$1,250	4 months
Buffer strips and conservation tillage.	$1,000	1 year
Fencing and conservation tillage.	$750	1 year
Buffer strips, fencing, and conservation tillage.	$1,500	1 year

Figure 6.5 *Example of graphical methods for application of planning constraints.*

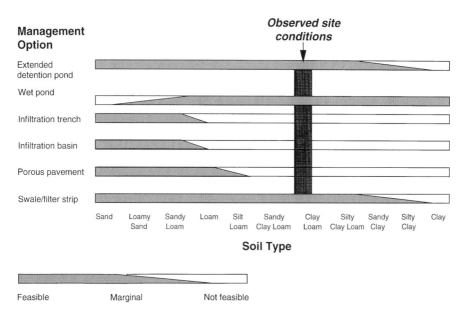

Figure 6.6 *Example of gradient bars used in the application of planning constraints (adapted from Schueler 1987).*

	Steep Slope	High Water Table	Close to Bedrock	Close to Foundations	Space	Maximum Depth	Restricted Land Uses	High Sediment Input	Thermal Impacts
Extended detention pond	●	●	◐	●	○	●	●	◐	●
Wet pond	●	●	◐	●	○	○	●	◐	○
Infiltration trench	○	○	○	○	●	○	●	○	●
Infiltration basin	◐	○	○	◐	◐	○	●	○	●
Porous pavement	○	○	○	○	○	○	○	○	●
Water quality inlet	●	●	○	○	●	○	○	○	●
Grassed swale	○	○	◐	◐	●	●	○	○	●
Filter strip	◐	◐	◐	◐	●	●	◐	○	●

○ Not usually suited for conditions
◐ Design amendments can improve feasibility
● Usually feasible

Figure 6.7 Example of feasibility matrix used in the application of planning constraints (adapted from Schueler 1987).

Design Option	Achieves Required BOD Standard	Achieves Required SS Standard	Achieves Required TP Standard	Fits Available Space	Materials Readily Available
Detention Pond	☺ marginal, +5%	☺ acceptable, -22%	☹ unacceptable, +65%	☺ acceptable, -13%	☺ acceptable
Infiltration Trench	☹ unacceptable, +43%	☺ acceptable, -27%	☹ unacceptable, +58%	☺ acceptable, -19%	☺ marginal
Porous Pavement	☺ acceptable, -6%	☺ acceptable, -13%	☺ marginal, +4%	☺ marginal, +2%	☺ marginal
Grassed Swale	☺ acceptable, -9%	☺ acceptable, -31%	☺ marginal, +3%	☺ unacceptable, +5%	☺ acceptable
Do Nothing	☹ unacceptable, +66%	☹ unacceptable, +27%	☹ unacceptable, +16%	☺ acceptable, NA	☺ acceptable

☺ = acceptable, ☺ = marginal, ☹ = unacceptable

Note: Percentages indicate amount exceeding target (e.g., +5% means 5% higher than desired).

Figure 6.8   Example of performance matrix used in the application of planning constraints.

the level of data quality and analytical capability possible in a simple assessment is found to be disappointing or in some way inadequate for the purposes of the planning initiative. In such a case, an outcome might be recommendations for the collection of additional data and/or a commitment to more detailed, and probably more costly, analysis.

Sometimes the output of a simple analysis is simply the realization that the true sources of use impairments are not what they were originally imagined to be, or that the scope of the initiative needs to be altered. New or previously underestimated use impairments may emerge during the course of the assessment, or the scale of a particular problem or source may turn out to be far greater than originally assumed.

In a recent watershed plan in Ottawa, Ontario, for example, bathing beach closures were thought to be caused by discharges of bacteria from storm and combined sewers, even during dry weather. The planners believed that the continuous nature of the loads (which had been measured in field studies) pointed to illegal cross-connections between storm and sanitary sewers. These cross-connections would have allowed untreated sewage to flow from residential sanitary lines into storm sewers, and then into the Rideau River, without benefit of treatment or disinfection. During the Rideau River Stormwater Study in the early 1980s, it was discovered that nesting waterfowl and pigeons—present in large numbers along the river, particularly on and under bridges—were contributing significant fecal loads to near-shore waters. This finding led to more detailed data collection and revised terms of reference and, ultimately, to a different suite of management actions than might have been contemplated at the outset of the study.

Finally, most planning initiatives extend beyond the present day into some imagined future condition. So another possible outcome of the assessment is that present and future problems and sources differ. This information is useful for long-term planning purposes and can be updated as more information about watershed problems and sources becomes available with time. Table 6.7 summarizes some possible outcomes of a simple assessment process.

6.5   AN EXAMPLE OF THE APPLICATION OF SIMPLE ASSESSMENT PROCEDURES

Tables 6.8 and 6.9 summarize the findings of a hypothetical simple assessment process for a watershed that is currently experiencing impaired swimming opportunities as a result of bacterial contamination and algae blooms at local beaches. It is thought that the algae blooms are directly related to elevated phosphorus concentrations in the river. Table 6.8 presents the findings for the present case; Table 6.9 presents predictions for a time 25 years into the future.

Tables 6.8 and 6.9 do not recap basic information about system characteristics nor state where data on these elements were obtained. It can be assumed that a mixture of state and federal agency sources, private sources (for instance,

Table 6.7   Some Possible Outcomes of a Simple Assessment Process

Time	Conditions meet desired targets	Conditions do not meet desired targets	Conditions marginal	Conditions uncertain	Conditions not as expected
Present	No action required.	Implement controls immediately.	Implement controls as available; plan for additional controls.	Additional data collection and analysis if warranted. Revise scope if warranted.	Additional data collection and analysis if warranted. Revise scope if warranted.
Future	No action required.	Begin planning for future controls and financing.	Begin planning for future controls and financing.	Additional data collection and analysis if warranted. Revise scope if warranted. Update projections as data become available.	Additional data collection and analysis if warranted. Revise scope if warranted. Update projections as data become available.

industrial records or septic system installation records), key informant interviews, and simple observation were used to generate the information given in the tables.

For this example, it can further be assumed that the planning horizon is 25 years, over which period 30% growth in population (with associated demand increases) is expected. The goal of the process is to restore the river to "swimmability" throughout the period of May to September.

Tables 6.8 and 6.9 obviously do not incorporate all elements of the watershed inventory, study scoping, the development and screening of alternatives, and the generation of output. They do, however, illustrate the following key features of a simple assessment:

- Focus on a small number of impaired uses
- Use of a small number of indicators with which to evaluate improvements in impaired uses
- Comprehensive inventory of sources
- Identification of key sources
- Step-wise elimination of infeasible options using systematic evaluation techniques
- Use of present and future scenarios to capture likely trends over time
- Specific outputs, including recommendations for immediate action, for deferred action, and for additional data collection and analysis

Table 6.8 An Example of the Outputs of a Simple Assessment Process (Present Case)

Potential Sources	Impact on Fecal Coliform Loadings	Impact on Phosphorus Loadings	Available Controls	Feasible Controls	Preferred Approach
Sewage Treatment Plant	Plant close to capacity; bypass common during wet weather; significant source even in dry weather.	Estimated to contribute 15% to total loadings to the river.	Add capacity; correct inflow and infiltration; disconnect roof leaders; repair CSO regulators; add filtration; add P removal capability.	Correct inflow and infiltration; repair CSO regulators; add filtration; add P removal.	Correct inflow and infiltration; repair CSO regulators; add P removal. Reassess problem in 5 years.
Industrial: Brewery	No effect.	Minimal effect.	No action required.	No action required.	No action required.
Industrial: Electroplaters	No effect.	No effect.	No action required.	No action required.	No action required.
Storm Sewer Discharges	Contribute high loads during wet weather.	Estimated to contribute 5% of total loadings to the river.	Detention ponds; wet ponds; porous pavement; grassed swales; stoop-and-scoop ordinances.	Detention ponds; wet ponds; stoop-and-scoop ordinances.	Recommend installation of 4 detention ponds to serve the 4 main sewered areas; implement and enforce stoop-and-scoop ordinances.
Combined Sewer Overflows	Contribute high loads during wet weather.	Minimal effect.	Build storage tanks; correct inflow and infiltration; repair CSO regulators; disconnect roof leaders.	Correct inflow and infiltration; repair CSO regulators; disconnect roof leaders.	Try correction of inflow and infiltration with repair of CSO regulators first. Reassess in 5 years to determine need for roof leader disconnection.
Uncontrolled Runoff from State Forest	No effect.	No effect.	No action required.	No action required.	No action required.

Source	Effect on water quality	Loading contribution	Potential controls		Recommended action
Agricultural Runoff: Spencer Creek Sub-Basin	No significant effect.	Estimated to contribute 60% of total loadings to river.	Milkhouse wash water controls; conservation tillage; buffer strips; intercropping; improved manure storage; contour plowing.	Milkhouse wash water controls; buffer strips; intercropping; improved manure storage.	First implement milkhouse wash water controls and improved manure storage. Encourage use of buffer strips and intercropping through extension agents. Reevaluate progress in 5 years.
Agricultural Runoff: Hastings Creek Sub-Basin	Intensive livestock operations create a continuous source of bacteria that is significantly increased in wet weather.	Estimated to contribute 15% of total loadings to river.	Milkhouse wash water controls; conservation tillage; buffer strips; inter-cropping; improved manure storage; contour plowing	Milkhouse wash water controls; improved manure storage.	Implement milkhouse wash water controls and improved manure storage immediately.
Quarry Instream Resuspension	No effect. Minor effect.	No effect. Minor effect.	No action required. Unknown: additional data collection required.	No action required. Unknown: additional data collection required.	No action required. Unknown: additional data collection required.
Leaking Septic Systems Along Shoreline	Continuous inputs of fecal coliforms during wet and dry weather.	Estimated to contribute 5% of total phosphorus loadings to the river.	Install sewers and central treatment plant; install sewers and several local treatment facilities; repair/replace deteriorating septic systems.	Repair/replace deteriorating septic systems.	Immediately begin to repair/replace deteriorating septic systems (may require enhanced enforcement of ordinances and/or financial subsidies).

Table 6.9 An Example of the Outputs of a Simple Assessment Process (Present Plus 25 Years: Future Case)

Potential Sources	Impact on Fecal Coliform Loadings	Impact on Phosphorus Loadings	Available Controls	Feasible Controls	Preferred Approach
Sewage Treatment Plant	Old plant replaced in year 15. Current facility operating well with bypass rare. All effluent is disinfected. Bacterial impacts now minimal.	Phosphorus removal added in planned plant replacement in year 15; current plant effluent estimated to contribute very low loadings of P (< 5%).	No action likely to be required at this time.	No action likely to be required at this time.	No action likely to be required at this time.
Industrial: Brewery Industrial: Electroplaters	No effect. No effect.	Minimal effect. No effect.	No action required. No action required.	No action required. No action required.	No action required. No action required.
Storm Sewer Discharges	Continue to contribute high loads during wet weather; development has increased % imperviousness and volume of runoff; significant wet-weather source of bacteria.	Estimated to contribute 25% of total loadings to the river.	Detention ponds; wet ponds; porous pavement; grassed swales; stoop-and-scoop ordinances.	Detention ponds; wet ponds; stoop-and-scoop ordinances.	Recommend installation of 6 additional detention ponds in areas of new development, staged over years 10–25; implement and enforce stoop-and-scoop ordinances.

Source		Effect			
Combined Sewer Overflows	Impact much reduced with gradual separation of sewers. CSO bypass occurs only 2–3 times a year and is low in volume.	Minimal effect.	No action likely to be required at this time.	No action likely to be required at this time.	No action likely to be required at this time.
Uncontrolled Runoff from State Forest	No effect.	No effect.	No action required.	No action required.	No action required.
Agricultural Runoff: Spencer Creek Sub-Basin	Livestock operations have expanded into Spencer Creek basin. This basin now contributes significant bacteria loadings to the stream in wet weather.	Earlier controls have achieved significant reductions in P loadings. Current loadings are estimated at 40% due to increased farming intensity.	Milkhouse wash water controls; conservation tillage; buffer strips; intercropping; improved manure storage; contour plowing.	Buffer strips; conservation tillage; intercropping; improved manure storage.	Implement improved manure storage. Encourage use of buffer strips, conservation tillage, and intercropping through extension agents. Reevaluate progress in 5 years.
Agricultural Runoff: Hastings Creek Sub-Basin	Intensive livestock operations continue to create a continuous source of bacteria that is significantly increased in wet weather.	Estimated to contribute 35% of total loadings to river.	Milkhouse wash water controls; conservation tillage; buffer strips; intercropping; improved manure storage; contour plowing.	Buffer strips; conservation tillage; intercropping; improved manure storage.	Implement improved manure storage. Encourage use of buffer strips, conservation tillage, and intercropping through extension agents. Reevaluate progress in 5 years.

Table 6.9 (*Continued*)

Potential Sources	Impact on Fecal Coliform Loadings	Impact on Phosphorus Loadings	Available Controls	Feasible Controls	Preferred Approach
Quarry Instream Resuspension	No effect. Minor effect.	No effect. Minor effect.	No action required. Unknown: additional data collection required; review as studies become available.	No action required. Unknown: additional data collection required; review as studies become available.	No action required. Unknown: additional data collection required; review as studies become available.
Leaking Septic Systems Along Shoreline	Continuous inputs of fecal coliforms during wet and dry weather; population density greatly increased; many older systems now in poor repair and contributing continuous loadings of bacteria.	Estimated to contribute 15% of total phosphorus loadings to the river.	Install sewers and central treatment plant; install sewers and several local treatment facilities; repair/replace deteriorating septic systems.	Install sewers and central treatment plant; install sewers and several local treatment facilities.	Install sewers and several local treatment facilities in villages of Pinedale, Smithville, and Wilson Heights.

The simple analysis illustrated in Tables 6.8 and 6.9 does not yield detailed predictions about pollutant loadings from specific sources, nor does it attempt to make accurate predictions of receiving water quality. Desktop methods are used to generate rough, even order-of-magnitude, estimates that allow the analyst to determine which sources are most urgently in need of attention and which can safely be ignored. The analysis of options is similarly simple, avoiding detailed prediction of expected performance levels or loading reductions. A long list of options is identified, feasible options are chosen from that list (probably on the basis of cost and rough estimates of pollutant removal), and, in many cases, one or more preferred options are identified for further action.

This level of analysis can be completed with available data, using desktop methods, in a very short period of time, perhaps two to three weeks. Yet its benefits are immediately obvious. By clarifying the roles of the different sources of the use impairment, it helps to focus public and agency attention where remedial actions will do the most good. In some cases, this will mean a request for resources to undertake more detailed assessments of specific system components (see Chapter 7) or of the system overall. In either case, this simple and inexpensive analysis will provide a useful foundation for dialogue and future action.

REFERENCES

Baltimore Regional Planning Council (BRPC). 1986. *Technical Summary: Jones Falls Urban Stormwater Runoff Project.* Baltimore, Md.: BRPC.

Chapra, S. C. 1979. Applying phosphorus loading models to embayments. *Limnology and Oceanography* 24(1): 168.

Dillon, Peter J., and F. H. Rigler. 1974. A test of a simple nutrient budget model predicting the phosphorus concentration in lake water. *J. Fish Res. Bd. Can.* 31: 1771.

Haith, D. A., and L. L. Shoemaker. 1987. Generalized watershed loading functions for stream flow nutrients. *Water Resources Bull.* 23(3): 371–478.

Hydrologic Engineering Center. 1975. *Urban Stormwater Runoff—STORM.* Davis, Calif.: U.S. Army Corps of Engineers.

Metcalf and Eddy Inc. 1991. *Wastewater Engineering: Treatment, Disposal, Reuse.* 3d ed. New York: McGraw-Hill.

Metropolitan Washington Council of Governments (MWCOG). 1983. *Urban Runoff in the Washington Metropolitan Area—Final Report.* Washington, D.C.: Area Urban Runoff Project.

Mills, W. B., et al. 1982. *Water Quality Assessment: A Screening Procedure for Toxic and Conventional Pollutants.* US EPA Report 600/6-82-004a and b. Athens, Ga.: U.S. Environmental Protection Agency.

Mills, W. B., et al. 1985. *Water Quality Assessment: A Screening Procedure for Toxic and Conventional Pollutants.* rev. ed. US EPA Report 600/6-85-002a and b. Athens, Ga.: U.S. Environmental Protection Agency.

Novotny, V., and H. Olem. 1994. *Water Quality: Prevention, Identification, and Management of Diffuse Pollution.* New York: Van Nostrand Reinhold.

Occoquan Watershed Monitoring Lab (OWML). 1983. *Final Contract Report: Washington Area NURP Project.* Prepared for Metropolitan Washington Council of Governments, Manassas, Virginia.

Ontario Ministry of the Environment (MOE). 1987. *Technical Guidelines for Preparing a Pollution Control Plan.* Report of the Urban Drainage Policy Implementation Committee, Technical Sub-committee No. 2. Toronto: Ontario Ministry of the Environment.

Sartor, J. D., G. B. Boyd, and F. J. Agardy. 1974. Water pollution aspects of street surface contamination. *J. WPCF* 46: 458–465.

Schueler, Thomas R. 1987. *Controlling Urban Runoff: A Practical Manual for Planning and Designing BMPs.* Washington, D.C.: Washington Metropolitan Water Resources Planning Board.

Shelley, P. E., and D. R. Gaboury. 1986. Estimation of pollution from highway runoff—initial results. In *Urban Runoff Quality*, edited by B. Urbonas and L. A. Roesner. New York: American Society of Civil Engineering.

Sullivan, W. D. 1978. *Evaluation of the magnitude and Significance of Pollution Loadings from Urban Stormwater Runoff in Ontario.* COA Report #81. Toronto: Ontario Ministry of the Environment.

Thomann, R. V., and J. A. Mueller, 1987. *Principles of Surface Water Quality Modeling and Control.* New York: Harper & Row.

U.S. Environmental Protection Agency (US EPA). 1976. *Area-Wide Assessment Procedures Manual*, Vols. 1, 2, and 3. US EPA Report No. 600/9-76-014. Washington, D.C.: US EPA.

U.S. Environmental Protection Agency (US EPA). 1983. *Results of the Nationwide Urban Runoff Program*, Vol. I. Final Report. Washington, D.C.: US EPA, Water Planning Division.

Vollenweider, R. A. 1976. Input-output models with special reference to the phosphorus loading concept in limnology. *Schweiz. Z. Hydrol.* 37: 53.

Wischmeier, W. H., and D. D. Smith. 1978. *Predicting Rainfall Erosion Losses—A Guide to Conservation Planning.* USDA Handbook 537. Washington, D.C.: Government Printing Office.

Zison, S. W. 1980. *Sediment-Pollutant Relationships in Runoff From Selected Agricultural, Suburban, and Urban Watershed. A Statistical Correlation Study.* US EPA Report 600/3-80-022, Tetra-Tech, Inc., Lafayette, Calif. Athens, Ga.: Environmental Research Laboratory.

7

Detailed Assessment Methods

Every step described in Chapter 6 for a simple watershed assessment has its parallel in a detailed watershed assessment. The difference between the two, as discussed previously, lies in the level of detail used, the quality (and often age) of data, and thus in the precision and accuracy of outputs. In a detailed assessment, a commitment has usually been made to collect additional data, at additional cost and over an extended period of time. Detailed assessments often, perhaps usually, make use of computer simulation models, which in themselves take time to develop, test, and apply. While the duration of a simple assessment may be weeks or, at the most, months, a detailed assessment process usually takes months or even years to complete.

Detailed assessments may be required when one or more of the following conditions is present:

- Pollutant sources are highly variable in quantity and quality and thus demand a more detailed understanding of their behavior in space and/or time.

- The dynamics of the receiving water are complex, for instance, in a large river or lake, requiring assessment of impacts in two or three dimensions as well as over time.

- Available data are inadequate to characterize sources or predict the impacts of management measures on the receiving environment (as in the case shown in Tables 6.8 and 6.9 with suspected in-stream resuspension of bacteria and phosphorus from river sediments).

- There is a need to optimize the behavior of a complex system such as a storm or sanitary sewer system, or a sewage treatment plant, where several unit processes or steps may be involved between influent and effluent.

Whereas a simple assessment relies on rules-of-thumb and simplified equations, a detailed assessment may (but does not always) use data and algorithms tailored to the site of interest, as well as two- or three-dimensional analysis of physical and chemical properties. The "but" arises because, as discussed in Section 6.1, in some of the computer simulation models described later in this chapter, the predictive equations upon which the models are based are in fact quite simplistic. The analyst may not, however, realize this, because those simple equations may be solved for many locations in a basin, and for every hour or half-hour of model time. In other words, the "complexity" of the model comes not from its intrinsic structure, but from the computational demands of hundreds or thousands of recalculations.

It is absolutely essential that the user understand the strengths and weaknesses of any models to be used in watershed assessment. High cost and high "complexity" does not guarantee high accuracy. As SWMM guru Bill James has often commented, "All models are wrong. Some are useful. The trick is to know why, and how."

7.1 THE DETAILED WATERSHED INVENTORY

As described in Chapter 2 and Section 6.1.1, the purpose of a watershed inventory is to develop an understanding of the many physical, chemical, and biological features and processes that affect water movement in a system. Data on features alone, such as water flows or quality, are of use only in describing the current condition. They cannot be used to predict changes in future land use or population situations or under the influence of potential management actions. To this end, it is also necessary to understand watershed processes and to use that understanding to inform predictions of watershed conditions in different situations.

A watershed inventory serves two purposes:

1. To determine the nature and extent of use impairments in the basin, and
2. To identify the causes of existing use impairments.

As such, the inventory provides the foundation for scoping the watershed study—that is, for deciding which problems are most urgent and which sources contribute most to those problems.

7.1.1 Using Existing Information

Most detailed assessments begin in the same way as simple assessments, with compilation of existing data (see Tables 6.1 and 6.2). As discussed in Section 6.1, these data are seldom sufficiently detailed or site-specific for the purposes of most planning initiatives. In a simple assessment, the analyst could nevertheless proceed, using simplifying assumptions or modifying data from neighboring systems. In a detailed assessment, there is likely an expectation that the analysis will yield:

- Identification and detailed characterization of specific sources
- Quantitative evidence in regard to the performance of different management alternatives
- Elucidation of processes, and thus cause-and-effect relationships within the basin (e.g., proof that algal blooms are primarily linked to high phosphorus levels)
- Detailed and quantitative projections about the impact of specific remedial measures on in-stream hydrology, water quality, and biological systems

The fundamental premise of a detailed assessment is that it will provide definitive answers to key planning questions, and although it may be repeated periodically, there is no implication that there will be "follow-up" data collection or analysis. The detailed assessment aims to collect and analyze all the data necessary to answer the questions that are posed at present. Thus, it is usually necessary to proceed beyond existing data sources to the collection of new data.

7.1.2 Filling Data Gaps

For a complex computer simulation model with lengthy input data requirements, existing data are seldom adequate for a detailed analysis. A typical dynamic simulation model requires input data (for instance meteorological conditions, stormwater quality and quantity, and effluent flows) on a one- or two-hour time step; that is, input data files that contain real or realistic values for variables such as hourly (or two-hourly) solar radiation, rainfall, and storm flows. In addition, a typical dynamic model requires definition of rate constants and limiting factors for the wide range of processes that are to be reflected in the simulation. The need to collect such a huge amount of data is often a daunting prospect. Indeed, it is primarily in attempting to fill gaps in the watershed inventory that the water manager becomes aware of the incremental costs and time associated with a detailed assessment. (Data collection can, in fact, be so costly and time-consuming that it limits the computer simulation methods that can be used. This problem is discussed further in later sections of this chapter.)

Whereas a simple assessment uses mostly off-the-shelf information, however incomplete or dated, a detailed assessment usually involves extensive field work to complete and/or update databases on some or all of the following:

- Use impairments and water conflicts (for instance, using user surveys, aerial mapping, satellite imagery, or in-depth analysis of agency records)
- Hydrology (often requiring the collection of new or more detailed stream-flow data, including stream cross sections, analysis of critical low and high flow levels, lake bathymetry, aquifer flows, and similar data)
- Water quality (which may incorporate both routine sampling, perhaps weekly or monthly, to determine "average" conditions at one or more depths and at multiple stations, and intensive sampling during wet weather to determine the shape and height of the hydrograph and associated pol-lutographs; detailed data collection may also be targeted at extending the number of parameters for which data are available—for instance, to add a range of industrial organic substances in addition to "conventional" pollutants like solids and phosphorus)
- The rates of physical, chemical, and biological processes in the system and the factors that influence those rates
- Biological systems (mapping of aquatic community structure, including algae, rooted plants, invertebrates, fish, and amphibians; terrestrial communities, including woody and herbaceous plants, mammals, birds, reptiles, and insects; habitat types; and the presence of rare or endangered systems)
- Population and land use (using agency records, census information, aerial mapping, satellite imagery, and "ground truthing" to confirm interpretations of remote data)
- Facilities and infrastructure (including detailed characterization of the behavior of sources in space and time, such as video surveillance of sewer systems to determine the condition of the physical fabric, intensive sampling of effluent quality and quantity, collection of data on bypass frequency and volume, and detailed projections of the size and nature of facilities proposed for the future)

7.1.3   Sampling Design Considerations

Which data should be collected, and where, and how frequently, will depend on the questions being asked. Accurately mapping current conditions in a watershed is essentially a problem of sampling design: choosing an appropriate set of samples in space and time that will reflect actual conditions with reasonable accuracy, but without unnecessary cost or effort. As stated in Section 2.7.3, clear formulation of the hypothesis to be tested provides valuable guidance in sampling design. A clear statement of hypothesis can allow the application of statistical tools in the development of sampling strategies and the analysis of analytical results.

Very often, the hypothesis is something like "Current average river concentrations for parameter X exceed desired target values." To test this hypothesis, the analyst must first determine the "average" concentration of parameter X and then apply a statistical test (for instance, Student's test) to determine whether, at a given confidence level, the hypothesis should be supported or rejected.

The average determined from field measurements is simply an estimate; the "true" average value (based on a complete knowledge of the system) could be higher or lower than this average. To quantify this "confidence interval," the analyst must have some information about the variability of the system. A good understanding of the range of variability, and the factors that affect it, is fundamental to the development of cause-and-effect relationships and, thus, to the testing of alternative management approaches. It can also help the analyst determine likely high and low conditions. In many cases the collection of additional data for a detailed screening process therefore focuses on characterization of variability (as opposed to average conditions) in flow or quality.

It is beyond the scope of this book to describe detailed sample collection procedures for specific situations. The interested reader is referred to Ostler and Holley (1997) for additional guidance. The following discussion is intended merely to outline some major considerations in designing data collection programs for time-variable systems.

The central problem of any sampling program, whether or not related to water, is to obtain a reasonably accurate picture of a complex system from a small set of observations. The fewer the observations, the greater the chance that we have guessed incorrectly as to the true condition of the system. For example, if there are a thousand cars in a parking lot and we wish to determine the relative proportions of different makes of car, we could select a small number of cars at random and use this sample to draw conclusions about the whole. It can readily be imagined that if we choose a hundred cars at random, we could make a pretty good guess about the relative proportions of makes in the parking lot. But if we choose only three cars (after all, three is a small number and would not require much effort or time for sampling), we would very likely draw incorrect conclusions about the system overall.

The same is generally true of water and wastewater sampling: the larger the number of samples, the more accurate the estimate. (There is, however, some risk in drawing conclusions from very large samples in that higher variability—an artifact of the sampling process—may be apparent in the sample than in the true condition.) A single sample is never enough to characterize a time-variable stream. At minimum, a sampling program should include at least three samples, representing low, mid-range or typical, and high conditions. A better approach is to use available information (however limited) about the variability of the system to inform the development of a sampling plan. Figure 7.1 illustrates some problems in field sampling of time-variable systems. It shows the differing conclusions that can be drawn from different sampling frequencies—differences of more than 100% from the lowest to the highest estimate.

More heterogeneous systems generally require more detailed—and therefore more costly—characterization. A good characterization provides information on:

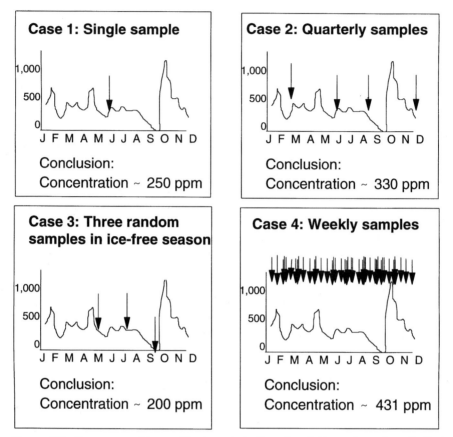

Figure 7.1 *Common problems with characterization of variable systems (hypothetical parameter measured in parts per million).*

- Probable maximum and minimum values
- Level of variation at different scales (e.g., for time: minutes, hours, days, weeks, months, seasons, years, multiyear cycles)
- Potential correlations with other variables (e.g., peak pollutant loadings in spring could be associated with snowmelt and runoff; multiyear flow cycles could be related to larger meteorological phenomena)
- Insight into cause-and-effect relationships

Some industrial effluents (for instance, those from petroleum refineries) are relatively constant through time. Sampling of these sources may require only a few days' effort to map normal operating ranges. In other industries (for instance, organic chemical manufacturing) effluent quality is highly variable because different products may be made every day of the week. Characterization of a highly variable effluent may require months of sampling to capture

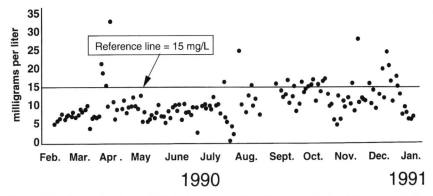

Figure 7.2 *Example of variability in industrial effluent: suspended solids concentrations in the final discharge from the Rio Algom (Stanleigh) mine, Ontario (adapted from MOEE 1992).*

the full range of products, and thus effluent quality, from this source. The same may be true even of municipal sewage treatment plant effluents, depending on the nature of the sewer system and its users. Systems with primarily residential users will have reasonably constant flow and quality (unless there are substantial inputs from stormwater via combined sewers or from inflow/infiltration). On the other hand, systems that receive industrial wastewater flows (very common in larger cities) may have highly variable quality, with occasional high pulses of solids, BOD, metals, or toxic organic substances, depending on the behavior of the contributing industries. Treatment plant bypass is similarly unpredictable and difficult to characterize without detailed sampling. Figure 7.2 illustrates the variability of effluent from a mining operation.

Ostler and Holley (1997) provide the following method for estimating the number of samples required to properly characterize the quality of a wastewater discharge or continuous waste generation process where several preliminary analysis results have already been obtained. The same approach can be used for characterizing a lake or river zone that is thought to be reasonably homogeneous through space but variable over time. (Note that in a body of water this zone does not necessarily extend to the bottom of the water column, but merely to a level where conditions cease to be homogeneous—for instance, of different temperature and, therefore, density.)

$$N = (t^2 s^2)/(RT - \bar{x})^2 \qquad (7.1)$$

where: N = required number of samples

t = t value taken from Students' t table (consult statistics text)

s^2 = variance = (standard deviation)^2 (see Eq. 7.2)

RT = regulatory threshold (target or standard)

$\bar{x}$ = mean

The variance, s^2, is the square of the standard deviation. The standard deviation, s, may be calculated as:

$$s = \sqrt{\frac{\Sigma(x_i - \bar{x})^2}{n - 1}}$$

where: n = number of observations
         x_i = value of individual observation
         $\bar{x}$ = mean of the sample = $(\Sigma x_i)/n$

This equation can be simplified for computational purposes as:

$$s = \sqrt{\frac{\Sigma x_i^2 - n\bar{x}^2}{n - 1}}$$

or

$$s = \sqrt{\frac{n\Sigma x_i^2 - (\Sigma x_i)^2}{n(n - 1)}}$$

It will be seen that equation 7.1 is very sensitive to changes in variance. In other words, if the system is known to be highly variable, more samples will be required for adequate characterization than if it is relatively constant through time. This point is illustrated in Table 7.1.

Example:
    Given the following five preliminary sample results from a wastewater stream, how many samples would be required to determine whether the stream meets a regulatory threshold of 6.0 mg/L?

Results of Preliminary Analysis (example drawn from Ostler and Holley 1997)

Individual measurement	Square of measurement
3.1 mg/L	9.6
4.2 mg/L	17.6
6.5 mg/L	42.3
5.2 mg/L	27
4.8 mg/L	23

Using these values, the standard deviation can be calculated as:

$$s = \sqrt{\frac{n\Sigma x_i^2 - (\Sigma x_i)^2}{n(n-1)}}$$

$$s = \sqrt{\frac{(5)(119.5) - (23.8)^2}{[5(5-1)]}} = \sqrt{\frac{(597.5 - 566.44)}{20}} = \sqrt{1.55}$$

Therefore the variance of this data set is simply the square of s or 1.55.

For 4 degrees of freedom ($= n-1 = 4$), and a 90% confidence level (i.e., a 5% chance that the characterization is too low, and a 5% chance that the characterization is too high $= 10\%$ chance of having made the wrong characterization), Student's t value (obtained from statistical tables) is 2.132.

We can therefore calculate the required number of samples as

$$N = (t^2 s^2)/(RT - \bar{x})^2$$

or

$$N = [(2.132)^2 1.55]/(6.0 - 4.8)^2 = 4.89$$

so about five samples are required to determine with 90% confidence whether the stream in question meets the regulatory standard of 6.0 mg/L. Table 7.1 illustrates how this number of samples would change at higher and lower variances (note that because the variance is a square, differences in standard deviation need not be large to create large changes in variance).

Note that the method proposed by Ostler and Holley requires that sample values be normally distributed ("bell curve" distribution). If, as is frequently the case, values are log-normally distributed (in other words, require a logarithmic transformation to obtain a normal distribution) or are distributed in some other way, it may be difficult to apply statistical tests and sample size estimation methods like this.

Table 7.1   Influence of System Variability on Required Number of Samples

Observed System Variance (s^2)	Required Number of Samples
1.55	5
1.0	3
5.0	16
25	79
100	316

7.1.4 Sampling Methods

Although also beyond the scope of this book, the problem of which sampling method to use is important to note. Basically, samples can be of two types: discrete or composite. In an effort to reduce the costs associated with laboratory analysis, water managers sometimes mix aliquots (subsamples) of different discrete samples to create a "composite" sample. In this way, they believe, an average value can be obtained over the sampling period. Although it is true that composite samples blend, or average, the results of several individual samples, the blended sample does not reveal any of the variability that would have been present in the discrete measurements. Depending on the goals of the study, this may or may not present a problem. If only "average" water quality is required, composite samples may be a cost-effective sampling method. But if it is important to know the true range of results—the maximum and minimum values (for instance, to evaluate the risk of toxicity to aquatic biota)—then a composite sample will be inadequate for the purpose. The analyst must be cautious in trading off sampling and analytical cost against the information to be gained from the sampling effort.

Flow-weighted sampling is sometimes conducted during periods of rapid change in flow or quality, for instance, during wet weather events. If significant variation is anticipated, the frequency of sampling should be increased until the system is stable again. Metcalf and Eddy (1991) recommend intervals as short as 10 to 15 minutes for variable wastewater flows; MOE (1987) recommends intervals of 0.5 to 2 hours for in-stream sampling during wet weather. If flow is monitored simultaneously, the flow and concentration data can be combined to yield loading data for each time interval. Simultaneous monitoring of sources and receiving water during periods of high variability will provide very useful insights into the relationships between the two. Marsalek (1976) provides a helpful discussion of required equipment and sampling techniques appropriate for sampling intermittent sources.

7.1.5 Understanding Watershed Processes

Understanding the rates of watershed processes, and the factors that influence them, is more difficult than simply mapping existing conditions. Chemical and biological processes are largely controlled by the thermodynamics of fundamental reactions; physical processes are strongly influenced by the physics of water, air, and solids.

Analysts attempting to understand the dominant processes affecting water flow and quality within a watershed should aim to answer the following questions for each parameter under study:

Sources

1. What are the primary mechanisms by which materials are transported from sources to the receiving water (e.g., advection, diffusion, volatilization, atmospheric deposition, etc.)?

2. What are the primary mechanisms (if any) by which source materials are transformed between the source and the receiver (e.g., sedimentation, resuspension, adsorption, die-off, etc.)?

3. Which of these mechanisms has a significant effect on the material? Which has a trivial effect (and thus can safely be neglected)?

4. What is the time scale at which important mechanisms operate (e.g., seconds, minutes, hours, days, weeks, months, seasons, years, multiyear cycles)?

5. What is the spatial scale at which important mechanisms operate (e.g., microns, millimeters, centimeters, meters, kilometers, etc.)?

6. What equations can be used to represent these processes for the purpose of predicting system response within altered management scenarios?

7. What factors affect the rate at which these processes occur?

8. What data are available to substantiate the operation of processes that are thought to be important, and the rate at which these processes occur (e.g., rainfall, runoff, and infiltration rates, temporal variation in industrial effluent flow and quality, erosion rates, etc.)?

9. What data remain to be collected? What sampling frequency and spatial distribution would be appropriate for such a data collection program?

Receiving Waters

1. What are the primary mechanisms by which materials are transported within the receiving water (e.g., advection, diffusion, volatilization, atmospheric deposition, etc.)?

2. What are the primary mechanisms (if any) by which source materials are transformed within the receiver (e.g., sedimentation, resuspension, adsorption, die-off, etc.)?

3. Which of these mechanisms has a significant effect on the material? Which has a trivial effect (and thus can safely be neglected)?

4. What is the time scale at which important mechanisms operate (e.g., seconds, minutes, hours, days, weeks, months, seasons, years, multiyear cycles)?

5. What is the spatial scale at which important mechanisms operate (e.g., microns, millimeters, centimeters, meters, kilometers, etc.)?

6. What equations can be used to represent these processes for the purpose of predicting system response within altered management scenarios?

7. What factors affect the rate at which these processes occur?

8. What data are available to substantiate the operation of processes that are thought to be important, and the rate at which these processes occur (e.g., stream hydrographs, patterns of species diversity, annual cycling of algal populations, fish spawning patterns and timing, etc.)

9. What data remain to be collected? What sampling frequency and spatial distribution would be appropriate for such a data collection program?

In most cases, a knowledge of watershed processes is used to generate a one-time or step-wise accounting of mass, number, or quality in the system. (If the accounting is repeated many times—for instance, once per model hour, the result is a more or less continuous approximation of watershed conditions.)

In a detailed assessment, development of the watershed inventory therefore cannot be separated completely from the requirements of the methodology that will be used to evaluate the performance of different management alternatives. In other words, once a commitment has been made to conduct a detailed assessment, data collection and development of a methodology for testing alternatives often proceed more or less simultaneously. Indeed, a computer simulation model may be selected simply *because* it requires only data that are readily available or easily collected.

A good example of this problem can be found in the prediction of runoff and soil erosion from agricultural fields. Ideally, the watershed manager may want to simulate conditions on an hourly basis, 365 days of the year. Such a simulation would demand very detailed knowledge of existing conditions and dominant processes—on an hourly basis. Yet data like these are scarce and would be prohibitively expensive to collect. So in many cases the manager will instead opt for a model that simulates "average annual" conditions (thus losing information about system maxima and minima), or perhaps one that simulates "event" conditions—that is, those associated with a single rainfall event (thus losing information about other events or soil saturation conditions resulting from previous storms). In either case, the analyst has made a decision to trade off the information to be obtained from detailed modeling against the cost of collecting the data necessary to drive a detailed model.

The physical processes most affecting water flow—runoff, evaporation, and infiltration—have been discussed in Chapter 2. The following are the most important processes governing the transformation of water quality in aquatic systems.

The general equation for mass balance in any given volume of water or sediment, or in any area of land (for plants and animals) is:

$$\text{Rate of mass increase} = \text{rate of mass entering} - \text{rate of mass leaving}$$
$$+ \text{rate of mass created internally}$$
$$- \text{rate of mass lost internally}$$

with each term expressed in units of mass per unit time. (If numbers of plants or animals are used, the units are number per unit time.) For the purposes of the following discussion, water quality constituents are used to illustrate the general theory. Metcalf and Eddy (1991) describe the processes affecting the rates of change of water quality constituents as discussed in the following paragraphs.

Transport processes are of two kinds, (1) advection, or transport of material by the flow of water into or out of a control volume, and (2) diffusion, or transport of the constituent by turbulence in the water.

For *advection* into or out of an infinitesimally small control volume, the following equations apply:

Rate of mass increase in the control volume

$$= \frac{\delta C}{\delta t} \, dx \, dy \, dz \tag{7.3}$$

Rate of mass entering the control volume

$$= CU dy \, dz \tag{7.4}$$

Rate of mass leaving the control volume

$$= \left(C + \frac{\delta C}{\delta x} \, dx \right) \left(U + \frac{\delta U}{\delta x} \, dx \right) dy \, dz \tag{7.5}$$

where: C = mass concentration of constituent (mass per unit volume)
U = water velocity in x direction (distance per unit time)
dx, dy, dz = dimensions of the control volume in x, y, z directions
t = time

With the addition of flow components in the y and z directions, this set of equations can be simplified and substituted to yield:

$$\frac{\delta C}{\delta t} = -U \frac{\delta C}{\delta x} - C \frac{\delta U}{\delta x} - V \frac{\delta C}{\delta y} - C \frac{\delta V}{\delta y} - W \frac{\delta C}{\delta z} - C \frac{\delta W}{\delta z} \tag{7.6}$$

where V and W are the velocity components in the y and z directions.

Using the equation of continuity from fluid mechanics, this equation can be further reduced to the effect of advection on concentration changes with time, as follows:

$$\frac{\delta C}{\delta t} = -U \frac{\delta C}{\delta x} - V \frac{\delta C}{\delta y} - W \frac{\delta C}{\delta z} \tag{7.7}$$

Diffusion is caused by small velocity variations acting in conjunction with concentration gradients. If advection is equivalent to stirring in a cup of coffee, diffusion is the gradual mixing that occurs without the stirring force. The corresponding equations for diffusive transport are as follows:

Rate of mass increase in the control volume

$$= \frac{\delta C}{\delta t} \, dx \, dy \, dz \tag{7.8}$$

Rate of mass entering the control volume

$$= -E_x \frac{\delta C}{\delta x} \, dy \, dz \tag{7.9}$$

Rate of mass leaving the control volume

$$= -\left[E_x \frac{\delta C}{\delta x} + \frac{\delta}{\delta x} \left(E_x \frac{\delta C}{\delta x} \right) dx \right] dy \, dz \tag{7.10}$$

where: E_x = diffusion coefficient (or "diffusivity")

With the addition of flow components in the y and z directions, this set of equations can be simplified and substituted to yield:

$$\frac{\delta C}{\delta t} = \frac{\delta}{\delta x} \left[E_x \frac{\delta C}{\delta x} \right] + \frac{\delta}{\delta y} \left[E_x \frac{\delta C}{\delta y} \right] + \frac{\delta}{\delta z} \left[E_x \frac{\delta C}{\delta z} \right] \tag{7.11}$$

Transformation processes depend on the underlying chemical reactions and thus are constituent-dependent. Usually, transformation occurs independently of transport, and thus the effects of transformation and transport are additive.

Some of the most important transformation processes in water quality are:

1. The *oxidation of BOD* (i.e., consumption of available oxygen by oxygen-demanding compounds), a first-order process in which the oxidation rate is proportional to the amount of BOD present. The oxidation of carbonaceous BOD can be expressed as:

$$r_C = -K_C L_C \tag{7.12}$$

and that of nitrogenous BOD as:

$$r_N = -K_N L_N \tag{7.13}$$

where: r_C and r_N = rates of carbonaceous and nitrogenous (respectively) BOD loss per unit time per unit volume of water
  L_C, L_N = carbonaceous and nitrogenous (respectively) BOD concentrations (mass per unit volume)
  K_C, K_N = rate constants for carbonaceous and nitrogenous (respectively) BOD oxidation (expressed as units of time^{-1})

2. *Surface reaeration* occurs when the dissolved oxygen concentration of water is below saturation and oxygen from the atmosphere enters the water surface. This flux is proportional to the oxygen deficit and can be calculated as:

$$r_R = k_R \frac{A}{V} (C_s - C) = \frac{k_R}{H} (C_S - C) = K_2(C_S - C) \tag{7.14}$$

where: r_R = rate of oxygen gain due to reaeration per unit time
  per unit volume of water
  k_R = reaeration flux rate (distance per unit time)
  A = free surface area of the control volume
  V = control volume
  C_S = saturation concentration of dissolved oxygen
  C = dissolved oxygen concentration
  H = depth of control volume
  K_2 = surface reaeration rate (1/time)

Although k_R is more closely linked to the physics of oxygen transfer, K_2 is used more frequently. O'Connor and Dobbins (1958) and Metcalf and Eddy (1991) provide formulas for K_2. Methods for field measurement of reaeration coefficients are described in the literature; they usually employ release/recapture of volatile gases, which can be assayed in sample volumes of water.

3. *Sediment oxygen demand* is an important consideration in systems with high loadings of organic matter that may settle out and decay—often the case downstream of sewage treatment plant effluent discharges, for example. For a control volume above bottom sediments, the sediment oxygen demand can be calculated by:

$$r_S = \frac{k_S}{H} \tag{7.15}$$

where: r_S = rate of oxygen consumption due to sediment oxygen demand
  (mass per unit time per unit volume)
  k_S = sediment oxygen demand uptake rate (mass per
  unit area per unit time)
  H = depth of control volume

Like reaeration rate, sediment oxygen demand can be measured directly in the field using a closed chamber (sometimes termed a "respirometer") to separate a control volume of water and its associated bottom sediments from overlying water. Continuous measurement of dissolved oxygen concentration within the chamber yields an estimate of oxygen depletion over time. Metcalf and

Eddy (1991) suggest values for r_S ranging from 2–10 g/m^2 near municipal wastewater outfalls, to 0.05–0.1 g/m^2 for mineral soils.

4. *Settling of suspended solids* can be represented by the simple relationship:

$$r_D = \frac{w}{H} \, S \tag{7.16}$$

where: r_D = settling rate (mass per unit time per unit volume)
  S = suspended solids concentration (mass per unit volume)
  w = average settling velocity of the solids (distance per unit time)
  H = depth of control volume

Settling velocity is difficult to measure but may be in the order of 0.0001 cm/sec for turbid natural waters or mixed primary/secondary wastewater effluent.

5. *Bacterial die-off* is highly variable in natural waters and is affected by a wide range of factors including temperature, salinity, incident light, available nutrients, and the growth characteristics of the species under consideration. The general equation for bacterial die-off is first-order:

$$r_B = -K_B C_B \tag{7.17}$$

where: r_B = bacterial die-off rate (number of bacteria per unit time
      per unit volume)
  K_B = die-off constant (1/time)
  C_B = bacterial density (number per unit volume)

The die-off constant, K_B, is determined empirically and is used to capture species-specific response to the various aforementioned factors. These rates are highly variable, even for the same species in the same water body. Metcalf and Eddy (1991) report ranges of K_B from 0.12 to 26 per day for freshwater systems, with much more rapid die-off in seawater.

6. *Adsorption to solids* was discussed in Chapter 2 as an important mechanism by which water quality is altered as water passes through porous media such as soils. Here again, removal rates are highly variable, depending on the characteristics of the solid and the chemical constituent affected. A more complete discussion of adsorption kinetics can be found in the literature on groundwater contamination.

7. *Volatilization* is an important process for certain constituents—for instance, methyl mercury in open waters (such as sometimes occurs in impoundments behind water control structures), and for volatile organic compounds (such as benzene) in general. The relationship here is similar to that in surface reaeration (discussed earlier), except that material is transferred out of the water surface rather than into it, and the atmospheric saturation concentration of

most compounds (the equivalent value to C_S in the equation for surface reaeration) is usually close to zero because the partial pressure of the material is usually close to zero. Here again, the relationship follows first-order kinetics, with the challenge being the determination of the volatilization constant, K_V:

$$r_V = -K_V C \tag{7.18}$$

where: r_V = volatilization rate (mass per unit volume)
K_V = volatilization constant (1/time)
C = concentration of substance in water (mass per unit volume)

If estimates of the surface reaeration rate rate, K_2, are available, K_V can be approximated by relationship to that value:

$$K_V \approx K_2 \frac{D_o}{D_C} \tag{7.19}$$

where: K_2 = surface reaeration rate (1/time)
D_o = diffusion coefficient for oxygen in water (area/time)
D_c = diffusion coefficient for substance of interest in water (area/time)

8. *Photosynthesis and respiration* are critical processes in many systems where the growth of rooted aquatic plants and/or algae is a prominent force. These processes have a fundamental effect on dissolved oxygen concentrations and are themselves affected by concentrations of nutrients, especially phosphorus and nitrogen, but also trace elements such as iron. They are also affected by biological processes such as crowding, tissue aging and death, and variable growth rates. Specialized models are available for the prediction of plant growth with associated dissolved oxygen effects (see, for example, Lopez-Ivich 1996), although this area of endeavor is still in its infancy. The majority of available models generate crude estimates of biomass accumulation in response to changing nutrient, light, and temperature conditions. A few (such as that discussed by Lopez-Ivich) provide species-specific representations of the stoichiometry of photosynthesis and respiration, species-specific growth, deterioration, death, and washout rates, and the presence of additional considerations such as "luxury" uptake of excess nutrients for storage in plant tissue. A full discussion of these processes, and how they may be accounted for in the testing of watershed management alternatives, is beyond the scope of this book.

This discussion has introduced some of the most important processes that must be understood and accounted for in compiling a watershed inventory. Most of these processes are in fact highly complex and variable through time and space. The foregoing discussion suggested a fairly consistent approach to representing transformation processes primarily as simple first-order removal mechanisms. The same approach can be used for other processes not discussed here, such as photolysis, with the same concerns, such as determination of removal rates. This approach unquestionably simplifies what can in fact be an extremely complex physico-chemical system, but it is widespread in the literature and forms the foundation of most water quality simulation models. It is based on the conservation of mass equation, a second-order partial differential equation that is difficult to solve in three dimensions but that can be simplified in various ways.

The general equation, incorporating both transport and transformation processes, is:

$$\frac{\delta C}{\delta t} = -U \frac{\delta C}{\delta x} - V \frac{\delta C}{\delta y} - W \frac{\delta C}{\delta z} + \frac{\delta}{\delta x} \left[E_x \frac{\delta C}{\delta x} \right] + \frac{\delta}{\delta y} \left[E_y \frac{\delta C}{\delta y} \right]$$

$$+ \frac{\delta}{\delta z} \left[E_z \frac{\delta C}{\delta z} \right] + \Sigma r_i + \Sigma I_j \tag{7.20}$$

where: I = input rate from external sources (mass per unit time per unit volume)
i = transformation process index
j = input identification index
E_x = diffusion coefficient in the x direction
E_y = diffusion coefficient in the y direction
E_z = diffusion coefficient in the z direction
U = water velocity in x direction
V = water velocity in y direction
W = water velocity in z direction
r_i = rate of transformation (production or consumption)

In summary, the challenge of compiling a watershed inventory for a detailed assessment is to collect enough information to answer the fundamental management questions under study, without wasting scarce human or fiscal resources. Usually, this means:

1. An evaluation of the current condition of the watershed ecosystem (components)
2. Identification of the physical, chemical, and biological mechanisms that influence water quality and flow in the watershed
3. An assessment of the quality and quantity of available data on watershed components (both sources and receivers) and mechanisms
4. Preliminary selection of a methodology for evaluating the performance of alternative management strategies
5. Comparison of the data requirements of the methodology with available data

6. Development of a field program to collect new data or supplement missing or outdated data

7. Evaluation of the resource requirements of the proposed field program

8. If adequate resources are available, execution of the program followed by scenario testing (Section 7.3.7) as appropriate

9. If adequate resources are not available, reevaluation of the proposed scenario testing methodology (or selection of a different methodology) with a view to simplifying the representation of the watershed system in space or time, so as to reduce required input data

10. Development and evaluation of a revised field program

11. Continued iteration of the methodology-data requirements-data required-resource availability analysis until an acceptable balance of resource expenditure and management information has been reached

7.2 SCOPING

7.2.1 Defining the Problem

Section 6.2 outlined the basic methods by which the most urgent problems in a watershed can be determined:

1. Through a listing of existing (or perceived) use impairments and their causes

2. By comparing existing receiving water or sediment quality with published national or state/provincial targets

3. Through key informant interviews and media reports

Scoping for a detailed assessment can begin in the same way, with the proviso that detailed analysis allows the luxury of site-specific comparisons, collection of extensive new data, and longer time for surveys, interviews, and data reduction. Table 7.2 gives some examples of how scoping techniques may differ for a detailed assessment as compared with a simple assessment.

The use of current, detailed, and site-specific data allows the detailed assessment to develop very precise statements about the timing and extent of use impairments and other potential factors in scoping. This precision is necessary to support the detailed questions posed to, and answers required of, the planning process.

The difficulty with detailed analysis is that the analyst may be faced with an overwhelming amount of data, so that the answers to even simple questions are not always easy to find. The three questions central to scoping are:

1. Which uses are currently impaired?

2. Of these, which are the two or three that are most important to the community?

Table 7.2 Comparison of Scoping Techniques for Detailed and Simple Watershed Planning Initiatives

Simple Assessment	Detailed Assessment
Agency staff brainstorm list of use impairments.	Detailed survey of water users to determine nature and quantify extent of impairments.
Literature review of typical effluent characteristics of municipal sewer-using industries.	Detailed survey of the quality and quantity of effluent from specific sewer users and the variability of that effluent over time.
Drive-by survey of watershed land uses.	Analysis of satellite imagery and/or aerial photographs to yield detailed and specific mapping of watershed land uses.
Agency staff compare typical water quality with standards.	Extensive collection of new data tailored to the needs of this study; statistical analysis on a reach-by-reach basis to determine frequency and extent of water quality problems.
Evening public meeting to hear public input on scoping issues reports.	Series of workshops on specific scoping elements, such as use of aboriginal knowledge, status of endangered species, emerging technologies.
Typical, average, or regional data, some or all several years out of date.	A preponderance of new, site-specific data, whose collection was tailored to the needs of the specific planning initiative.

3. For each of these two or three use impairments, what are the best indicator parameters of a healthy condition? What specific (even numerical) targets should be met, to consider the impaired use as having been restored (e.g., total phosphorus <0.02 mg/L)?

It is easy to lose sight of these three "beacons" as more and more data are collected. This problem does not exist in a simple analysis. For example, a simple analysis may involve a few key informant interviews and a review of historical documents and media reports, concluding that most uses are unimpaired but that there is a problem with "swimmability" in the summer. In a detailed analysis, existing data may be supplemented with intensive sampling at specific bathing beaches, demonstrating that sometimes desirable conditions are met and sometimes they are not. What should the analyst conclude from this?

Figure 7.3 illustrates how the relative contribution of a single "important" source can change, depending on the season, weather, or even time of day.

Figure 7.3 illustrates the need to develop reliable and specific indicators of progress toward desired watershed conditions—the evaluation criteria discussed in Section 5.5.2. The higher costs and staff requirements of a detailed assessment therefore demand care in defining problems, in identifying planning constraints, and in selecting appropriate indicators of progress.

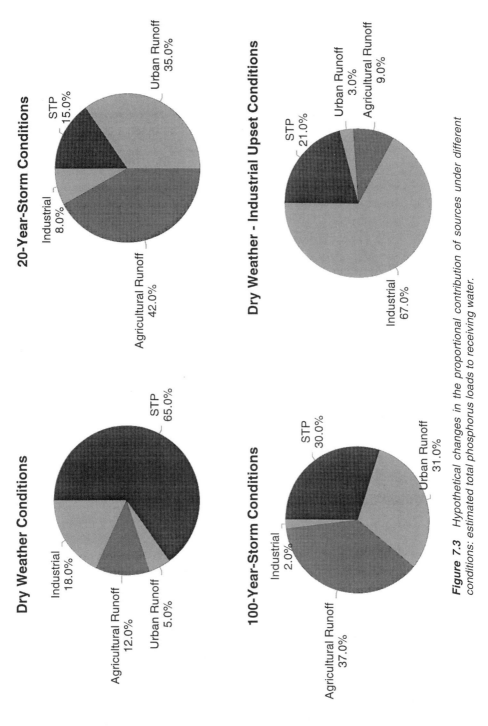

Dry Weather Conditions

STP 65.0%

Industrial 18.0%

Agricultural Runoff 12.0%

Urban Runoff 5.0%

20-Year-Storm Conditions

STP 15.0%

Urban Runoff 35.0%

Industrial 8.0%

Agricultural Runoff 42.0%

100-Year-Storm Conditions

STP 30.0%

Urban Runoff 31.0%

Industrial 2.0%

Agricultural Runoff 37.0%

Dry Weather - Industrial Upset Conditions

STP 21.0%

Urban Runoff 3.0%

Agricultural Runoff 9.0%

Industrial 67.0%

Figure 7.3 Hypothetical changes in the proportional contribution of sources under different conditions: estimated total phosphorus loads to receiving water.

Given the vast range of problems and sources that may be found in any given planning area, it is difficult to generalize about scoping methodologies. The following points may be helpful in understanding the different challenges involved in scoping a detailed study as compared with a simple study.

Scoping is usually considered to comprise two separate tasks:

1. Boundary setting, and
2. Focusing.

Boundary setting refers to the need to place limits on the time period, spatial extent, and populations that will be evaluated in the planning initiative. *Focusing* means choosing the two or three most important issues or problems, which will form the focus of most of the sampling and analytical effort.

In a simple analysis, boundary setting is usually crude and general, and the justification for the choice of boundaries may be simple consensus. For instance, agency staff undertaking a simple analysis may simply decide on a 20-year time frame, because that is the expected life of the local sewage works, and to use the watershed boundaries as a geographic boundary (for analytical simplicity), excluding groundwater considerations (because data are too scarce and analytical techniques too difficult), and focusing only on human populations (again because of concerns about potential resource expenditure if many species were assessed).

By contrast, a detailed analysis usually involves careful and detailed scoping, as described in the following paragraphs.

Boundary Setting—Time   Time boundaries (the planning horizon) may be determined through a process of infrastructure evaluation (expected facility life), surveys of public opinion (determined by in-person or mailed surveys, public meetings, workshops, etc.), and an iterative series of boundary-setting proposals that are successively reviewed and revised by stakeholders. Planning may be conducted over several time periods, for instance an immediate future (1 to 2 years post-project), a short-term future (3 to 5 years post-project), and a long-term future (10 to 50 years post-project or even longer). The differing requirements of these planning levels demand different types of information, both at the inventory stage and in the development and testing of alternatives.

Many major projects will involve construction activities and the use of heavy machinery. This construction phase will have its own special impacts and concerns (for example, dust, noise, and vibration) that are separate and different from those that will occur during system operation or closeout. It is important to understand how long such a construction phase might last and the nature of the impacts that may occur in this stage.

A different set of concerns is likely to arise during the operational phase, the time following completion of construction when the system or device is operating normally. The operational phase is usually much longer than the construction phase—perhaps 25 years, as compared with 3 or 4 years for construction.

Finally, the analyst must understand the possible implications of "closeout" or decommissioning of the project. When the operational life of a structure or facility is finished, some action has to be taken to close it down and seal it off from the environment. In some cases, this will mean dismantling it; in others, it may mean capping an underground structure. Whatever the necessary action, the analyst will need to understand what is involved with closeout, how long it will take, and whether ongoing care is required. If perpetual care is needed, as it would be in the case of a landfill or abandoned mine, the post-closeout phase, with its associated impacts, could continue for tens or hundreds of years.

In certain cases different time boundaries may be set for different environmental compartments or media; this is seldom required in a simple analysis. For example, planners may decide to look at surface water quality and flows over a 20-year time horizon, but groundwater quality and flows (especially where there has been groundwater contamination) over 100 years or longer. In such a situation, predictive simulations over the longer term would focus only on the compartment of interest, in this case groundwater.

Choosing one or more planning horizons takes work and thought but is seldom a highly controversial issue. The exception may be where contamination by persistent, toxic, or radioactive substances currently exists or is predicted for the future. In such instances people who hold more protective, resource-conservative values may prefer a very long planning horizon, anticipating the potential for cumulative or intergenerational effects. Others may hold conflicting values, perhaps emphasizing careful expenditure of funds or anticipating the development of better treatment technologies in the distant future. Such a conflict may arise, for example, in regard to deep-water disposal of tailings from mining and milling operations. Deep-water disposal offers a way to "bury" tailings where they will not cause acid mine drainage—in the deepest waters of a lake or reservoir. Yet if these materials are especially toxic, as in the case of uranium mine tailings, downstream water users may prefer an analysis that extends over hundreds or thousands of years to be sure that predicted releases from deep waters are low or nil.

Boundary Setting—Space The task of setting spatial boundaries on an analysis can be both more difficult and more controversial than setting time boundaries. For one thing, impacts of current or proposed management actions may have both local effects (such as disturbance of fish habitat) and very widespread effects (such as changes in local and regional traffic patterns). It may not be clear whether it is preferable to adopt a very local, intensive scope or a much larger, and perhaps more general, scope. As with time boundaries, it is possible to conduct an analysis at several scales. A very local scope (a radius of less than a kilometer around the proposed site of an activity) could be used for situations where there is a single major source or where intensive activity is planned for the future. A larger scope (for instance, the watershed boundaries) could be used to evaluate alternative management scenarios for the whole basin. And if major changes are anticipated within the basin, it may be necessary to adopt a scope that extends beyond the watershed boundaries; this may be especially true for groundwater flow and surface traffic patterns. Different levels of detail, and even different watershed components, can be examined at each scale of analysis.

It should perhaps be added here that in urban centers, sewer systems in fact override natural drainage patterns and may thus supersede natural watershed boundaries in scoping. In other words, the analyst may need to use the "sewershed" boundaries rather than those of the natural watershed.

Boundary Setting—Populations   A third important element in boundary setting relates to the choice of which species of plants and animals will be included in the analysis. If the analyst is seeking to determine current and future impacts of alternative management actions, the question arises, "Impacts on whom?" It seems to be implicit in most planning initiatives that the species of most, perhaps sole, concern is the human species. But this is by no means always the case. Fish species in particular may form an important part of the analysis, as may macroinvertebrate assemblages (see, for example, the case of watershed planning for the Latrobe River basin, discussed in Section 3.4.2; Grayson et al. 1994).

There are potentially thousands of plant and animal species in an area to be assessed, including not only mammals and birds but insect, amphibian, and reptile species that may be of special interest or value. Domestic animals such as livestock and house pets may also be contributors to, and affected by, water problems in the basin.

The task of setting boundaries on the populations to be assessed may be the most complex and the most controversial of any boundary-setting exercise. Even if there is consensus that the human population is the only one of interest, there may be conflict in regard to *which* humans are affected, and how. Furthermore, within what appears to be a single user group (for instance, recreational boaters) there may be conflict between stillwater boaters, such as sailors and canoeists, and users of motor craft. If resources permit, a detailed analysis should include all significant users. If these populations—or, in the case of nonhuman animals and plants, advocates for those populations—are allowed to speak for themselves, the scoping exercise becomes more straightforward. Problems, and conflict, arise when one group, such as agency staff, attempt to make these kinds of scoping decisions without input from the community. The techniques described in Chapter 4 are especially helpful in developing scoping guidelines that are acceptable to all water users.

Focusing   Implicit in Chapters 3 and 4, and in the foregoing discussion in this chapter, is the notion that community needs and values should drive the scoping exercise. That is, there is no single correct outcome toward which planners should strive. Rather, an appropriate scope for the detailed study will emerge only after discussion, preferably ongoing, with the community affected by the plan. This premise reflects Bishop's (1970) sense of watershed planning as an mechanism for social change—for moving from the current, unacceptable condition to some condition that is desired by the community.

Focusing is the logical extension of that concept. Focusing recognizes the problem of too much data to analyze meaningfully and demands that the analyst examine the issues that are truly key to the community. These are usually strongly felt issues and, thus, factors that may, if treated cavalierly, adversely affect the implementation success of a management plan. Focusing asks the community to decide which problems it wants solved first and then directs the planners to expend time and resources on those problems, with the result that secondary issues are treated in much less depth or may even be ignored.

In a detailed assessment, with the luxury of time and money to spend on data collection, careful focusing should emerge as a matter of course: the key issues in the basin will be mentioned again and again by those involved in the process. Key informant interviews (discussed in the following paragraphs) provide an excellent mechanism for developing a clear focus for the planning initiative. It must be emphasized, however, that focusing is not a scientific process. That is, no amount of data analysis will reveal with certainty the key foci of public concern. Data analysis will certainly help to reveal areas where beneficial uses are, or soon may be, impaired, but it will not show which issues are considered by the public to be most serious or most urgent. Focusing is therefore not an analytical exercise, but rather a consensus-building process. The community must decide which problems to fix first and then agree to live with that decision.

The key informant interview has many applications in watershed management—for instance, in the quantification of intangible benefits (see Chapter 8). Interviews with positional, reputational, and decisional leaders (Section 4.3.2) are almost essential in a detailed analysis. Table 4.1 illustrated some possible outcomes of a watershed planning initiative, depending on the nature of goal setting, which includes focusing. The table illustrates Bishop's (1970) important point that the ways in which goals are set, and the clarity of goals, have a direct impact on the nature of change that occurs. Inclusion of community leaders in the focusing process is central to collaborative goal setting.

Key informant interviews should include people from all sides of an issue (in this case, people who are likely to support the project and those who are opposed to it, as was the case in policy profiling, discussed in Section 4.3.4). Appropriate key informants may include government representatives, public interest group staff, local residents' association leaders, and professional staff from local industries or consulting operations.

In the interview, the analyst should be open and receptive, stating that he or she is looking for advice on scoping and would welcome new ideas and viewpoints. Then the key informant can be invited to discuss the important issues in the basin, boundary-setting considerations, sources of data, and suggestions for additional contacts. The larger the number of key informants interviewed, the more comprehensive the analyst's understanding of issues will be and the stronger the rationale for boundary setting and focusing.

Conflicts may certainly arise in the process of scoping. They can be dealt with in various ways, including:

1. *Technocratic decision.* The agency undertaking the study simply makes a decision and proceeds with the analysis. Those not satisfied with the decision are likely to withdraw their support from the planning process and may become vocal opponents.

2. *Conflict resolution.* (Alternative Dispute Resolution). Conflict resolution (Section 4.4.7) can be used to find equitable solutions to disagreements on boundary setting, scoping, or the impacts or benefits of proposed management actions. The technique has broad application and has been used in the resolution of aboriginal land claims, forest management issues, social and economic impacts of dams and impoundments, and similar controversies. However, it will be successful only if participants can be encouraged to abandon entrenched positions and think creatively about alternative solutions. This may not be possible if there are longstanding disputes and a history of failed negotiations in the basin. In such instances, new ground may be found using circle techniques.

3. *Circle processes.* As described in Section 4.4.5, circle processes allow the exchange of information, including values and priorities, without a formal meeting structure or chair. The self-regulating nature of circle processes makes them less daunting to groups that consider themselves marginalized by the process and may thus reveal attitudes and solutions that would not emerge within a more rigid and hierarchical format.

4. *Other processes:* Section 4.4.5 describes a variety of other small-group processes that may be helpful in resolving conflicts. Adaptive Environmental Assessment and Management (Section 3.4.2), for instance, uses a series of workshops to develop boundaries, foci, and methodologies for the assessment process.

Also part of scoping is the identification of planning constraints and the development of evaluation criteria. These elements were discussed in Sections 5.5 and 6.3 and are not treated at length in this section. The central considerations relating to planning constraints and evaluation criteria in a detailed analysis are as follows:

Constraints must be:

- Exhaustive; inclusive of all factors that may place practical limits on the planning exercise.
- Specific and quantitative; if possible, expressed as numerical limits so as to permit a "pass-fail" evaluation.
- Endorsed by the decision-making group and by the wider community.

Evaluation criteria must be:

- Able to capture the vision of the community in terms of a desired future condition.
- Useful in separating alternatives; in other words, a criterion on which all options score equally is of little use as a screening tool.
- Explicit and specific: measurable in some objective way, or at least permitting qualitative ranking, such that participants in the decision-making process can agree on the performance of each option.

7.2.2 Identifying the Sources

A detailed watershed inventory will quickly reveal a range of pollutant sources and should suggest their relative contributions to the problem of interest. The problem of characterizing variable sources and receivers was discussed in Section 7.1.3. As in a simple assessment, the goal of source identification and prioritization is to determine where remedial efforts will have the greatest impact. A source analysis such as that shown in Figure 6.3 can form the basis of such a decision.

Figure 7.3 illustrates the problem of identifying sources when the "problem" appears to vary from time to time and place to place. Table 7.3 illustrates how clear scoping can assist in the identification of sources.

As observed in the simple analysis, and as shown in Table 7.3, there are frequently several sources of a single problem. Careful scoping, including attention to the true interests of the plan, can help to narrow the field of investigation. For instance, the flooding case presented in Table 7.3 has one primary source, rapid urbanization in the basin, but may have many solutions. The analyst must be very clear about the goals of the study: in this case, whether to reduce flood *damages* rather than the depth of flooding. Reducing the depth of flooding would demand land management measures upstream, because existing flood control structures are considered adequate. Given the extent of flooding observed, changes in land management practices would have to be extensive, and thus costly, to be effective. If, on the other hand, the goal of the study is to reduce the impact of flooding—the dollar costs, the injuries and death, the social disruption—then other "sources" of the problem (in this case, building on the floodplain) could be examined along with changes in land management practices.

7.2.3 Evaluating the Relative Importance of Sources

Source evaluation has seven major steps:

1. Explicit identification of the issue to be resolved, usually expressed as a use impairment.

Table 7.3 Source Identification Based on Scoping and Indicators

Issue	Indicator(s) of Progress and Targets	Available Data Reveal …	Sources and Significance
Flood damage in Windsor Heights and Walkerton in spite of flood protection structures	Flood levels less than 100-year storm (1% risk) should not overtop levees. Human casualties, flood losses ($), and number of residences flooded may also be used as interim indicators	Direct losses include damage to buildings, contents, roads, and vehicles, currently estimated at $4.5 million/year. Social disruption including lost productivity, business losses, diminished quality of life, etc., is valued at a further $2.5 million/year. Human injury currently affects approx. 20 people/year at a health-care cost of $1.5 million/year. Properties most at risk of flooding were constructed in the floodplain during the 1920s and 1930s.	1. Intensive and rapid urban development in the basin have increased peak flows and shortened the time to peak flows. Some flows could be intercepted by land management practices such as reforestation, porous pavement, on-site storage, and so on. This source is considered to be the only major factor in creating flood flows. 2. Land use controls prohibiting development on the flood plain could reduce the risk and extent of flood damages and should be investigated as an adjunct to land management practices. 3. Existing flood protection structures are in good repair and should not be removed or replaced. Existing structures are not considered to be a source of the flooding problem.
Closure of Heron Point and Pine Point swimming beaches	Fecal coliform counts should be less than 100/100 mL within 24 hours of the end of a rainfall event.	Beach contamination occurs only during and shortly after rainfall events. The ratio of fecal coliform to fecal streptococcus bacteria suggests a nonhuman source of bacteria. Highest in-stream bacterial densities	1. Dry weather sources (e.g., sewage treatment plant, industry) contribute minimal loads that do not vary significantly over time. 2. Combined sewer overflows occur only 2–3 times a year; therefore, are only occasional contributors.

are measured near Municipal Drain #4, which drains feedlots owned by Best Beef Inc. and Irving Hofstetter and family.

3. Stormwater outfalls are located downstream of the beaches.
4. Inspection of feedlots reveals a faulty manure storage tank on the property of Best Beef Inc. Immediate abatement action is warranted.

| Rapid siltation rates at Thornton Reservoir | Suspended solids should be less than 15mg/L at all times in the water entering the reservoir. Secchi disk depth should be at least 1.5 m except during spring high-flow periods. | Suspended solids levels entering Thornton Reservoir range between 37 and 778 mg/L. The highest levels are observed during spring snowmelt and runoff periods. The reservoir experiences algal blooms in summer, which may contribute to turbidity and sediment accumulation. Land use upstream of the reservoir is primarily new urban (residential) development, with 37% of land surface now exposed in new construction. | 1. New construction at the southeast corner of Davis and South Streets has exposed almost 5 ha of land to erosion. Source is considered to be a major contributor to in-stream sediment loads in wet weather.
2. Cropland upstream of the reservoir is planted in corn with 50 m deep buffer strips. This source is not considered to be significant.
3. In-stream bank erosion is significant in fast-moving reaches in the till plain upstream of the reservoir. This is considered to be a source of moderate importance, especially at high flows.
4. Truck-washing operations at McKinley Transport discharge wash water into storm drains that, in turn, discharge to the river upstream of the reservoir. This is considered to be a source of minor importance, because it is limited to dry weather. |

2. Explicit identification of the target(s) to be met (i.e., the point at which the problem will be considered "solved" or the use restored).

3. Explicit identification of measures of progress toward desired targets, sometimes termed "indicator variables."

4. An inventory of all sources within the basin that affect the levels of these indicator variables.

5. Description of the behavior of these sources in space and time.

6. Ranking of these sources in terms of their overall contribution to the problem (usually expressed as total mass, for pollutants, or volume, for flows, contributed or as a percentage of the total mass or volume). If a source is highly variable through time, accurate calculation of total contribution may require subcalculations for individual events (e.g., rainfall events, industrial spills, etc.), even on an hour-by-hour basis using flow-proportional weighting.

7. Elimination of trivial sources (usually those contributing less than 5% of the total load) from further consideration.

These steps parallel those followed in a simple analysis, varying only in that the extent of detail involved is much greater. In particular, Step 6, ranking of sources, is a complex task if done thoroughly, taking into account the time- and space-variable nature of sources and receivers.

Much will hinge on the outcome of a detailed analysis: expenditures of time and money, impacts on people and structures, long-term changes in life-style and taxation. It therefore behooves the analyst to undertake a detailed analysis of each suspected source and its behavior in space and time. Often, this will demand both careful sampling and the use of an effective computer simulation model to extrapolate from available observations to a wider range of operating conditions. The computer model is first calibrated (tuned) to a set of observations, then validated (without further tuning) by the analyst's trying to predict a different set of observed conditions. The predictive error of that validation run (the difference between observed and predicted values) gives some indication of how well the model will perform in simulating future management scenarios.

Figure 7.4 illustrates the results of a hypothetical calibration and validation sequence for a continuous hydrologic model. Calibration was performed on data obtained from one storm event (Figure 7.4a), and the "tuned" model was then applied without further changes in input variables to simulate a second storm event (Figure 7.4b).

For the purposes of understanding the relative importance of sources, a good simulation model can provide the analyst with a way of exploring how sources contribute to a problem under different meteorological, flow, or operating conditions. Although available data alone give some insight into the behavior of each source, a predictive model allows the analyst to pull together data on several sources simultaneously and investigate their relative influence on in-stream conditions.

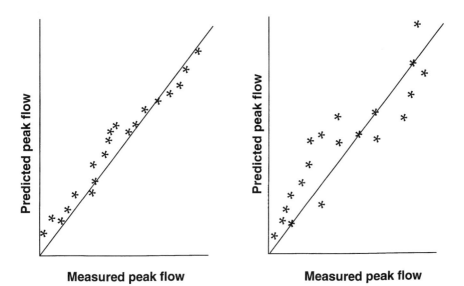

Fig 7.4a Calibration run - Storm #1   **Fig. 7.4b Validation run - Storm #2**

Figure 7.4 *Example calibration and validation sequence for a continuous hydrologic model (after James 1992).*

In a sense, a simulation model provides the basis for developing an understanding of current cause-and-effect relationships in the basin. It can reveal how the relative contributions of various sources change during a major rainfall event, for example, or can allow the analyst to examine source behavior during high-risk–low-frequency events such as hurricanes or major industrial upsets.

Chapter 3 discussed the need for systematic comparison using a predetermined set of planning constraints and criteria. These criteria, in particular, become critical in the evaluation of sources in the basin, and in subsequent testing of management options (Section 7.3). It is therefore recommended that planning constraints and evaluation criteria be reviewed carefully and endorsed by the decision-making body before further analysis of remedial actions begins.

7.3   DEVELOPING AND SCREENING MANAGEMENT ALTERNATIVES

7.3.1   Developing Alternatives

As in a simple analysis, the development of management alternatives in a detailed analysis should flow naturally from an inventory of sources. As discussed in Section 6.3.1, the steps to be followed are:

- Identify the sources that contribute most to the problem under study.

- For each nontrivial source (those contributing, say, more than 5% of the total load), identify at least one management option (if possible) in each of the following categories:

  "Do nothing"—retaining status quo

  Structural measures—built technologies or structures

  Vegetative approaches

  "Best management practices" (nonstructural measures)

Ideas for management options can be obtained from the literature, from local experts such as agency staff, consultants, or university researchers, or can be formulated de novo by stakeholders or decision makers. The various techniques described in Chapters 4 and 5 are helpful in structuring idea-generation processes for maximum productivity.

A more challenging task is evaluating the performance of different management options. The outcome of the detailed analysis is likely to be real dollars spent on real, tangible improvements, not just completion of a paper exercise. Each step of option screening must therefore be carefully justified, including a full description of anticipated predictive accuracy. Statements of relative effectiveness ("better than," "not as good as," etc.) will not provide the justification necessary to support decisions about resource allocation and expenditure, which must often be approved by elected officials who are accountable to the larger population. For these reasons, computer models are now routinely used in detailed watershed studies and are required for the preparation of certain discharge permit applications under the U.S. Clean Water Act.

Although the time and effort needed to build and calibrate a good simulation model may be daunting, the rewards can be tremendous. Some of these benefits include:

- Insight into basin processes and cause-and-effect relationships between sources and receiving water impacts

- Compilation of an extensive database that may reveal other issues or trends not previously recognized

- A powerful, flexible tool for the comparison of management alternatives and for extending the understanding of the system to periods for which data are unavailable.

Section 7.3.2 describes some approaches to simulating existing conditions and predicting the impacts of management changes on a current or future system.

7.3.2 Types of Simulation Models

The mid -to late 1970s and early 1980s saw the development of dozens, per-haps hundreds, of sophisticated computer simulation models for use in water management. These models were costly to develop and costly to run, requir-ing the use of minutes or hours of processing time on a mainframe computer. Through the later 1980s some of these models were converted for use on desk-top microcomputers, although most remain Fortran-based and, to a large extent, inconsistent with newer software technology. Many models were developed for a given task and/or geographic location and may have limited application in other systems. The huge range of models now available may be confusing to the inexperienced user, perhaps especially because there is seldom a clear "right" choice. The following discussion may be helpful in assisting water managers in selecting a model appropriate for their needs. Figure 7.5 illustrates how dif-ferent models may be used at different stages of the planning process.

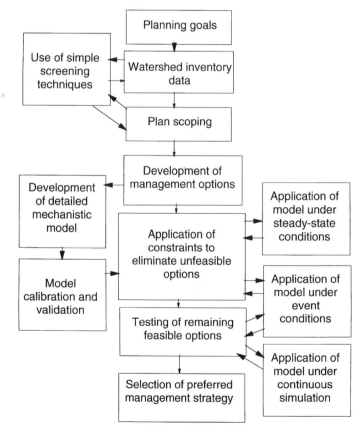

Figure 7.5 *Use of computer simulation models in the development of watershed man-agement plans.*

Models can be of various types:

1. Simple statistical relationships (such as the frequencies of violation, as shown in Table 6.6, or predictions based on the expected frequency distribution of flows or pollutant loads).
2. Complex statistical or empirical models incorporating nonlinear processes or time-variable rates.
3. Unit area loads (e.g., expected sediment loss per hectare from a given land use).
4. Simple relationships incorporating interaction between basic processes such as slope, runoff, and roughness (e.g., the Universal Soil Loss Equation).
5. Simple steady-state deterministic models (which are likely to be based on the same relationships as in (3) but which incorporate more detailed hydrologic functions, allowing prediction at multiple locations in the system under study.
6. Dynamic deterministic models incorporating step-wise changes in input and generating step-wise changes in output. The time step employed in a dynamic model may vary from a few minutes to days or weeks, but is typically on the order of one to two hours. These models usually incorporate more detailed and comprehensive representations of meteorological, hydrologic, chemical, and biological phenomena rather than lump complex processes into a handful of algorithms.

Figure 7.6 illustrates the components of a typical dynamic, deterministic

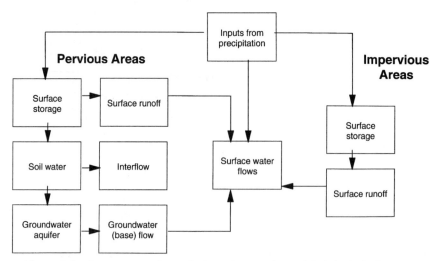

Figure 7.6   *Components of a typical continuous, deterministic hydrologic model.*

hydrologic model. It is important to note that models like this simulate flows, runoff, and infiltration processes in a detailed fashion. By contrast, simulation of water quality is often very crude and weak. For instance, sediment buildup and wash-off may be estimated by the simple relationships described in Section 6.3.3. The load of a given pollutant (e.g., total phosphorus or total zinc) may then be estimated simply as a proportion of total sediment load. This approach (discussed in regard to the use of potency factors in buildup and wash-off simulation) was also described in Section 6.3.3. This is a good example of the point made in Chapter 6 that even the most computationally complex models may reduce complex natural processes to very simplistic representations. Failure to understand the weaknesses, not just the strengths, of deterministic models may lead the analyst to make naive decisions or to misrepresent findings to the decision-making body and the wider public.

Table 7.4 summarizes the characteristics of major water quality simulation models.

There are many other simulation models available, and those outlined in Table 7.4 are continually revised and improved. Novotny and Olem (1994) describe some other types, including stochastic and neural network models. The interested reader is encouraged to contact model authors or responsible agencies for current information on the models in Table 7.4.

Models are not always used singly. It is not uncommon to find several models, perhaps of different types, linked together in the modeling process. (Indeed, several of the models described in Table 7.4 incorporate submodels or model linkages.) An example of a linked model system is shown in Figure 7.7. Clearly, the challenges involved in building, calibrating, and validating a model increase substantially when more than one model is in use. Nevertheless, use of linked models allows the analyst to tailor simulation tools to the particular task at hand. In the work of Lopez-Ivich (1996), for example, the US EPA channel model WASP5, which contains good two-dimensional deterministic hydrologic and water quality routines, was linked to the plant growth simulation model ECOL, which contains more sophisticated algorithms representing plant growth and nutrient utilization dynamics for three different aquatic plant species. In this way, Lopez-Ivich was able to reduce computational complexity while making optimum use of each model's strengths.

7.3.3 Selection of an Appropriate Modeling Methodology

Given the enormous range of modeling approaches possible, the analyst is faced with a difficult decision: which model will provide the desired level of information within the desired time frame and budget? Table 7.4 illustrates that the choices to be made include the following:

1. *Dimensionality:* To simulate the system in one, two, or three dimensions?
2. *Time period:* To simulate the system for a single rainfall event (in that

Table 7.4 Characteristics of Major Water Quality Simulation Models (from James 1992 and Novotny and Olem 1994)

Model Name	Primary Application	Mode of Operation	Comments
Stormwater Management Model (SWMM) (US EPA)	Simulation of urban runoff quantity and quality, including processes in storm and combined sewer systems.	Event or continuous; time step can be minutes or hours.	1. Can be run at different levels of detail: very detailed for design purposes, or at a coarser level for planning purposes. Pollutant concentrations are simulated using potency factors. 2. Real storm events are modeled on the basis of rainfall and pollutant inputs and system characterization to predict responses in the form of flow and pollutant concentrations.
STORM (US EPA)	Simulation of rainfall-runoff-water quality in urban and rural catchments.	Event or continuous; fixed time step of one hour.	1. Similar to SWMM, which has largely supplanted it. 2. Suitable mainly for large catchments. 3. Estimates runoff by one of three simple methods, a formula similar to the Rational Formula, the U.S. Soil Conservation Service Curve Number Technique (developed for small rural catchments), or a combination of these two methods, whereby the first is used for impervious surfaces and the second for pervious surfaces.
MOUSE (Danish Hydraulic Institute)	Simulation of complex piped stormwater/sewer systems.	Continuous; three computation levels available, depending on user needs.	1. Comes with a rain database with historical series of rainfall events and a full-screen menu for editing input data, executing model runs, and graphing results. 2. Uses either a time-area method approach or a more detailed estimation based on a kinematic wave model. 3. Includes separate models for runoff, pipes and networks, and combined sewer overflows.
MIKE-11 (Danish Hydraulic Institute)	Simulation of unsteady-state one-dimensional flows, transport, and biological-chemical reactions.	Continuous unsteady-state in one dimension.	1. Contains separate modules for discharge and water levels in rivers and floodplains, sediment transport, transport-dispersion, and water quality (related to the transport-dispersion module). 2. Provides an interactive menu system for data management, execution of model runs, and processing of results.

Model	Description	Type	Notes
MIKE-21 (Danish Hydraulic Institute)	Simulation of unsteady-state two-dimensional flows with vertically homogeneous flow, transport, and biological-chemical reactions.	Continuous unsteady-state in two dimensions (uses a rectangular grid covering the area of interest).	1. Contains three "stages": a hydrodynamic stage that simulates water level variations and flows in response to various forcing functions in lakes, estuaries, bays, and coastal areas; a transport-dispersion stage simulating the spreading of a substance in an aquatic environment (conservative or nonconservative pollutants, inorganics and organics, salt, heat, suspended sediment, dissolved oxygen, inorganic phosphorus, nitrogen, etc.; and a water quality stage, which is an advanced eutrophication model describing nutrient cycling and phytoplankton growth, including impacts on dissolved oxygen levels.
Hydrologic Simulation Program—Fortran (HSP-F) (Bicknell et al. 1993)	A modular system of models incorporating both watershed and channel simulation of hydrology and water quality.	Dynamic and continuous simulation of fate and transport overland and in one-dimensional channels.	1. This model allows integrated simulation of land and soil erosion processes with instream hydraulic and sediment-chemical interactions. 2. Incorporates a wide range of watershed processes that can be combined or ignored to suit the needs of the use. 3. Data requirements can be extensive. Continuous simulation requires continuous input of data to drive the simulations. 4. Model documentation is weak, and the model is frequently revised. Application will likely require experienced users and sufficient time for model building and testing.
Enhanced Stream Water Quality Model (QUAL2E) (US EPA)	Simulation of several water quality constituents in a branching stream system.	Steady-state, steady-flow simulation of hydrology, heat balance, and materials balance for pollutant(s).	1. A model that has seen wide use in water management around the world. It has to some extent been supplanted by continuous models such as WASP and HSP-F but is still popular for water quality simulation at steady state. 2. Application may require an experienced user for model building and calibration (more than 100 separate inpute variables are required). 3. A version (QUAL2E-UNCAS) incorporating uncertainty analysis for the evaluation of predictive error is available.

Table 7.4 (Continued)

Model Name	Primary Application	Mode of Operation	Comments
Water Quality Analysis Simulation Program (WASP5) (US EPA)	Simulation of contaminant fate and transport in surface waters.	Dynamic and continuous or steady-state simulation of water quality in one, two, or three dimensions.	1. A powerful tool for the simulation of water quality constituents such as dissolved oxygen, BOD, nutrients, phytoplankton-related eutrophication, bacteria, and heavy metal contamination. 2. Incorporates six transport mechanisms (advection and dispersion in the water column; advection and dispersion in pore water; settling, resuspension, and sedimentation of up to three classes of solids; evaporation or precipitation. 3. The user must specify the proportion of flow routed through various environmental compartments—a difficult task. 4. Data requirements are extensive, and calibration may require an experienced user.
Source Loading and Management Model (SLAMM) (Pitt 1995, 1998)	Simulation of urban runoff quality and quantity.	Dynamic and continuous simulation of mass balances for particulate and dissolved pollutants and runoff flow volumes.	1. SLAMM is intended as a planning-level tool, not a detailed urban runoff model. Its strengths lie in its detailed algorithms for buildup and wash-off of street dirt and the incorporation of small storm hydrology. 2. Can be linked to other models such as HSP-F to improve the simulation of urban runoff quality and quantity. 3. Requires information about the runoff controls to be evaluated and the development characteristics of the study area, including data such as infiltration rate, area and cross-sectional shape of basins, and the sizes and types of outlet structures
Areal Nonpoint Source Watershed Environment Response Simulation (ANSWERS) (Purdue University)	Simulation of flows and soil erosion in agricultural watersheds.	Event-oriented, distributed parameter model. A single storm hyetograph drives the model.	1. Uses a square grid system (1–4 ha grid area). Within each grid element, the model simulates various processes, including interception, infiltration, surface storage, surface flow, subsurface drainage, sediment detachment, and sediment transport. Each grid element becomes a source for an adjacent elements. 2. Nutrient concentrations are simulated using statistical relationships between sediment load and pollutant concentration. 3. Cannot be used for the simulation of snowmelt or for the movement of pesticides.

Model	Purpose	Approach	Notes
Agricultural Nonpoint Source Pollution Model (AGNPS) (USDA)	Simulation of nutrient concentrations and sediment load for the comparison of agricultural "best management practices" on agricultural watersheds.	Continuous or event-oriented. Flow and pollutant routing uses a unit hydrograph (lumped parameter) approach.	1. Uses a square grid system with grid elements grouped according to subwatershed drainage. 2. Uses a modified version of the Universal Soil Loss Equation to predict soil loss in five different particle size classes. 3. Requires extensive input data, but most of the required data are obtainable from topographic and soil maps, meteorological data, field observations, and similar published material.
Agricultural Runoff Management Model (ARM) (Donigian 1978)	Simulation of runoff, snow accumulation and snowmelt, sediment, pesticides and nutrient loadings from surface and subsurface sources.	Dynamic and continuous simulation of fate and transport overland and in one-dimensional channels	1. A version of HSP-F that can be run as a separate model or included as a module of HSP-F. 2. Like HSP-F, requires extensive input data and an experienced user. Calibration may be difficult.
Chemicals, Runoff, and Erosion from Agricultural Management Systems (CREAMS) (USDA)	Simulation of soil loss and nutrient export from agricultural watersheds.	A field-scale model using separate submodels for hydrology, erosion, and chemistry.	1. Hydrology is based either on the SCS Runoff Curve Number model (when only daily rainfall data area available) or the Green-Ampt equation (when hourly data are available). 2. The soil erosion model incorporates soil detachment, transport, and deposition (Universal Soil Loss Equation is used to model detachment). 3. The model can simulate up to 20 water quality constituents at any one time. 4. Can simulate user-defined management practices, including a range of agricultural "best management practices."

233

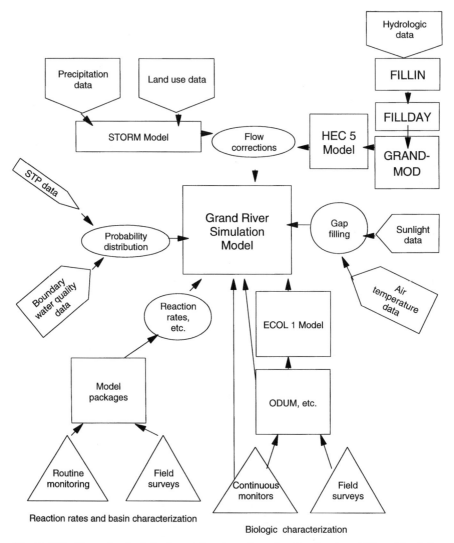

Figure 7.7 *Example of a linked model system: flow chart for the Grand River Simulation Model (GRSM) (after Willson 1981).*

case, which event?), for several individual but not necessarily contiguous events, or continuously?

3. *State:* To simulate the system at "steady state" (a single set of conditions) or "unsteady state" (flow and other conditions varying as they naturally would)?

The analyst must answer a series of other questions as well:

1. What parameters will be used to make decisions; i.e., what parameters must be modeled?
2. What kinds of information will be useful in making decisions (e.g., load reductions from land sources? in-stream water quality? average contributions from a given source, etc.)?
3. Over what time period (days, hours, seasons, years) would predictions be helpful? In other words, is it enough to know that average annual phosphorus will decline by 0.2 ppm in Scenario 4? Or is it necessary to know the maximum and minimum phosphorus concentrations that are likely to occur during a particular rainstorm?
4. What computer models or other predictive tools are available for the parameters that must be evaluated? Do available models predict the variable(s) of interest in a realistic fashion? (i.e., are the algorithms upon which the model is based sufficiently realistic to represent the system, and its response to management intervention, accurately?)
5. What are the input data requirements for each model that could be used?
6. What data have already been collected for the watershed system of interest?
7. What hardware/software does each model require? Is the required system available? Is sufficient memory, hard disk space, etc. available? Is an experienced user available to operate the model or to advise a novice user?
8. Of all the available models for the time period and parameters of interest, which will be easiest to use and/or require input data closest to that which is available and/or be most realistic or accurate in its predictions?

James (1992) has the following suggestions to make regarding model selection:

1. Identify all the hydraulically and hydrologically significant elements and processes in the basin so that the model(s) selected can be shown to include all relevant processes.
2. Develop study objectives that relate objective functions to design options.
3. Identify important performance criteria so that minimum modeling effort can be used to select the best option.
4. Analyze systematic and random error in field data to reduce model effort.
5. Review available models.
6. Assess study resources, including time, money, and staff.
7. Compile and review existing data. Review existing technical expertise. Choose the minimum-cost model to deliver the required accuracy for the desired objective functions.

8. Use a simplified data set to verify each identified process.
9. Produce objective functions for each hydrologic element, including design options. Draw a system schematic.
10. Estimate parameters. Discretize homogeneous subspaces.
11. Conduct a sensitivity analysis: produce a family of plots illustrating model response to changes in input variables. Use extreme high and low values for input variables, and assess impact on predicted values.
12. Calibrate the model using both simplified and full data sets. Choose optimum values for the most sensitive parameters. Simulate the system over the maximum possible number of rainfall events.
13. Validate the model against an independent data set.
14. Interpret the output: discuss the magnitude and direction of probable errors. Plot ranges, showing output and expected confidence bands.
15. Document the results. Identify the model and version used, archive input, output, and source coding.

Figure 7.8 illustrates this process graphically.

There is an old saying in computing: "Garbage in, garbage out." A model is only useful as a planning tool if the user understands how, and how well, it represents the natural world. To gain this understanding, the analyst must go through four steps:

1. *Model verification*—Determining whether the algorithms used by the model are appropriate for the system, or should be replaced with others more suitable.
2. *Model calibration*—Tuning the model so that it simulates a set of observed conditions with a high degree of accuracy.
3. *Model validation*—See how well the tuned model predicts a different data set, without changing any input variables or rate kinetics.
4. *Sensitivity analysis*—To determine which input variables most influence model output and must therefore be chosen with great care.

The first of these steps, verification, is usually straightforward, if somewhat subjective. The rest will be discussed in the following sections.

7.3.4 Model Calibration

Calibration is usually the most challenging step in model testing, particularly with a complex continuous model. Bicknell et al. (1993) define calibration as "a test of a model with known input and output information that is used to adjust or estimate factors for which data are not available." The more complex the model, the more numerous the adjustments that must be made, and the more complex the input data set. Both input data and variable parameters (such as

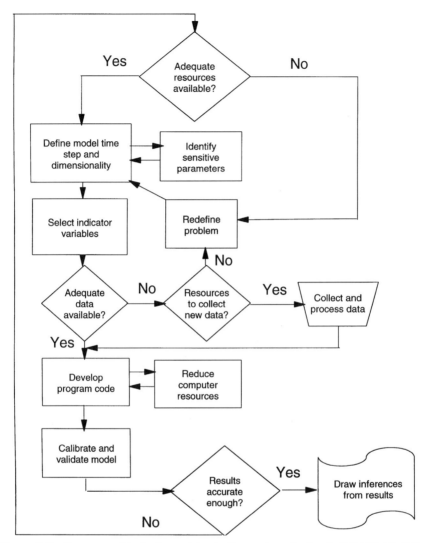

Figure 7.8 *Sequence of tasks in model selection and application (adapted from James 1992).*

diffusion or removal rates) can be in error. The model may therefore contain dozens or even hundreds of error sources that singly or in combination can make the model system fail to replicate the real system. Figure 7.9 illustrates some of these sources of error.

The challenge in calibration is to determine the probable sources of error and to eliminate them systematically, leaving (it is to be hoped) a model with only very minor deviation from observed processes. This task requires a combination of system understanding, common sense, and random luck.

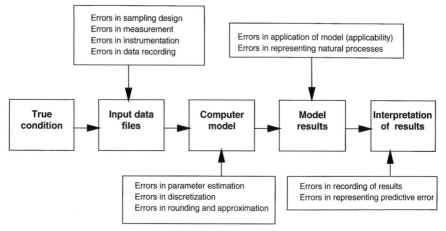

Figure 7.9   *Sources of error in model building and application.*

Models may be calibrated manually by changing one parameter at a time and trying to understand how it affects the model overall or by using an automated (computer-based) system. Automatic calibration is much faster than manual calibration. (Al-Abed (1997) estimates that manual calibration requires five times more time than computer-assisted calibration.) Most automatic calibration programs operate on the basis of optimizing some objective function using maximum likelihood theory or a least-squares approach for parameter estimation. Although more efficient than manual calibration, automatic calibration may produce parameter values that are "optimal" but that make little sense in terms of the mechanics of the system. By contrast, manual calibration is tedious, but provides good insight into the behavior and credibility of individual parameter values.

A typical calibration sequence works like this:

1. The model code is compiled, and a model run is executed using either a simplified data set or a full data set of observed input conditions.

2. Observed data are compared with (usually plotted against) model predictions. (This step may be done manually or with computer assistance.)

3. The differences between observed and predicted data are evaluated, and a decision is made as to whether the calibration is complete or should continue. (This is seldom a clear-cut decision and may require considerable subjective judgment.)

4. This run-compare-evaluate sequence is repeated until the differences between observed and predicted data are within acceptable ranges.

5. The model is considered complete and ready for validation. No further adjustments are made to model parameters.

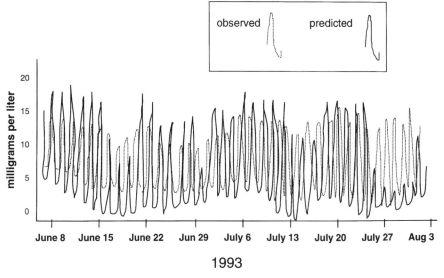

Figure 7.10 Results of a partial calibration of the ECOL model.

Example of a Partial Calibration   Figure 7.10 illustrates the results of a partial calibration of ECOL, a model for the simulation of three species of aquatic plants and algae in temperate river systems. It shows a situation common in manual calibration: certain oddities in the simulated results suggest areas for further parameter adjustment or model revision. In the model illustrated in Figure 7.10, diurnal variations in dissolved oxygen are generally captured by the model for the first six weeks or so of model time. Starting at about the third week in July of the year being simulated, the model fails to reproduce observed oxygen peaks. This gives the modeler several pieces of information. First, the fact that the model is reproducing a diurnal oxygen cycle with an appropriate timing indicates that the basic mechanisms of photosynthesis are adequately modeled. Second, the fact that the general relationship between simulated and observed data changes over a short period of time suggests that some process or source originating at that time has been omitted from the model. Third, the magnitude and timing of the shortfall gives the modeler clues as to what might be missing.

In this case observed oxygen levels continue, on a diurnal cycle, throughout the problem period, so the modeler can conclude that plant growth (and thus photosynthesis) is responsible for the observed peaks. The large difference between observed and simulated oxygen maxima suggests that some major plant population (or increase in the photosynthetic rate of a smaller number of plants) is causing high oxygen maxima to continue throughout July. The model simulates the die-off of one species, the attached alga *Cladophora*, in response to rising temperatures, while two rooted aquatic plant species, *Potamogeton* and *Myriophyllum*, begin to take over habitat formerly occupied by *Cladophora*. In fact, simulated biomass curves clearly show this transition taking place, consistent with field observations. So three possibilities remain:

1. There was more biomass in the stream than was apparent from field observations.
2. The growth kinetics of the rooted aquatic plants are not correctly represented; in other words, the model is not growing the plants fast enough to generate the observed levels of oxygen.
3. There is some other source of photosynthesis in the stream that has been omitted from the model.

Subsequent calibration demonstrated that changes to the growth kinetics of the two rooted aquatic species produced only modest changes in predicted oxygen levels. Field data were reviewed and judged to be accurate and robust. There remained the possibility that some other source of photosynthesis had been overlooked. Consultation with field naturalists revealed that large numbers of insect larvae emerge from aquatic systems in late June in this system. In the natural system, these larvae feed off diatoms and other algae. When they leave the system, this "grazing pressure" is greatly reduced and algal growth can increase rapidly—creating the high dissolved oxygen levels observed in the stream.

Calibration offers the analyst the opportunity to learn how well the model system reproduces the real system, and to explore its strengths and weaknesses. These insights are important in understanding model output, especially during scenario testing (Chapter 11).

7.3.5   Model Validation

A calibrated model has been carefully tuned to a particular set of observed conditions. In Figure 7.10 these were the summer months of 1993 in the lower Grand River, Ontario. The user of the model has, however, no clear idea of how the model would perform if asked to simulate, say, the winter of 1957 in the same system or the summer of 1993 in a different system. Similarly, the user cannot know how accurately the calibrated model would reflect conditions that may occur in the future. (It is perhaps wise to add here that model output *looks* official—it looks like fact—and the incautious user can be tempted to communicate it as fact when it is, in reality, only guesswork. Model predictions should *always* be shown with some estimate of error or uncertainty, to give the user information about how to interpret them. The AEAM process described in Section 3.4.2 has as a cardinal rule that participants must never make a hard copy of the output, so that they can never treat the output as fact.)

The modeler can learn more about how a model would perform under other circumstances by testing it, without further adjustment, on a different data set. This process is called validation. Validation results give the user an understanding of the probable predictive error in the model, and thus its probable reliability in simulating future conditions with and without management intervention. The data set chosen for validation purposes should be similar in most respects to that used for calibration. The most common approach in validating continuous simulation models such as WASP or SWMM is to use data from a different year.

7.3.6   Sensitivity Analysis

The final step in model building is sensitivity analysis. This step allows the analyst to determine which parameters are sensitive, and which relatively insensitive, to changes. Sensitivity in itself can provide valuable insight into the model's adequacy in representing the natural system. For instance, the ECOL model whose results were shown in Figure 7.10 was very sensitive to changes in species-specific temperature optima (the temperature at which growth is best). The modeler must judge whether this is a reasonable finding (which, in this case, it was) or not. If the response of the model to a parameter seems odd or unrealistic, recalibration may be required.

Sensitivity can be judged qualitatively ("very sensitive," "fairly sensitive," "insensitive," etc.) or as a percentage change in output, given a percentage change in input. Sometimes these results are plotted in graphical form, as illustrated in Figure 7.11.

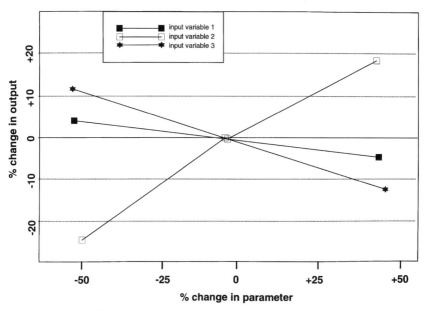

Figure 7.11   *Example plot of sensitivity analysis.*

In sensitivity analysis plots like Figure 7.11, parameters that have relatively little influence on model output (parameters for which the model could be said to be insensitive) should plot as more or less horizontal lines: very large changes in the parameter create very small changes in the model objective function (for instance, predicted mean annual flow). Parameters that have a major influence on model output (those to which the model is sensitive), by contrast, will appear as steeply sloping lines (large changes in model output as a result of small to large changes in input values).

Sensitivity analysis is a useful adjunct in understanding model results and in anticipating predictive error in scenario testing. It can also point to areas where the model is deficient mechanistically or where lumped parameters should be separated into individual values.

7.3.7 Scenario Testing

Once an appropriate model has been selected, calibrated, and validated, it is available for evaluating alternative management approaches in the basin. Scenario testing is usually the foundation for plan development and resource expenditure. Generally speaking, individual management options are tested first, and their performance evaluated against the decision criteria. Those that do not perform at an adequate level are rejected. When all options have been tested in this way, individual options are combined into management strategies, which may be "thematic" in nature. That is, one strategy may, for example, encompass urban management options, while another groups agricultural management options. Mixed strategies may also be developed. Typically, the analyst will evaluate high-cost, high-benefit strategies, mid-cost, "typical" management strategies, and least-cost, lower-benefit strategies, so that decision makers have a range of costs and performance to evaluate in arriving at a best management approach.

The topic of scenario testing and selection of a preferred management strategy is discussed in greater detail in Chapter 11.

REFERENCES

Al-Abed, Nassim. 1997. Modelling Water Quality and Quantity on a Watershed Scale Using Automatic Calibration and GIS. Ph.D. thesis, University of Guelph.

Bicknell, B. R., J. C. Imhoff, J. L. Kittle, Jr., A. S. Donigian, Jr., and R. C. Johanson. 1993. *Hydrological Simulation Program—Fortran: User's manual for Release 10.* Environmental Research Laboratory, Report #EPA/600/R-93/174. Athens, Ga.: US EPA.

Bishop, Bruce. 1970. *Public Participation in Planning: A Multi-Media Course.* IWR Report 70-7. Fort Belvoir, Va.: U.S. Army Engineers Institute for Water Resources.

Donigian, A. S., and H. H. Davis. 1978. *User's Manual for Agricultural Runoff Management (ARM) Model.* Athens, Ga.: US EPA.

Grayson, R. B., J. M. Dooland, and T. Blake. 1994. Application of AEAM (Adaptive Environmental Assessment and Management) to water quality in the Latrobe River catachment. *J. Envir. Management* 41:245–258

James, William. 1992. *Workbook for Stormwater Management Modeling.* Guelph, Ont.: CHI.

Lopez-Ivich, Karina. 1996. Feasibility of modeling phosphorus dynamics in stormwater wetland. Master's thesis, University of Guelph.

Marsalek, Jiri. 1976. *Instrumentation for Field Studies of Urban Runoff.* COA Report No. 42. Toronto: Ontario Ministry of the Environment.

Metcalf and Eddy Inc. 1991. *Wastewater Engineering: Treatment, Disposal, Reuse.* 3d ed. New York: McGraw-Hill.

Novotny, V., and H. Olem. 1994. *Water Quality: Prevention, Identification, and Management of Diffuse Pollution.* New York: Van Nostrand Reinhold.

O'Connor, D. J., and W. E. Dobbins. 1958. Mechanisms of reaeration in natural streams. *Transactions of the American Society of Civil Engineers* 123:641–666.

Ontario Ministry of the Environment. 1987. *Technical Guidelines for Preparing a Pollution Control Plan.* Report of the Urban Drainage Policy Implementation Committee, Technical Sub-committee No. 2. Toronto: Ontario Ministry of the Environment.

Ontario Ministry of the Environment and Energy (MOEE). 1992. *Status Report: The Metal Mining Sector Effluent Monitoring Data for the Period February 1, 1990 to January 31, 1991.* Toronto: MISA Program, MOEE.

Ostler, Neal K., and Patrick K. Holley, eds. 1997. *Sampling and Analysis.* Vol. 4, Environmental Technology Series. Upper Saddle River, N.J.: Prentice-Hall.

Pitt, Robert. 1998. Unique features of the source loading and management model (SLAMM). In *Advances in Modeling the Management of Stormwater Impacts*, Vol. 6. Edited by William James. Guelph, Ontario: Computational Hydraulics Inc.

Pitt, R., and J. Voorhees. 1995. *Source Loading and Management Model (SLAMM).* National Conference on Urban Runoff Management: Enhancing Urban Watershed Management at the Local, County, and State Levels. U.S. Environmental Protection Agency Report No. EPA/625/R-95/003. Cincinnati, Ohio: U.S. Environmental Protection Agency, Center for Environmental Research Information.

Willson, Keith. 1981. *Grand River Simulation Model.* Grand River Basin Water Management Study Technical Report Series, Report #30. Cambridge, Ontario: Grand River Conservation Authority.

8

Costing and Financing

Cost is a central consideration in the development of most watershed management strategies. Water managers are usually operating under some sort of fiscal constraints, so there is only a certain amount of money to be spent on watershed projects. Difficult issues also arise in the costing of individual measures, in the quantification of benefits, and in the selection of discounting rates and analytical methods for economic considerations. Sometimes it is important to include uncertainty—probability—in cost analyses, adding another complication to the study.

This chapter reviews fundamental principles of cost estimation, comparison of alternatives on the basis of costs, application of uncertainty in economic analyses, and similar issues relevant to watershed planning and management.

8.1 SCOPE AND MEASURES

8.1.1 Concepts of Cost

What should be included in an economic analysis of watershed projects? Obvious components are capital costs—the labor and materials costs required to build or implement water management measures—and operating and maintenance costs for those structures or measures. "Costs" might also be taken to include items that are sometimes called "disbenefits," or impacts—for instance, impacts on the environment. These costs are much more difficult to quantify than the costs of lumber and concrete, because they often include intangible impacts such as reduced beauty—impacts that are hard to express in dollar values. Section 7.5 discusses some approaches for quantifying intangibles.

Several types of cost may be of interest:

Fixed Costs—costs that are not affected by changes in production or activity, such as taxes, insurance, salaries for management and administrative staff, license fees, and interest on borrowed capital.

Variable Costs—costs that vary with the quantity of output or other measures of activity, such as the costs of labor and material to produce a product or service. Variable costs often constitute the primary difference between two alternatives, because their fixed costs will be similar.

Incremental Costs—the additional costs or revenues that will result from increasing the output of a system by one or more units. Incremental costs can be estimated only if the analyst can identify a "normal operating range" beyond which incremental costs might be expected to apply. An example of incremental costs may be those associated with hiring summer students to conduct a special one-time-only field investigation; such labor would not be considered part of routine operations.

Costs can also be categorized by their frequency of recurrence. This consideration is important in developing cash flows for long-term projects, where money will flow into, and perhaps from, projects over many years. These categories include:

Recurring Costs—costs that are repetitive, occurring in a foreseeable cycle. Recurring costs include fixed costs such as rent and insurance.

Nonrecurring Costs—costs that occur only once or infrequently, at intervals that are difficult to predict over the long term. Nonrecurring costs include those associated with the purchase of major equipment or real estate to develop a new facility (rather than to operate an existing facility).

Costs may also be categorized as direct or indirect. This distinction becomes important in calculating the true "cost" of an alternative, including not only the obvious, direct costs, but also less obvious costs that are nevertheless attributable to the project. Direct and indirect costs may be defined as follows:

Direct Costs—costs that can be measured and that are obviously attributable to a particular product or work activity (e.g., lumber required to build a piece of furniture and the labor required to build that item).

Indirect Costs—costs that are shared and thus difficult to attribute—for instance, the cost of tools that are shared among several workers, general supplies such as lubricants, equipment maintenance costs, heating and lighting of a shared facility, and so on. Indirect costs are usually allocated as a proportion of direct labor hours, direct labor dollars, or direct material dollars, but other methods of allocation may be used. Sometimes indirect costs are referred to as "overhead" costs.

A variety of other cost concepts are relevant to watershed planning and management. Some of these relate to accounting issues, such as keeping track of the "value" of an asset as it depreciates over time. Among these are:

Cash Cost—a cash cost involves a cash payment. Cash costs are those that are used in calculations of future expenditures. Usually, only cash costs are used in making economic decisions about water management. A cash cost can be differentiated from the following types of costs.

Book Cost—the estimated cost of an asset, including any depreciation.

Sunk Cost—a cost that is behind you, that no longer figures in current planning or analysis. Examples of sunk costs are the costs of defunct structures or machinery. Sunk cost can also be thought of as the difference between the book cost (estimated value) of an asset and the price that a buyer will pay for it.

Opportunity Cost—the cost of losing an opportunity to use the money elsewhere. An example of an opportunity cost can be found in the development of a landfill site: the opportunity costs are those of foregoing the chance to develop the land for residential or industrial purposes (e.g., from the municipality's perspective, taxes that would have accrued to the municipality, etc.).

The concept of "life cycle costs" refers to the fact that costs are incurred over the entire life of a project, from need definition through preliminary design, detailed design, construction, operation, production of goods and services, transportation and delivery, operation of facilities, retirement of structures, and postretirement activities such as long-term monitoring. The true cost of a structure therefore legitimately includes not only the costs of building and maintaining it, but also the costs of designing it, operating all aspects, and retiring it from service. The greatest potential for cost errors lies in the early stages of the life cycle, for instance, at the conceptual design phase. Undetected errors at this stage, carried through the life of the facility, can have enormous cost implications. By comparison, errors in costing the retirement phase will likely have a much smaller impact on total project cost. The clear lesson here is to pay particular attention to cost analysis for the early stages in a project's life cycle, and to have cost estimates for these (and other) stages carefully reviewed by objective analysts to assure their completeness and accuracy.

A third concept of interest is the break-even point. Although more usual in manufacturing contexts, the break-even point also has application in watershed management. It is a simple and familiar concept: that the revenues from a project should equal or exceed its costs, or the project is not viable:

$$\text{Profit or loss} = \text{total revenue} - \text{total costs}$$

This can be thought of in terms of demand for some product or service—for instance, water supply. At a given demand level, revenues will be sufficient to recoup costs. At higher demand levels, the operator of the utility will realize a profit. At lower demand, costs will exceed revenues and a net loss will result. If the unit costs and revenues for the product (in this case, water) can be determined, it is a simple task to solve the equation for the break-even demand point.

8.1.2   The Time Value of Money

At the heart of economic analysis is the notion that the value of money changes with time. Two factors operate on the value of money over time: interest, paid by a bank or other custodian of an investment, and inflation, the phenomenon by which prices for goods and services are not constant over time but fluctuate in response to a variety of social and economic conditions.

Interest has been part of human society for more than 4,000 years and is recorded in documents of ancient societies such as Babylon. In a sense, interest is a fee paid by the borrower or user of money (or some other commodity) for the privilege of that use. Legitimate interest rates in history have ranged between a few percent and 25%; higher interest rates were considered socially unacceptable "usury." The Bible (Exodus 22:25) prohibits usury. Over the past two decades, interest rates in Canada and the United States have ranged as high as 18% and as low as 2%, depending on the time and the type of investment. Often, higher interest rates are paid for riskier investments.

Simple interest is interest paid as a percentage of the original loan. For example, a loan of $1,000 at 10% interest over five years would require the following interest payments:

Year 1   10% of $1,000 = $100
Year 2   10% of $1,000 = $100
Year 3   10% of $1,000 = $100
Year 4   10% of $1,000 = $100
Year 5   10% of $1,000 = $100

Total interest paid would therefore be $500. Simple interest is almost never used in modern business or government.

Compound interest is the more commonly used form. In compound interest, the "fee" is paid on the original loan amount (the principal) plus any accumulated interest charges. Using the preceding example, a schedule of compound interest payments would be:

Year 1   10% of $1,000 = $100
(At the end of Year 1, principal plus interest = $1,000 + $100 = $1,100)
Year 2   10% of $1,100 = $110

(At the end of Year 2, principal plus interest = $1,100 + $110 = $1,210)

Year 3   10% of $1,210 = $121

(At the end of Year 3, principal plus interest = $1,210 + $121 = $1,331)

Year 4   10% of $1,331 = $133.10

(At the end of Year 4, principal plus interest = $1,331 + $133.10 = $1,464.10)

Year 5   10% of $1,464.10 = $146.41

(At the end of Year 5, principal plus interest = $1,464.10 + $146.41 = $1610.51)

So in the case of compound interest, the total interest charges paid on the loan would be $610.51, as compared with $500 using simple interest. This concept can be expressed by an exponential relationship, as follows:

$$F = P(1 + i)^n \tag{8.1}$$

where: F = Future value, the future amount payable on a loaned amount, P
         P = Present value, the principal of the loan, the amount borrowed (or loaned)
         i = the interest rate expressed in percentage
         n = number of compounding intervals (if compounded once a year, this would be number of years)

This simple relationship is the foundation of all time-value-of-money calculations. It can be used to calculate the amount payable on a loan after n years at $i\%$ interest, or the amount earned by an investment over n years at 1% interest (in other words, from the perspective of the borrower or the lender).

Inverting the relationship yields the formula to obtain the present value of a sum, given its future value:

$$P = F\left[\frac{1}{(1 + i)^n}\right] \tag{8.2}$$

This is a helpful formula if you know the amount that will be needed at some point in the future and wish to find out how much you must invest today to ensure that the necessary funds are in hand at the time they are needed.

Other useful formulas can be derived from these basic equations. The four most commonly described relate present and future values to annual payments. Their application is obvious, because most financing requires periodic payments into or out of some fund, not a simple lump sum payment. These four equations are:

$$F = A \left[\frac{(1 + i)^n - 1}{i} \right] \tag{8.3}$$

where: A = the individual payment amount.

This formula allows the user to find the future value of a series of uniform payments (for instance, monthly payments into a bank account: how much money will have accumulated after five years?).

The inverse is also possible to calculate (that is, the individual payment amount required to accrue a total of F dollars at the end of the period of investment). The amount inside the brackets is sometimes termed the "sinking fund" factor:

$$A = F \left[\frac{i}{(1 + i)^n - 1} \right] \tag{8.4}$$

Finally, we can calculate the present value (the value in today's dollars) of a series of uniform payments. This calculation may be of interest if the investor wishes to withdraw a certain amount in each of several years and wants to know how much money must be invested now to support those withdrawals:

$$P = A \left[\frac{(1 + i)^n - 1}{i(1 + i)^n} \right] \tag{8.5}$$

Again, the inverse may be calculated to determine the value of a series of uniform payments required to repay a loan of present value P:

$$A = P \left[\frac{i(1 + i)^n}{(1 + i)^n - 1} \right] \tag{8.6}$$

These concepts of present value, future value, and payment amount (sometimes termed annual value, where there is a single annual payment or revenue to be considered) are ubiquitous throughout economic analysis. Indeed, the fundamental task of economic comparison of alternatives is to reduce all options to a common basis of comparison, which can be present value, future value, or payment amount (annual value).

Goodman (1984) presents an economic analysis of two engineering options for an electric power project. The two options under consideration are a 1,500,000-kilowatt coal-fired thermal power plant and a 400,000-kilowatt hydroelectric power project. The scenario presented in this example is typical of (although probably simpler than) most public sector projects. One option, the hydropower project, will not require fuel or other similar variable costs. By contrast, although the thermal plant will incur fuel costs, it is larger and will thus satisfy a larger demand. Either project may be financed by private investors or from government sources; private investment costs are higher because higher interest rates must be paid in the private sector. There are other differences between the two projects as well. The hydropower project is estimated to have a 50-year life span, while the thermal plant is expected to last only 30 years. For planning purposes, it is assumed that both plants will operate at an "average plant factor" (ratio of average output to rated capacity) of 55%. This assumption may prove to be false over the long term—for instance, if operation of the hydropower plant can be made more efficient, or if the efficiency of the thermal plant declines as it ages and equipment deteriorates.

Table 8.1 summarizes the economic aspects of the decision to be made in this example, assuming that funding comes from private sources. Table 8.2 provides the same analysis, assuming government sponsorship of the project.

Which of these options is "best"? The answer lies in:

- Choosing a *common time period* (decision horizon) over which to compare the two options
- Choosing an *appropriate discount rate* (applicable interest rate, or blended rate reflecting both interest and inflation over the planning period)
- Choosing a *common basis of comparison*, for example, total cost expressed in present-day dollars, or annual cost over the planning horizon; various measures are possible

The following paragraphs describe the analyst's approach to each of these steps for the example under discussion.

Choosing a Planning Horizon The choice of a time period over which to evaluate economic aspects may be determined by land use planning needs (e.g., regional planning forecasts extend 50 years into the future), by the expected useful life of a major structure (such as a dam), or by other social, economic, or political considerations. There is seldom a single correct approach to choosing a planning horizon, although planning in the public sector seldom extends over less than 5 or more than 100 years. Typical water resources planning horizons fall in the 20- to 50-year range, because this is the horizon over which population growth forecasts and basin hydrology can be predicted with some degree of accuracy. Forecasts over very short time periods are rare, because major public planning exercises usually involve the expenditure of large sums of money and public agencies must consider financing logistics, environmental impacts, and benefits over a much longer term. The analyst is well advised to seek the opinion of the community and interested agencies in the choice of a planning period and to ensure that all interests and concerns will be captured by the time period chosen. (For example, cumulative environmental impacts such

Table 8.1 Annual Cost of Electric Power Under Two Engineering Options, Assuming Private Sponsorship (After Goodman 1984)

Project Cost or Revenue	Percent	Dollars per Net Kilowatt of Production (assumes 55% average annual plant plant factor)
400,000 kW Hydroelectric Plant		
I. Plant investment (capital costs)—annual equivalent		$1,000.00
II. Annual fixed costs:		
A. Fixed charges		
Cost of money	10.50	
Depreciation (10.5% 50-year sinking fund)	0.07	
Insurance	0.10	
Taxes (federal, state, local)	5.00	
TOTAL FIXED CHARGES	15.67	156.70
B. Fixed operating costs		
Operation and maintenance @ 5 mills/kWh		2.50
Administration and general @ 35% of		
operation and maintenance costs		0.88
TOTAL FIXED OPERATING COSTS		3.38
TOTAL ANNUAL FIXED COSTS (IIA + IIB)		160.08
III. Annual production—variable costs:		
Energy costs (fuel)		0.00
Operation and maintenance		0.00
Energy costs—total variable operating costs		0.00
1,500,000 kW Coal-Fired Thermal Plant		
I. Plant investment (capital costs)—annual equivalent		$700.00
II. Annual fixed costs:		
A. Fixed charges		
Cost of money	10.50	
Depreciation (10.5% 30-year sinking fund)	0.55	
Insurance	0.25	
Taxes (federal, state, local)	5.00	
TOTAL FIXED CHARGES	16.30	114.10
B. Annual carrying costs of fuel inventory		1.50
C. Fixed operating costs		
Operation and maintenance @ 1.83 mills/kWh		8.82
Administration and general @ 35% of		
operation and maintenance costs		2.95
TOTAL FIXED OPERATING COSTS		11.77
TOTAL ANNUAL FIXED COSTS (IIA + IIB + IIC)		127.37
III. Annual production—variable costs:		
Energy costs (fuel) @ 9500 Btu/kWh @		
$1.10/10^6$ Btu × 1000 mills/dollar		10.45 mills/net kWh
Operation and maintenance		1.55 mills/net kWh
Energy costs—total variable operating costs		12.11 mills/net kwh

(*Note:* 1,000 mills = $1.00)

Table 8.2 Annual Cost of Electric Power Under Two Engineering Options Assuming Government Sponsorship (After Goodman 1984)

Project Cost or Revenue	Percent	Dollars per net kilowatt of production (assumes 55% average annual plant factor)
400,000 kW Hydroelectric Plant		
I. Plant investment (capital costs)—annual equivalent		$1,000.00
II. Annual fixed costs:		
A. Fixed charges		
Cost of money	7.50	
Depreciation (7.5% 50-year sinking fund)	0.25	
Insurance	0.08	
Taxes (federal, state, local)	—	
TOTAL FIXED CHARGES	7.33	73.30
B. Fixed operating costs		
Operation and maintenance @ 5 mills/kWh		2.50
Administration and general @ 35% of		
operation and maintenance costs		0.88
TOTAL FIXED OPERATING COSTS		3.38
TOTAL ANNUAL FIXED COSTS (IIA + IIB)		76.68
III. Annual production—variable costs:		
Energy costs—total variable operating costs		0.00
1,500,000 kW Coal-Fired Thermal Plant		
I. Plant investment (capital costs)—annual equivalent		$700.00
II. Annual fixed costs:		
A. Fixed charges		
Cost of money	7.00	
Depreciation (7% 30-year sinking fund)	1.06	
Insurance	0.20	
Taxes (federal, state, local)	—	
TOTAL FIXED CHARGES	8.26	57.82
B. Annual carrying costs of fuel inventory		1.00
C. Fixed operating costs		
Operation and maintenance @ 1.83 mills/kWh		8.82
Administration and general @ 35% of		
operation and maintenance costs		2.95
TOTAL FIXED OPERATING COSTS		11.77
TOTAL ANNUAL FIXED COSTS (IIA + IIB + IIC)		70.59
III. Annual production—variable costs:		
Energy costs (fuel) @ 9500 Btu/kWh @ $1.10/$10^6$ Btu × 1000 mills/dollar		10.45 mills/net kWh
Operation and maintenance		1.66 mills/net kWh
Energy costs—total variable operating costs		12.11 mills/net kWh

(*Note:* 1,000 mills = $1.00)

as reservoir siltation would be underestimated if the planning period were only 5 years; 20 or 50 years would be a better choice.)

In the present example, the analyst has elected to evaluate costs on an annual basis, thus eliminating the need to choose a common planning period. This decision is discussed further in a following section, "Choosing a Common Basis of Comparison."

Choosing an Appropriate Discount Rate   As with the choice of a suitable planning period, there is no clear answer to the problem of choosing an appropriate discount rate—the interest rate to be used in Eqs. 8.1 through 8.6. This rate can be a blended rate, reflecting both the positive action of interest accumulation and the negative action of inflation. Typical discount rates for major public projects probably range between 6% and 12%, depending on prevailing economic conditions and the expected source of funding (private sector funding usually entails higher interest rates). The choice of an appropriate discount rate may be the most controversial element in the economic analysis of major public projects. (Private projects, such as those undertaken by major corporations, generally employ discount rates and minimum attractive rates of return set by company policy and endorsed by shareholders.) As in choosing a planning period, the analyst should seek the approval of community and agency representatives before proceeding to adopt a particular discount rate. Sensitivity analysis, in which several analyses are conducted using high and low interest rates, can give insight into the impact of prevailing interest rates on economic viability.

In the present example, the analyst has chosen a blended (interest + inflation) rate of 10.50%, termed the "cost of money," for projects under private sponsorship. For government sponsorship, lower interest rates would apply, but applicable rates may differ depending on the type of project. The analyst has chosen a rate of 7.5% for the hydroelectric project, and 7% for the thermal plant. Appropriate rates for government-sponsored public projects can be determined through consultation with interested agencies.

Choosing a Common Basis of Comparison   The economic performance of an option can be measured in various ways. The most obvious of these is its gross or net cost expressed in present-day dollars; this is termed its "present worth" or "present value," so comparing alternatives on the basis of present value is called comparison using the *present-worth* or *present-value method*.

Similarly, alternatives can be compared on the basis of their future value. This may be a good approach if revenues are expected over the life of the project and the analyst wishes to determine which option has the larger value at the end of the project. This type of comparison is termed the *future-worth* or *future-value method*.

Any series of cash flows can be reduced to a present or future value or converted into a uniform series of annual payments—the series that would yield a value equivalent to that represented by the lump sum. Essentially, this method, termed the *annual-worth* or *annual-value method*, reduces a series of nonuniform expenditures and receipts to an equivalent uniform series of cash flows.

Options can also be compared on the basis of their *internal rate of return*, a measure of their rate of growth over the planning period, or on their *benefit-cost ratio*, a measure that compares the total value of "benefits," including revenues, to total project costs.

Each method of comparison has advantages and disadvantages. The present-worth and future-worth methods, for example, are easily understood and easily calculated. However, they do not provide insight into cash flow exigencies that may arise over the project life, and they may require the analyst to work with a very long planning period to accommodate the various life spans of different project structures. Annual worth is also intuitive and obviates the need for selection of a common planning period (the year is the time unit on which all options will be compared), but gives no sense of the financial impact of the project at the outset, when financing issues may be crucial to success. Internal rate of return is critical in most private sector analyses, where economic performance may make or break a project, but it is relatively difficult to calculate and less easily understood than present or annual worth. Internal rate of return is used less often in public projects, which are usually intended to provide some necessary public service and for which revenues may be a small or nonexistent component of the project cash flows. Benefit-cost analysis is required by some environmental and planning legislation, but it continues to be controversial because of the difficulty of quantifying intangible costs and benefits and the problem of deciding what to include in, and exclude from, the analysis.

In the present example, the analyst decided to examine annual costs, including both the annualized portion of capital (construction) costs and the routine operation and maintenance costs that can be expected each year. This eliminates the need to choose a common planning period, because both projects can be evaluated on an "annualized basis" using the following calculations.

Total project costs include both capital costs and operating costs, and thus are incurred at irregular intervals over the entire life of the project. Each individual expense can be converted to present-day dollars using Eq. 8.2, and these "present values" then summed over the project lifetime to create a total cost expressed in present-day dollars. For example, expected operating expenses in year 12 can be converted to a present value, and expected expenses at close-out or decommissioning can also be converted to a present value. But the two calculations would involve different values for n in Eq. 8.2, to reflect the different times at which they occur. The calculation for year-12 expenses would set $n = 12$, and that for close-out costs would set $n = 30$ or 50 (the expected project life span), depending on which option was under consideration. Revenues can be calculated in the same way. If both revenues and costs are included in the calculation, costs should be shown as negative values and revenues as positive values.

The present values of individual costs (and revenues) are then summed to create a net project cost expressed in terms of present-day dollars. This present value or individual cash flow values can then be converted to an annual equivalent over the life of the facility, using Eq. 8.6.

If the analyst had instead decided to use total project costs as a basis of comparison, a decision would have to be made about an appropriate time period for comparison. Because the life spans of the two projects differ, we cannot simply use either 30 or 50 years. Several other approaches are available. The easiest is probably to choose the lowest common multiple of years, 150 years in this case, and assume that each project would be replaced during that time as the useful life of the structure ended. In this case, the thermal plant would have to be replaced five times (30-year life × 5 replacements = 150 years) and the hydropower project three times (50-year life × 3 replacements = 150 years). The analyst would then examine expected cash flows over the 150-year period, including the capital costs involved in replacing these major structures several times.

The lowest-common-multiple approach is often used when the expected life span of an option is less than the desired planning period. This frequently occurs in major public projects, where government agencies may wish to plan over 50 or 100 years, but project structures such as sewage treatment plants may have expected useful lives of only 20 or 25 years.

The example illustrated in Tables 8.1 and 8.2 shows how apparently disparate management options can be reduced to a common basis of comparison. In summary, Table 8.1 shows that, under private sponsorship, the thermal plant is overall a cheaper option despite its higher fuel and maintenance costs, on a per-kilowatt-hour basis. Under government sponsorship (Table 8.2), the difference between the two options becomes much smaller, so the decision may be driven by factors other than economics (for instance, social impacts or aesthetics).

8.1.3 Equivalency and the Problem of Inconsistent Ranking

In theory, a series of payments and receipts can be expressed as a present worth, a future worth, or an annual worth, with each form of expression equivalent to the others. This notion of equivalence is fundamental to economic analysis. Table 8.3 illustrates this idea for four different repayment schemes.

In each of the schemes illustrated in Table 8.3, $8,000 in principal will be repaid. In each, 10% interest will apply. And in each, the payments will be made over a four-year period. The differences between the total payments arises because of the *timing* of payments and, thus, the interest payable on the outstanding balance. In theory, therefore, the analyst should be able to propose any of these four scenarios as equivalent to the others. The same is true in economic analysis of major public projects: the analyst should be able to represent the economic value of the project in any of several different, equivalent, forms.

In general, this notion will be found to be true. If rate-of-return methods are used, however, alternatives may sometimes appear to rank differently using present-, future-, or annual-worth methods than with the rate-of-return method.

The internal rate of return for an alternative is the percentage growth of that alternative over the period of study; in this sense, it is rather like an interest rate

Table 8.3   Four Repayment Schemes for a Loan of $8,000 at 10% Interest (After De Garmo et al. 1997)

Nature of Scheme	Total Interest Accrued	Total Principal Repaid	Total End-of-Year Payments Made
1. Pay $2,000 (principal + interest due) at end of each year.	$2,000	$8,000	$10,000
2. Pay interest due at the end of each year, and repay all of principal at the end of year 4.	$3,200	$8,000	$11,200
3. Pay in four equal end-of-year installments	$2,096	$8,000	$10,096
4. Pay principal and interest in one payment at the end of year 4.	$3,713	$8,000	$11,713

and is expressed as a percentage of the original investment. To use a simple example, an individual may wish to determine whether it was better to invest $1,000 in a bank deposit at 5% interest or to invest the same amount of money in a hot dog stand: which is the better investment? Analysis of the projected cash flows from the hot dog stand can be used to estimate the "internal rate of return" (IRR) for that investment. If the IRR is higher than 5%, the individual is better to invest in the hot dog stand than to put the money in the bank. If the IRR for the hot dog stand is less than 5%, the money is better in the bank.

Internal rate of return analysis is seldom used in major public projects. But where it is, it can lead to serious errors in ranking alternatives if the correct analytical procedure is not used. This is because IRR is a growth *rate*, not an absolute dollar value. Calculation of IRR assumes that any positive cash flows will be reinvested at the calculated rate of return, not at the prevailing interest rate, as would be the case in present-, future-, or annual-worth methods. Thus, an alternative might be ranked as "best" using the IRR method, but second or third using one of the other methods. This is termed the "inconsistent ranking problem."

To get around this problem of inconsistent ranking, engineering economic texts recommend incremental analysis of alternatives. Simply put, this means examining alternatives one at a time, taking the lowest-cost alternative first and moving progressively toward the highest-cost alternative. This allows the analyst to see the performance of the base (starting) alternative, and then of each incremental cost that would be incurred by moving to a higher-cost option. The recommended steps are as follows:

1. Rank the options in order of increasing first (capital) costs.
2. Take the lowest-cost alternative as the base case, or current "best" alternative. Evaluate the present, future, or annual worth of this option, or its IRR.
3. Determine whether the performance of this base option on the chosen

measure is adequate, or whether the money would be better left in the bank (but note that for major public projects, the "do nothing" option may not be viable, whether or not the economic performance of the base option is adequate; see Section 8.2). The option will be viable economically if its present (or future or annual) worth is positive or if its IRR exceeds some predetermined target, such as the prevailing bank interest rate.

4. Identify the next most expensive option. Calculate the performance of the *incremental* cash flows for this option using the same measure applied in (2) for the base case. That is, subtract the costs of the base case from the costs of the next most expensive option, and conduct the analysis only on these incremental costs (and revenues).

5. Determine whether the expenditure of the additional funds to achieve the second option is warranted. In other words, determine whether the present (or future or annual) worth of the incremental cash flows is positive or the IRR of the incremental cash flows exceeds some predetermined target, such as the prevailing bank interest rate.

6. If the expenditure of the incremental funds is deemed to be viable, the second alternative becomes the "current best" alternative. If the expenditure of incremental funds is not deemed to be viable, the original, base option remains the "current best" and the second alternative is discarded.

7. Repeat steps 4 through 6 until all alternatives have been examined.

8. Select as the preferred alternative the last one for which the incremental cash flow analysis was favorable.

This analytical sequence may appear cumbersome, but it provides a useful framework for the systematic comparison of alternatives and guards against the common error of inconsistent ranking. Incremental analysis must, for this reason, be used with rate-of-return methods; it is, however, optional with equivalent-worth methods.

8.2   COSTING MAJOR PUBLIC WORKS

8.2.1   Special Needs in the Economic Analysis of Major Public Works

As mentioned in Section 8.1, the economic analysis of major public works differs from that of private projects in several major aspects:

1. The capital investment required for major public projects is often very large, as compared with private projects.

2. The ratio of fixed to variable costs is often very high for major public projects.

3. Public utilities are required to provide service within established rate

schedules.

4. Major public works are subject to close public scrutiny and must be highly accountable to the public and to regulatory agencies. They must therefore keep abreast of, and incorporate, any technology that would reduce costs.

5. Major public projects generally aim for full-cost recovery (which can include setting aside funds for maintenance and replacement) but do not generally seek excess revenues (profit).

6. Public utilities usually have stable demand, and therefore income, as compared with private industries, which may experience fluctuating price, demand, or both.

7. Major public works are usually financed with 50% to 60% or more debt (borrowed) capital. A proportion this high would be unusual in the private sector.

8. Major public assets can involve longer write-off periods than privately owned structures or equipment.

The combination of long planning period, high public accountability, and focus on cost recovery but not profit means that economic analyses of major public projects often must include a number of elements that would not be required in a similar analysis of a private project. Among these elements are the following:

1. There is emphasis on the interests of the customer, as compared with the interests of the utility.

2. There is usually a need to "do something," so the "do nothing" alternative is generally not viable and may even be omitted from the analysis.

3. Administrative and supervision costs may be excluded from economic assessments if these costs will be incurred in any case.

4. The costs of money, depreciation, income taxes, and property taxes are usually expressed as a percentage of total capital invested.

Goodman (1984), De Garmo et al. (1997), and other engineering economic texts contain more detailed discussion of the basic principles of economic analysis and the application of those principles to major public projects.

8.2.2 Utility Rate Setting

Economic analysis of major public projects can include the need to estimate revenues, for instance, from public utilities. This is by no means a simple task, even for the utility involved. Rate setting is the subject of many major works. The following discussion is intended only as an overview of the problem.

The main challenge in rate setting is to include all the fixed, variable, direct, and indirect costs that are legitimately attributable to the provision of the service being sold. So, for example, development of a drinking water supply entails the costs of drilling and maintenance of wells, or systems to extract water from surface water sources; treatment and disinfection of the water; labor and materials; construction and maintenance of buildings in which to house workers, equipment, and supplies; construction and maintenance of a water distribution system; and similar costs. All of these costs must be factored into the price charged the customer, or the utility will end up subsidizing costs to some degree. It is not at all uncommon for prevailing water rates to fall short of the actual costs of service. The MISA Advisory Committee (1991) reported that, on average, Canadian water prices were about 65% of the real cost of providing water supply and sewerage. The difference was made up by government grants and subsidies that are invisible to the consumer but are often buried in the residential tax bill. The MISA Advisory Committee endorsed the notion of "full-cost" pricing that would eliminate grants and subsidies and make the full cost of service transparent to the customer. While this would undoubtedly increase water costs, the consumer could also be expected to realize a tax saving.

A typical rate-setting exercise uses the following calculation (annualized values implied) (De Garmo et al. 1997):

Rate base = total capital costs of the plants in service
  − accumulated depreciation + materials and supplies
  + fossil fuel inventory + working capital allowance
  − deferred income taxes − deferred investment tax credit
  + construction work in progress

Current thinking suggests that rates should also include provision for sinking funds to cover future maintenance and replacement costs. In Ontario, which is typical of the systems the MISA Advisory Committee was describing, no such provision had been made over many years, with the result that replacement costs for water supply and sewer systems are now approaching $50 billion. It may be far better to build these costs into rates when they are established.

Table 8.4 summarizes the components of current water and sewer charges in the city of Guelph, Ontario. Guelph meters residential water use.

Annual water and sewer charges in Guelph average about $200 a year per household. This is less than half the equivalent charges in the city of Kitchener, located less than 25 kilometers away. Kitchener currently faces severe water supply shortages and has undertaken an aggressive demand-management program to encourage water conservation. Among these demand-management measures is higher pricing for water and sewer services.

Table 8.4 Explanation of Current Residential Water and Sewer Charges, City of Guelph, Ontario

Item	Explanation
Actual water consumption	Actual water use @ present charge (1997) of $0.2416/m^3.
Meter charge	Monthly charge based on the size of customer's water meter. Helps defray the cost of operating residential meters and water treatment/distribution system.
Sewer surcharge	An additional charge that provides funds for sewage treatment plant operation and a portion of capital upgrades. Calculated as a percentage of total water charges (= water use + meter charge).
Water environmental surcharge	An additional charge that provides funds for the reconstruction of Guelph's primary water supply pipeline and associated environmental protection works. Assessed at 125% of the monthly meter charge.
Sprinkler charge	Customers having separate water services used for fire protection purposes are assessed an additional charge for those services.

8.3 BENEFIT-COST ANALYSIS

8.3.1 Overview

Benefit-cost analysis is a relatively simple method of evaluating the economic performance of a project. It was popularized by the U.S. Flood Control Act of 1936 and has since been widely used (and misused) in the assessment of large-scale public and private projects. It has particular value in the public sector, however, where profits are seldom sought, but where maximum benefits are desired for each dollar spent.

The calculation of benefit-cost ratios is a relatively simple matter. More controversial is the problem of what to include in, and exclude from, the analysis. This problem is dealt with later in this section.

The conventional benefit-cost ratio is calculated as the ratio of equivalent worth of benefits to equivalent worth of costs. Any of the three equivalent worth measures (present worth, future worth, or annual worth) may be used, but present worth and annual worth are by far the most common. The basic formula for conventional benefit-cost ratio is:

$$\frac{B}{C} = \frac{\text{benefits}}{\text{costs}} \tag{8.7}$$

For example, if present worth is the desired measure, Eq. 8.7 would look like:

$$\frac{B}{C} = \frac{PW_{\text{benefits}}}{PW_{\text{costs}}} \tag{8.8}$$

As discussed in Section 8.1, the term "costs" can include both capital (i.e., construction, investment) costs and operating and maintenance costs; thus:

$$\frac{B}{C} = \frac{PW_{\text{benefits}}}{PW_{CC} + PW_{OM}} \tag{8.9}$$

where: CC = capital costs

OM = operation and maintenance costs

According to the principle of equivalence, Eq. 8.9 can equally well be expressed in terms of annual worth:

$$\frac{B}{C} = \frac{AW_{\text{benefits}}}{AW_{CC} + AW_{OM}} \tag{8.10}$$

An alternative formula, termed the "modified benefit-cost ratio," is sometimes used and should generally yield the same results as the conventional formula. For a present-worth analysis, this formula can be written as:

$$\frac{B}{C} = \frac{PW_{\text{benefits}} - PW_{OM}}{PW_{CC}} \tag{8.11}$$

In this case the denominator of the equation contains only the present worth of project capital costs. The numerator contains the present worth of benefits, less the costs of operation and maintenance.

The most usual, but not the only, approach in benefit-cost analysis is to maximize the B/C ratio. Depending on the type of project and the goals and objectives of decision makers, it may sometimes be important to maximize or minimize components of the benefit-cost analysis. Some common targets are to:

1. *Maximize total benefits*—choose the alternative that has the highest present worth of total benefits, for instance, to emphasize the positive impact a project would have on a community.
2. *Maximize net benefits*—choose the alternative that has the highest net benefits (net benefits = total benefits – total costs), to draw attention to the net (incremental) benefit to the community.
3. *Minimize total costs*—choose the lowest-cost alternative that still demonstrates a B/C ratio > 1.0, perhaps because total available funds are limited.
4. *Minimize capital (investment) costs*—choose the alternative that has the lowest initial costs, but which still demonstrates a B/C ratio > 1.0, perhaps because a fixed sum is available for project construction.
5. *Minimize operating and maintenance costs*—choose the alternative that has the lowest operation and maintenance costs, but which still demon-

Table 8.5 Decision Making Under Different Benefit-Cost Analysis Scenarios (Values in Millions of Dollars)

	Alternative 1	Alternative 2	Alternative 3	Alternative 4
Capital Investment	50	65	42	42
Operating and Maintenance Costs	10	12	16	11
Total Costs	60	77	58	53
Total Benefits	72	81	67	59
Net Benefits	12	3	11	6
Benefit-Cost Ratio	1.20	1.05	1.16	1.11

strates a B/C ratio > 1.0, perhaps to minimize the annual financial load of a project.

6. *Maximize rate of return*—for some projects like major irrigation projects, which have significant expected revenues, cost recovery or even cost surpluses may be possible. In these cases, decision makers may opt for the alternative that shows the highest rate of return.

Table 8.5 illustrates how the outcome of a decision-making process might differ depending on the goal of the process.

Table 8.5 illustrates the following differences under different decision criteria:

1. *To maximize benefit-cost ratio*—choose Alternative 1.
2. *To maximize total benefits*—choose Alternative 2.
3. *To maximize net benefits*—choose Alternative 1.
4. *To minimize total costs*—choose Alternative 4.
5. *To minimize capital (investment) costs*—choose Alternative 3 or 4 (choice would likely depend on other factors).
6. *To minimize operating and maintenance costs*—choose Alternative 1.

As with other types of decision criteria, the analyst must make explicit the goals of the decision-making process and the targets that are to be met for a management alternative to be considered "acceptable." Table 8.5 reveals that any one of the four alternatives could be considered "acceptable" or "unacceptable," depending on the criterion used to judge performance.

An Example of Benefit-Cost Analysis The following example of a benefit-cost analysis is adapted from DeGarmo et al. (1997). It employs a conventional B/C ratio method for the analysis of a flood-control and power project on the White River in Missouri and Arkansas. The two management options under consideration were to:

Table 8.6 Annual Flood-Related Losses on Three Stretches of the White River

Item	Annual Value of Loss	Annual Loss per Acre of Improved Land in Floodplain	Annual Loss per Acre for Total Area in Floodplain
Crops	$1,951,714	$6.04	$1.55
Other farm losses	215,561	0.67	0.17
Railroads and highways	119,800	0.37	0.09
Levees	87,234	0.27	0.07
Other losses	168,326	0.52	0.13
Total losses	$2,542,635	$7.87	$2.01

1. Build a reservoir only.
2. Build a reservoir and associated channel improvements for flood protection.

Table 8.6 presents an overview of pre-project flood-related losses in the White River basin. Table 8.7 presents an analysis of costs and benefits for the reservoir-only option and the reservoir-plus-channel option.

This example illustrates that both options are viable (B/C ratio > 1.00) but the multipurpose project has significantly higher benefits, and thus a significantly higher B/C ratio, than the single-purpose dam. Significant benefits accrue in the Mississippi River basin, as compared with the target White River basin; flood losses (which would be reduced by either management option) are not shown in Table 8.7. Finally, the "benefits" incorporated in this analysis are only the obvious (and measurable) flood loss reductions (prevented flood losses) and the value of the power from the power generation facility. They do not include a wide range of other considerations such as increased earning power among the people employed in construction, spin-off benefits in industries affected by that buying (e.g., house construction, automobile manufacturing, etc.). This problem of what to include in, and exclude from, a benefit-cost analysis is discussed in Section 8.3.2.

8.3.2 The Problem of Scoping a Benefit-Cost Analysis

In theory, analysis of benefit-cost ratios is straightforward: a project is considered economically viable if benefits outweigh costs, or B/C ≥ 1.0. In practice, however, this analysis is often more complicated by several factors. As mentioned earlier, there is the problem of deciding what to include as "benefits" and as "costs." Benefits in particular can be problematic, because of the difficulty in attaching dollar values to intangible qualities such as aesthetics or quietude. This problem is addressed further in Section 8.5.

The problem of scoping is also a difficult one, even for the most experienced analyst. It is not uncommon for an agency to request a second benefit-cost anal-

Table 8.7 Estimated Costs, Annual Charges, and Annual Benefits for Proposed Table Rock Reservoir and Bull Shoals Channel Improvements (After DeGarmo et al. 1997)

Item	Table Rock Reservoir	Dam Plus Bull Shoals Channel Improvement
Capital costs, including clearing, excavation, associated structures, etc.	$20,447,000	$25,240,000
Powerhouse and equipment	6,700,000	6,650,000
Power transmission	3,400,000	4,387,000
Land	1,200,000	1,470,000
Highway relocations	2,700,000	140,000
Cemetery relocations	40,000	18,000
Damage to villages	6,000	94,500
Damage to misc. structures	7,000	500
Total construction costs	**$34,500,000**	**$38,000,000**
Federal contribution to construction costs	$34,500,000	$38,000,000
Interest during construction	1,811,300	1,995,000
Total federal contribution	36,311,300	29,995,000
Present value of federal properties	1,200	300
Total federal investment	**$36,312,500**	**$39,995,300**
Annual charges: interest, amortization, maintenance, operation	1,642,200	1,815,100
Annual Benefits		
Prevented direct flood losses in White River basin: present conditions	60,100	266,900
Prevented direct flood losses in White River basin: future conditions	19,000	84,200
Prevented indirect flood losses in White River basin	19,800	87,800
Property value enhancements in White River basin	7,700	34,000
Annual flood benefits in Mississippi River basin	220,000	980,000
Total annual flood benefits	**$326,600**	**$1,452,900**
Power value	1,415,600	1,493,400
Total annual benefits	**$1,742,200**	**$2,856,300**
Conventional B/C ratio	**1.06**	**1.57**

ysis if the first seems inadequately scoped. For example, a conservative benefit-cost analysis of an irrigation scheme might include only measurable benefits, such as increased crop revenues, and only capital and construction costs. An alternative analysis might comprise a range of benefits, including quantification of improved "quality of life" for the people who no longer have to haul water to the field, perhaps improved project aesthetics, and secondary or indirect benefits such as the sale of electricity for pumping. The alternative analysis might also include quantification of a range of environmental "costs" or benefits, such as altered water quality or flow regimes. The scoping of benefit-cost analysis is not a simple problem and should be discussed with the community and interested regulatory agencies. In one recent analysis of a major public-sector project, the first, more conservative, benefit-cost analysis showed that only $800,000 worth of benefits would accrue from more than $3 million in expenditures: clearly, the project was not viable. When this analysis was repeated by a different firm, using a wider net of secondary and indirect benefits, the benefit-cost ratio was found to be 3 : 1, and the project clearly justifiable.

The U.S. Water Resources Council (1983) provides a series of guidelines for the evaluation of the benefits of major public water projects. The types of benefits it considers, and proposed measures to evaluate these benefits, are summarized in Table 8.8.

8.3.3 Classification of Benefits and Costs

Other complications in benefit-cost analysis include the problem of whether to count certain types of benefit explicitly as benefits or as decreased costs. The salvage value of equipment is one such problem. It may be treated as a positive benefit, in the numerator of the equation, or as a reduced cost, in the denominator of the equation (normally, the latter is the case). De Garmo et al. (1997) have pointed out,

> An arbitrary decision as to the classification of a benefit or a cost has no effect on the acceptability of a project ... regardless of the classification of a cash flow item as an additional benefit or a reduced cost, the present worth of the project will be the same and the acceptability of the project based on the B/C ratio will be unaffected.

8.3.4 Impact of Discount Rate on Decision Outcome

As discussed in this and earlier chapters, the choice of a discount rate can be a very controversial issue in water management planning. The reasons for this relate to the timing of costs and benefits over the course of the project and can be easily illustrated by an example.

Assume that a major public project will extend over five years. Capital costs are, of course, highest at project initiation, but benefits (for example, flood control benefits, as shown in Table 8.7) will accrue over the life of the project. Operation and maintenance costs will also occur over the project life. The following simple example (developed by Reid Kreutzwiser, Department of Geography, University of Guelph) illustrates the effect of discount rate on the present

Table 8.8 Evaluation of Benefits for Major Public Water Projects

Activity	Type of Benefit	Measures
Municipal and industrial water supply	Clean, reliable water supply for residential and industrial customers	Direct measures: Demand curves for each category of water user (show true willingness to pay); historical data may be inadequate.
Agricultural water and drainage projects	Irrigation, drainage, flood protection, erosion control, sediment control	Direct measures: Cost of damage to or impaired productivity in crops, pasture, and range due to inundation, drought, sedimentation and erosion; net income per crop or per unit area.
Urban flood damage reduction	Flood protection, extended geographic area of operation	Direct measures: Cost of physical damages or losses, income losses, emergency response costs, increased costs of police, fire, military, or medical service.
Hydroelectric power generation	Generating capacity, avoidance of fossil fuel usage, environmental and possibly safety benefits	Indirect measure (alternative cost approach) is usual method of estimation, using equivalent thermal plant as alternative for costing (see Section 8.5).
Navigation	Reduced costs of transportation	Reduction in cost of using waterway, or difference between using this waterway and an alternative route; reduction in delivered price of products transported on the waterway.
Recreation	Swimming, boating, water skiing, fishing, water-enhanced recreation including camping, hiking, picnicking, hunting, bird-watching, photography, aesthetic enjoyment, sightseeing	Entry and use fees actually paid plus any unpaid (surplus) value to consumers. Indirect measures include contingency valuation, unit-day value, and travel-cost method; see Section 8.5.

worth of costs and benefits, and thus on the benefit-cost ratio. Values are in thousands of dollars.

	Year 0	Year 1	Year 2	Year 3	Year 4	Year 5	Total Without Discounting
Benefits	$0	$35	$50	$100	$100	$100	$385
Costs	$185	$50	$10	$10	$10	$10	$275

The following table shows the impact of discount rate on the benefit-cost ratio and thus on the "viability" of the project:

Discount Rate	Total Discounted Benefits	Total Discounted Costs	Benefit-Cost Ratio	Project Viable?
0%	385.0	275.0	1.40	Yes
2%	359.0	271.2	1.32	Yes
7%	300.6	262.5	1.14	Yes
12%	249.3	254.8	0.97	No

To obtain a discounted value for costs or benefits, the value for any given year is taken (e.g., costs of $10,000 in Year 2) and converted from that "future" value to a present worth using Eq. 8.2. Discounted benefits are then totaled, as are discounted costs, and the ratio of the two calculated.

The reason that the benefit-cost ratio (and thus the project viability) changes depending on discount rate is that the benefits and costs occur at different times throughout the project. Most of the costs occur at the beginning of the project, so the total discounted value of costs is not much different from the total of undiscounted costs ($254,800 discounted at 12% vs. $275,000 undiscounted). By contrast, the benefits accrue only after the project has been in place for some time—and a dollar five years in the future is worth much less than a dollar today. The difference between total discounted benefits ($249,300 at 12%) and undiscounted benefits ($385,000) is therefore much larger. As the discount rate increases, the difference between the two also increases.

This discussion serves to illustrate that, while useful in many applications, and indeed required by law in some, benefit-cost analysis is far from being an objective analytical tool. It requires careful application by the water resources analyst—and cautious interpretation by the user—to be a reasonable representation of project pros and cons.

8.4 ALLOCATION OF COSTS AMONG MULTIPURPOSE PROJECTS

The flood-control/power generation project described in Table 8.7 illustrated that the multipurpose project offered more benefits than either a dam alone or (not shown in the table) channel improvements alone. Table 8.7 shows the total costs and benefits accruing from the dam-plus-channel improvement project, but does not attempt to allocate costs and benefits to the two component projects. The most common method for making such allocations is called the "separable cost-remaining benefits" (SCRB) method.

The SCRB method allocates costs on the basis of "purposes"—that is, intended impacts of the project (e.g., irrigation, power generation, flood control, etc.). The separable costs of a purpose are those that are clearly linked to that purpose, and then can be calculated as the difference between the total costs of a multipurpose project and the costs of a project in which the purpose is excluded. In the example shown in Table 8.6, the separable costs for the channel improvement project are the costs of the dual-purpose project minus the costs of the dam. The purpose is also assigned a portion of the total joint costs of the multipurpose project; this allocation is made on the basis of remaining benefits (described in a subsequent paragraph). The total joint costs are the difference between the total costs of the multipurpose project and the sum of all the separable costs for all purposes.

The separable and joint costs allocated to any purpose cannot exceed the "adjusted benefits" or "justifiable costs" of that purpose, which are either the benefits attributable to the purpose by the most favorable method available or the benefits evaluated as the costs of the most economic alternative (usually a single-purpose project), whichever is least.

Finally, the "remaining benefits" for each purpose are the adjusted benefits (justifiable costs) minus the separable costs. The remaining benefits are used to allocate a portion of the joint costs to each purpose, in proportion to the amount of each purpose's remaining benefits.

Methods of cost allocation are not perfect and, like benefit-cost analysis, can be misapplied. Loughlin (1977) and the U.S. Water Resource Council's *Principles and Guidelines* (1983) provide an overview of cost-allocation methods, including illustrative examples, and explain some of the shortcomings of the SCRB and other methods.

8.5   QUANTIFYING INTANGIBLES

Implicit in the notion of quantifying benefits is the problem of how to attach dollar values to intangible considerations such as quality of life, aesthetics, silence, and so on. While it can be argued that no such quantification is legitimate (who can put a price on peace of mind, for example?), the fact is that such analyses are often part of water management planning and are routinely incorporated into benefit-cost analyses. An understanding of how they are developed is helpful in identifying weaknesses in the analysis and in developing new estimates.

Generally speaking, quantification of intangibles aims to determine the so-called consumer surplus—that is, the value to the consumer over the price that would normally be paid. There are several methods of estimating the consumer surplus. All are fraught with error, but each can be useful if carefully applied. Using a combination of methods may help to offset the weaknesses of an individual method.

8.5.1 Contingency Valuation: Consumer Surveys of Willingness-to-Pay

As discussed in Chapter 4, consumer surveys are useful tools in determining public opinion. Survey forms can include mailed, telephone, or in-person questionnaires, but because of the need for one-to-one interaction they tend to be costly to prepare and distribute. There is usually a need to analyze large numbers of questionnaires to obtain representative results; this also adds to the cost.

A well-designed survey can produce valuable insights into willingness-to-pay, for example. Willingness-to-pay is a measure of how highly a resource or asset is valued in the community. A survey may, for instance, ask individuals how much they would be willing to pay to retain an existing resource or to obtain it if it is currently absent. Survey design is a fine art, however, and should be undertaken by an expert. Improperly worded questions may result in answers that are difficult to interpret or that do not address the issue of interest. In terms of willingness-to-pay, for example, there is a risk that the person being interviewed will interpret the questions as an attempt to determine the amount that should be *charged* for access to a resource or asset. An example can be found in quantifying the value of bird-watching opportunities. A survey of public willingness-to-pay, if not correctly worded or conducted, could lead the interviewee to believe that in the future an admission fee will be charged for access to bird habitat. Properly conducted surveys of willingness-to-pay have been used in decision making relating to increased water treatment levels in major municipal centers, because they give an idea of how much extra the public would be willing to pay (for instance, in property taxes) to obtain a higher quality of drinking water.

Contingency valuation techniques can also include iterative bidding in a personal-interview context. The respondent is asked to respond "yes" or "no" to values suggested by the interviewer to obtain a specified increment in the desired resource. The highest price the respondent is willing to pay is the respondent's bid for the incremental improvement in the commodity.

8.5.2 The Travel-Cost Method

Some situations can benefit from application of the travel-cost method, which, simply put, evaluates how much an individual on average or at maximum will spend to visit a particular resource or asset, above the actual costs of gaining admission. A good example here can be found in tourism at national parks. The travel-cost method may identify the person who has traveled farthest to reach the park—for instance, a visitor from Europe—and evaluate the travel costs (airfare, car rental, gasoline, etc.) and time costs (value of the individual's time—for instance, on an hourly or daily basis) necessary for that person to reach the park. It may be found that the travel and time costs incurred by this traveler approach $2,000, while admission to the park is only $10. The consumer surplus would then be evaluated at $1,990. Tobias and Mendelsohn (1991) used the travel-cost method to assign a value to ecotourism in Costa Rica. They used multiple regression techniques to express visitation rate as a function of distance traveled, population density, and illiteracy rate, plus some random error. Their results suggest that travel by Costa Ricans to ecological reserves in that country has a consumer surplus value of about $116,200 US/year, or about $35 per visit, based on a typical travel cost of $0.15 per kilometer. These values translate to a present worth for the ecotourism industry of between $2.4 and $2.9 million US for domestic visitors only. Adding international visitors brings the total consumer surplus value to over $10 million US, or about $1,250 per hectare of ecological preserve. This is roughly 10 times the price that is currently paid for the acquisition of new land for ecological preserves in Costa Rica.

8.5.3 Hedonic Price Approach

The hedonic price approach can be used in cases where it can be proven that people have historically paid more to have a certain attribute or access available to them. The method requires a comparison of historical real estate prices of matched properties with and without the attribute. For example, the consumer surplus associated with ravine aesthetics may be estimated by comparing the recent sale prices of homes that backed on a ravine with those of similar homes that did not back on a ravine. The difference between the sale prices can be taken as the consumer surplus value of that resource.

8.5.4 Alternative Cost Approach

Where it is possible to identify another way to supply the resource or asset to be quantified, the cost of that alternative can be used to estimate the value of the consumer surplus. For instance, if existing fish habitat is to be destroyed by some management action that will result in excessive siltation, new habitat could be created elsewhere to replace that which is destroyed. The cost of developing the new habitat, including construction, vegetation, and maintenance, could be used to estimate the value of the habitat being destroyed. Alternatively, the value of a clean drinking water supply could be estimated as the cost of creating and maintaining an alternative water supply—for instance, through the use of trucked or bottled water.

8.5.5 Qualitative Techniques

A range of qualitative techniques can be used in estimating the value of intangibles. Typically, these techniques involve ranking or scoring, but not direct measurement. In this regard, consider the example of visits to a national park. Visitors may be asked to rate the reasons for their visits on a scale from 1 to 5. Categories evaluated may include considerations such as solitude, exercise opportunity, time with family, nature enjoyment, and so on. Taken over a large number of visitors, responses to a survey like this can reveal that most people visit the park to enjoy nature, indicating that park management may wish to extend visitor services in this area. If, by contrast, most people want "time with family," group camping or campfire activities may be promoted. While not—strictly speaking—quantification of intangibles, these methods are nevertheless a helpful adjunct to more quantitative methods and provide useful insight into public opinions and values.

8.6 INCORPORATING RISK AND UNCERTAINTY IN ECONOMIC ANALYSES

It is frequently the case that some components of the analysis will be subject to chance events, particularly when planning encompasses future conditions that cannot be known with certainty. The further into the future an event will happen, the less precise our estimates about it will be. These estimates may include considerations such as prevailing interest rates, utility rates, capital investment requirements, market share, inflation rates, and taxation rates. All of these factors will certainly vary with time, and all can in turn affect estimates of cash flow. Some are correlated with others; some are autocorrelated (that is, values at a given point in time are correlated with values that occurred some time earlier).

Two situations may occur. In the first, the possible outcomes are known, and their respective probabilities are also known. This condition is called decision making under risk. In the second situation, possible outcomes may be known but their probabilities of occurrence are not known. The analyst must therefore decide whether or not to incorporate uncertainty into the decision making (decision making under uncertainty) or whether to assume that some condition will occur with 100% probability.

Attitudes to risk and uncertainty vary with the population and with the sponsor agency. Individuals, for instance, may willingly accept the risk of heart and lung disease by smoking cigarettes, but adamantly oppose exposure to much lower health risks from nuclear power generation. Formal risk analysis, which incorporates estimates of probability with estimates of impact, are sometimes used in the evaluation of potential human health impacts caused by chemical pollutants, of high-impact low-frequency disasters such as earthquakes or nuclear accident, or to develop priorities for environmental remediation.

Various methods are employed to deal with decision making under risk and under uncertainty. These methods are reviewed in the following sections.

8.6.1 Decision Making Under Risk

Decision making under risk tends to be easier and more robust than decision making under uncertainty. The basic approach to incorporating risk is called "risk aggregation." It can be illustrated with a simple example.

Assume that a project will result in a series of expenditures and revenues

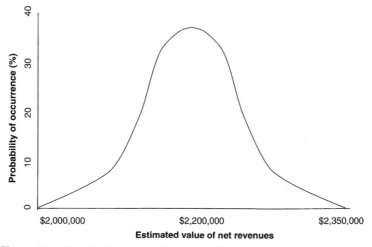

Figure 8.1 *Hypothetical probability distribution based on analyst's intuition.*

over time, but that the precise value of those cash flows cannot be known with certainty. The analyst probably has some insight into the relative likelihood of various levels of cash flow, however, and can translate that intuitive understanding into a rough probability distribution. Figure 8.1 illustrates this process.

Figure 8.1 simply records the analyst's "gut feel" for the project over time. For instance, the analyst may feel that a very pessimistic prediction might be $2 million in net revenues, while a very optimistic prediction would be $2,350,000. The most likely outcome is probably somewhere around $2,200,000. (Although this analysis may seem arbitrary, it should be borne in mind that an analyst with considerable experience in the watershed under study, and with good knowledge of water resources management techniques, is usually able to guess these outcomes with fair to good accuracy.)

We can then assign probabilities to these three conditions—for instance, 2% each to the low and high conditions and 35% to the "most likely" case. We can add two or three more high and low guesses to round out the range of probabilities.

This rough probability distribution can then be converted into a cumulative probability distribution (Figure 8.2), which in turn allows the analyst to interpolate probabilities for intermediate values of net revenue. From an intuitive feeling for the project, the analyst has created an approximate probability distribution that can be used in subsequent risk analyses.

Table 8.9 illustrates the values that can be estimated from this kind of analysis.

Recall that each of the entries in Table 8.9 is simply a guess by an experienced analyst. These values can, however, be used to estimate an overall present worth for the cash flow series, based on the premise that the expected present worth of a series of cash flows is equal to the sum of the present worths of the

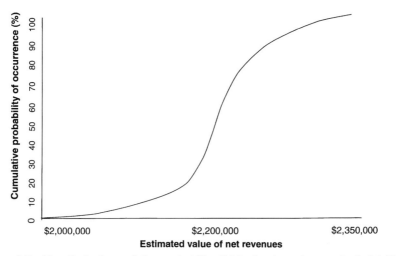

Figure 8.2 *Hypothetical cumulative probability distribution based on analyst's intuition.*

expected values of the individual cash flows. Recalling Eq. 8.2,

$$P = F \left[\frac{1}{(1 + i)^n} \right]$$

we can convert anticipated cash flows over time to a present worth. Similarly,

$$E(P) = \Sigma E(F) \left[\frac{1}{(1 + i)^n} \right] \tag{8.12}$$

where E is the expected value of present worth (P) or future worth (F).

Table 8.9 Estimated Cash Flows from a Hypothetical Water Management Project

Year	Source	Expected (Most Likely) Value	Possible Range of Outcomes	Standard Deviation
0	Capital costs	−$11 M	−$9.5 M– −$12 M	$500 K
1	Revenues minus expenses	$2.2 M	$2.05 M–$2.35 M	$50 K
2	Revenues minus expenses	$2.2 M	$1.9 M–$2.5 M	$100 K
3	Revenues minus expenses	$2.2 M	$1.9 M–$2.5 M	$100 K
4	Revenues minus expenses	$2.0 M	$1.7 M–$2.3 M	$100 K
5	Revenues minus expenses	$1.0 M	$700 K–$1.3 M	$100 K
5	Salvage value	$6.0 M	$4.8 M–$7.2 M	$400 K

(*Note:* M = million, K = thousand)

Table 8.10 Example Results of Present-Worth Calculations Using Estimates of Cash-Flow Values

Project	Expected Present Worth	Expected Variance
Alternative A	$20 M	$335,000
Alternative B	$38 M	$12,600,000

If two projects are being compared, the analyst can estimate pessimistic, optimistic, and expected (most likely) cash flow values for each. The outcome is often as shown in Table 8.10.

Given these results, the analyst can make an informed decision as to whether to accept the project with the higher potential worth (but much higher risk) or the one with lower worth but more certainty.

Analysis of this type makes uncertainty explicit in the analysis and thus allows the analyst and other participants in the planning exercise to make maximum use of available information. Uncertainty in several variables can be considered simultaneously, rather than independently, creating an analysis that is more realistic. Although it is computationally convenient to develop a solution under assumed certainty (e.g., interest rates will be 12%) rather than the more realistic risk (e.g., interest rates will lie between 8% and 14%, with the following probabilities for each interest rate ...), in real life probabilities are seldom known. An understanding of the probable "spread" of results can be just as important, or more important, than estimating the "expected" value. This point is made very clearly in the simple example in Table 8.10. Possibly most important, an explicit analysis of risk forces the analyst to confront unknowns one at a time, and to separate what is known from what is unknown or uncertain.

Several different methods can be used to assess problems under risk. These are briefly discussed in the following paragraphs.

Simulation (Monte Carlo) Approach The Monte Carlo simulation technique is often used in a complex situation with many variables. It can also be helpful in validating the results of analytical solutions, although it can introduce a random element that may not be present in those solutions.

The general approach is to develop a range of estimates for revenues and expenses and attach probabilities to each estimate. Random numbers are then assigned to the estimates, with the number of random values proportional to the probability assigned to the estimate. A computer program is then written that draws at random from the pool, assigning values to the calculation based on the random numbers drawn. Many calculations are performed in this way, usually hundreds or even thousands, and the analyst develops a probability distribution for present worth using the results. Table 8.11 illustrates how a simple Monte Carlo simulation might be set up.

Many decision techniques employ analytical solutions based on the following elements:

Table 8.11 Example Assignment of Random Numbers in a Monte Carlo Simulation

Expected Value of Present Worth (E)	Probability of (E)	Range of Random Numbers Assigned
$2 million	0.25	00–24
$3 million	0.50	25–74
$4 million	0.25	75–99

"States"—Designated as S_k States are the possible future conditions that may occur, numbered from 1 to k. Examples of states are different levels of drinking water demand (e.g., S_1 = 250 L/person/day; S_s = 300 L/person/day; S_3 = 350 L/person/day, etc.) or number of bulldozers required for a particular construction job (e.g., S_1 = 3 bulldozers; S_s = 4 bulldozers; S_3 = 5 bulldozers, etc.). States are components of an analysis over which the analyst has no control, and which are subject to chance occurrence.

Feasible Alternatives—Designated as A_j Feasible alternatives are the various management decisions one could make (e.g., supply enough drinking water for 300 L/person/day; buy four bulldozers, etc.). Each is the factual situation that would exist under a given management decision.

Outcomes—Designated by θ_{jk} Outcomes are the combinations of "state" and "alternative." For example, suppose that random chance has led to a situation where 250 L/person/day is required for drinking water supply, but the management decision that was made resulted in a supply sufficient for 350 L/person/day. Or suppose it turns out that you needed five bulldozers ("state"), but in fact you bought only three (the management alternative selected).

Value of Outcomes—Designated by θ_{jk} The value of each outcome is the measure of its cost or benefit. For example, if you decided to supply enough drinking water for 350 L/person/day but needed only 250 L/person/day, you have overdesigned your utility and paid more than you needed to; this outcome can be assigned a value, which may be either a net cost or a reduced revenue over what would have occurred had you designed the utility to exactly the right capacity. Similarly, if you bought three bulldozers but needed five, you would have to lease the additional bulldozers at extra cost. This outcome can also be assigned a value, which again would represent a net cost or reduced revenue over what would have occurred if you had bought exactly the right number of bulldozers.

Probability of Each State—Designated by p_k The various states, which are subject to chance occurrence, can be assigned probabilities based on the analyst's understanding of the problem and other sources (for streamflow, for example, long-term hydrologic records are a good source of information about the probability of different flow levels).

Table 8.12   Hypothetical Decision Model Under Risk

		P_1	P_2	P_3	P_4	$\cdots$	P_k
	S_k	S_1	S_2	S_3	S_4	$\cdots$	S_k
A_j							
A_1		$V(\theta_{11})$	$V(\theta_{12})$	$V(\theta_{13})$	$V(\theta_{14})$	$\cdots$	$V(\theta_{11})$
A_2		$V(\theta_{21})$	$V(\theta_{22})$	$V(\theta_{23})$	$V(\theta_{24})$	$\cdots$	$V(\theta_{11})$
A_3		$V(\theta_{31})$	$V(\theta_{32})$	$V(\theta_{33})$	$V(\theta_{34})$	$\cdots$	$V(\theta_{11})$
$\cdots$						$\cdots$	$\cdots$
A_j		$V(\theta_{j1})$	$V(\theta_{j2})$	$V(\theta_{j3})$	$V(\theta_{j4})$	$\cdots$	$V(\theta_{jk})$

We can set up a hypothetical decision problem in matrix form, as shown in Table 8.12.

Table 8.13 gives an example of the types of cost estimates that may be developed in a decision under risk.

We will examine this simple example in illustrating several different analytical methods for decision making under risk.

Assumed Certainty   The analyst can simplify the analysis (while still having gained something from the explicit statement of probabilities) by assuming that one state is 100% certain and the others have a probability of zero. In this case, the analysis becomes the same as if only a single state were considered. This approach may be used in cases where corporate policy dictates a certain outcome (for example, corporate policy may be to use a discount rate of 12.5%, regardless of actual prevailing interest rates). With this approach, Table 8.13 is altered, as shown in Table 8.14.

Expectation-Variance Principle   If the analyst wishes to retain the separate states and their respective probabilities in the analysis, a weighted sum of outcome values can be calculated, as follows:

$$E(A_j) = \Sigma V(\theta_{jk})p_k \tag{8.13}$$

Table 8.13   Hypothetical Costs (Values of Outcomes) Under Risk

Probability (p) of Interest Rate Occurring		$p = .30$	$p = .50$	$p = .20$
	Actual interest rate			
Management alternatives		8%	12.5%	14%
		Equivalent uniform annual costs (AW)		
Alternative 1		$2 million	$2.2 million	$2.4 million
Alternative 2		$2.2 million	$2.2 million	$2.3 million
Alternative 3		$2.05 million	$2.05 million	$2.05 million

Table 8.14 Hypothetical Costs (Values of Outcomes) Under Assumed Certainty

Probability (p) of interest rate occurring	$p = 0$	$p = 1.00$	$p = 0$
Actual interest rate			
Management alternatives	8%	12.5%	14%
	Equivalent Uniform Annual Costs (AW)		
Alternative 1	$2 million	$2.2 million	$2.4 million
Alternative 2	$2.2 million	$2.2 million	$2.3 million
Alternative 3	$2 million	$2.05 million	$2.05 million

or, effectively:

Probability (p) of interest rate occurring	$p = 1.0$
Actual interest rate	
Management alternatives	12.5%
	Annual Cost
Alternative 1	$2.2 million
Alternative 2	$2.2 million
Alternative 3	$2.05 million

Using this approach and the values in Table 8.12, we obtain the following expected annual worth in millions for Alternative 1:

$$E(PW_1) = (2.0)(0.3) + (2.2)(0.5) + (2.4)(0.2) = 2.18$$

The equivalent value for Alternative 2 is:

$$E(PW_2) = (2.2)(0.3) + (2.2)(0.5) + (2.3)(0.2) = 2.22$$

and for Alternative 3 it is:

$$E(PW_3) = (2.0)(0.3) + (2.05)(0.5) + (2.05)(0.2) = 2.035$$

If the cash flows for these alternatives are positive (that is, revenues), the best choice is Alternative 2, with an expected annual worth of $2.22 million. If the cash flows are negative (expenses, costs), the best choice is Alternative 3, with an expected net cost of $2.035 million. If two alternatives were tied, the analyst might opt for the one with the lowest variance.

Most Probable Future Principle   In a situation where one state has a much higher probability of occurrence than others, that state can be assumed to be certain, thus reducing the case to one of assumed certainty as discussed earlier and illustrated in Table 8.14. This approach should be applied only where probabilities are greatly different, with one probability very high. In a case such as that shown in Table 8.13, this technique would not likely be justified unless other factors, such as corporate policy, prompted an assumption of certainty.

Dominance   Where one alternative is always preferred regardless of the state that prevails, that alternative can be said to dominate the analysis. In Table 8.13, if the goal of the decision were to minimize costs (minimize the value of the outcome), Alternative 3 would be dominant because it has the lowest value in every state.

Aspiration-Level Principle   If there is an ideal outcome level (for instance, total costs of $2.1 million or less, or profits greater than $2.25 million), the analyst can evaluate each alternative in terms of its probability of meeting that threshold. This is done by summing the probabilities (not the outcome values) of outcomes that meet the criterion. Using the example in Table 8.13, with a desire to ensure that total costs are less than $2.1 million, we find that:

- Alternative 1 has a probability of 30% that costs will be less than $2.1 million (because only State 1, interest rate = 8%, meets this criterion, and that state has a probability of 0.30),
- Alternative 2 has a probability of 0 of meeting the criterion (all outcome values exceed the threshold), and
- Alternative 3 has a probability of 100% of meeting the criterion (all three states have outcome values ≤$2.1 million).

Under this criterion, Alternative 3 would be the best choice. If the values in Table 8.13 referred to revenues, and we wanted to maximize the likelihood of those revenues exceeding $2.25 million, we would find that:

- Alternatives 1 and 2 each have a probability of 20% of meeting this threshold (State 3 only, with a probability of 20%), and
- Alternative 3 has zero probability of meeting this threshold (all values less than $2.25 million).

So either Alternative 1 or 2 would be a reasonable choice; again, the selection may be made on the basis of lowest variability or some other factor.

8.6.2   Decision Making Under Uncertainty

Decision making under uncertainty is more difficult than decision making under risk, and the results are less satisfactory. Goodman (1984) suggests several steps before attempting to make decisions when several factors are unknown:

- Collecting more detailed data to reduce measurement error and improve understanding of the system
- Using more refined analytical techniques
- Increasing safety margins in design
- Selecting measures with better-known performance characteristics
- Reducing the irreversible or irretrievable commitments of resources
- Performing a sensitivity analysis of the estimated costs and benefits of alternative plans

Nevertheless, situations may arise in which the analyst has no information about the relative likelihood of various possible outcomes, but would like to incorporate in the analysis some consideration of uncertainty. The following methods are useful in decision making under uncertainty.

Equal Probability The simplest approach to detailing with uncertainty is to assume that all possible future states have equal probability. If we applied this notion to the values in Table 8.13, we find that the expected annual worth of Alternative 1 is:

$$E(PW_1) = (2.0)(0.33) + (2.2)(0.33) + (2.4)(0.33) = 2.178$$

The equivalent value for Alternative 2 is:

$$E(PW_2) = (2.2)(0.33) + (2.2)(0.33) + (2.3)(0.33) = 2.211$$

and for Alternative 3 it is:

$$E(PW_3) = (2.0)(0.33) + (2.05)(0.33) + (2.05)(0.33) = 2.013$$

So if we are minimizing costs, the best choice is Alternative 3, with an expected annual cost of $2.013 million. If we are maximizing revenues, the best choice is Alternative 2, with an expected annual worth of $2.211 million. If two alternatives were tied, the analyst might opt for the one with the lowest variance.

Maximin and Minimax Principles If some information about the range of outcomes is available, but the analyst cannot specify probabilities for each, it is possible to "hedge one's bets" by choosing the option that performs best under worst-case conditions, or that which performs worst under best-case conditions. Taking the values in Table 8.13 as the basis, if the goal is to maximize profits, the worst-case condition is probably that with the lowest interest rate, State 1. Examining only the column for State 1, we find that the alternative that performs best under those conditions is Alternative 2, at $2.2 million; the other two alternatives can be expected to generate only $2 million. So the best choice would be Alternative 2. The analyst has maximized the minimum revenues, hence the term *maximin*, as applied to analysis of this type.

If the goal of the project is to minimize costs, the worst case would likely be State 3, with the highest interest rate. Examining only that column, we find that the alternative that performs best (has the lowest costs) is Alternative 3, so that would be the preferred option. The analyst has minimized the maximum cost, hence the term *minimax* is applied to this type of analysis.

Maximin and minimax analyses are pessimistic in that they attempt to find the option that would perform best under the worst conditions. It is also possible, but less common, to examine the most optimistic case, finding the option that earns the most revenues under best-case conditions (maximax) or that which costs least under best-case conditions (minimin).

Hurwicz Principle   While the maximin and maximax methods allow the analyst to examine performance under the most pessimistic or most optimistic conditions, the Hurwicz principle allows consideration of intermediate levels of optimism or pessimism. The method employs an "index of optimism," designated by α, the value of which is chosen by the analyst based on an understanding of the problem and its context.

To apply the Hurwicz principle, the most optimistic performance of an alternative is multiplied by α and the most pessimistic by $(1 - \alpha)$. The two products are added for each alternative, and the largest (for revenues) or smallest (for costs) sum is chosen as the best alternative.

Using the examples in Table 8.13, and assuming that we are reasonably optimistic about the outcome of the project, we could assign an index of optimism of $\alpha = 0.70$. If we are seeking to maximize revenues, then the performance of each alternative can be calculated as:

Expected PW of Alternative 1 = $(0.7)(2.4) + (0.3)(2.0) = 1.68 + 0.6 = 2.28$

Expected PW of Alternative 2 = $(0.7)(2.3) + (0.3)(2.3) = 1.61 + 0.66 = 2.27$

Expected PW of Alternative 3 = $(0.7)(2.05) + (0.3)(2.0) = 1.435 + 0.6 = 2.035$

So the best choice is probably Alternative 1, although it is sufficiently close to Alternative 2 that the option should probably also be examined as viable.

8.6.3   Sequential Decisions and Decision Trees

Concepts of decision making under risk can be applied to situations in which sequential decisions must be made—for example, in a project that will be built in several stages. Decision trees represent this time sequence graphically and allow the analyst to take advantage of information as it becomes available over the course of the project. A common application of decision trees is found in decision making where new and uncertain technology is contemplated.

Assume that the problem under scrutiny is to decide whether to install traditional pump technology or a new type of pump. The relevant data for the two technologies appears in the following table:

Technology	Cost	Probability That Technology Works
Traditional	$15,000	100% ($p = 1.0$)
New	$12,000	65% ($p = 0.65$)

Traditional technology may be completely reliable, but the manufacturers of the new technology warn that it has only 65% reliability. If the new type of pump fails, it will have to be replaced with traditional technology, incurring "penalty" costs of, say, $3,000 resulting from added delays. The money for the new technology will be refunded, but the installer will have to pay the full price of the traditional technology as well as the penalty, for a total of $18,000. Depending on the cost of the new technology, it may nevertheless be worth taking the risk of failure. The situation can be analyzed as follows.

The decision to be made can be represented in a decision tree, as shown in Figure 8.3. There are three "branches" on the right side of the tree. In evaluating this decision, the correct approach is to work from right to left, reducing the branches to a single either-or choice. In this case, we can reduce the "works" and "fails" branches to a single value using the expectation-variance principle (essentially, calculating a weighted sum of the expected costs):

$$V_2 = (\$12,000)(0.65) + ((\$18,000)(0.35) = \$7,800 + \$6,300 = \$14,100$$

We can then compare the cost of Alternative 1, installing traditional technology ($15,000), with the cost of Alternative 2, installing new technology and taking the chance of its failing ($14,100; see previous calculation), and find that trying the new technology is probably justified.

We can extend this analysis to make better use of information in the future by examining the impact of adding a field test of the equipment. This approach is illustrated in Figure 8.4.

Figure 8.4 illustrates that installation of the traditional pump technology without testing remains the most expensive option ($15,000) and can be discarded. This leaves two alternatives to consider: Alternative 2, proceed to install the new technology immediately ($14,100), and Alternative 3, conduct a test and then decide which technology to install. The cost of conducting a preliminary test can be calculated as the weighted sum of the two options, plus the actual cost of the test:

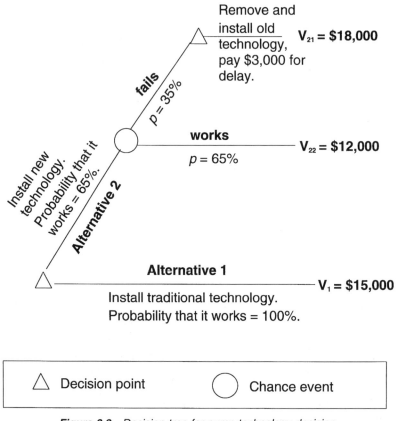

Remove and install old technology, pay $3,000 for delay. $V_{21} = \$18,000$

fails $p = 35\%$

works
$p = 65\%$ $V_{22} = \$12,000$

Install new technology.
Probability that it works = 65%.

Alternative 2

Alternative 1
$V_1 = \$15,000$
Install traditional technology.
Probability that it works = 100%.

△ Decision point	◯ Chance event

Figure 8.3 *Decision tree for pump technology decision.*

$$V_3 = (\$12,000)(0.65) + (\$15,000)(0.35) + T = \$13,050 + T$$

where T is the cost of the test.

Recall that the expected cost of Alternative 2 is $14,100. Figure 8.4 demonstrates that the company can spend up to $1,050 (= $14,100 − $13,050) on the test and Alternative 3 will still be as low in cost as Alternative 2. Therefore, if the testing can be done for $1,050 or less, the company should choose Alternative 3 and conduct a preliminary test before deciding which technology to install.

Decision trees capture a sequence of events, as well as the chance events that can influence outcomes over time, in a way that traditional economic analysis does not. They are particularly useful in evaluating high-impact, low-risk (low frequency) events such as extreme storm conditions, dam failure, and other "disasters". They allow the decision maker to keep track of many possible outcomes and incorporate the possibility of adding new information as time passes.

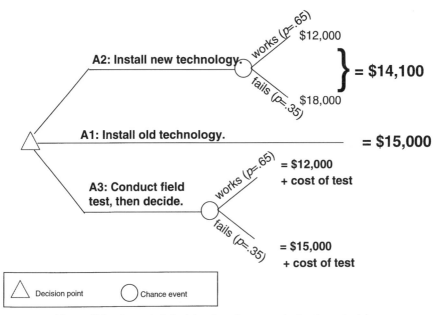

Figure 8.4 *Amended decision tree for pump technology decision.*

8.7 CAPITAL FINANCING (SOURCING)

The decision about where the money will come from is a difficult one, particularly in private-sector projects. In major public projects, financing is usually a combination of grants and loans from government and some independent debt financing.

There are two main sources of capital: debt capital (i.e., borrowed money) and equity capital (i.e., money already available in treasury). As anyone who has ever bought a house or a car knows, there are major differences in these sources.

Financing with debt capital binds the project proponent to paying interest to the lenders of the capital, and to repaying the capital at some fixed point in time. There may be conditions placed on the loan that restrict how or when the money can be used. Under some circumstances, the interest payable on a capital loan is an allowable deduction from income taxes. The most common type of debt financing in major public projects is through the sale of bonds.

Financing with equity incurs no additional interest costs. As a result, there are no "costs" of equity financing that can be used to offset income and reduce income tax payable. Equity capital can include funds raised through the sale of common or preferred stock, from retained earnings, or from depreciation funds. The following discussion is drawn largely from De Garmo et al. (1997).

8.7.1 Bond Capital

Probably the most common source of debt capital for major public works is bond capital. Bonds are also very common in private corporations. They differ from "stock" or "shares" in that a bond represents only indebtedness on the part of the borrower and does not confer any ownership in the company on the lender.

Bonds are long-term loans made by an investor (lender) to the borrower. The bond document specifies the interest rate to be paid, the term of the bond, and other repayment conditions. Bonds are usually issued in denominations of thousands or five thousands. The denomination of a bond is known as its "face value" or "par value," and the interest rate quoted on the bond is the "bond rate." The interest payable in any payment period is simply the face value multiplied by the bond rate for the period of interest. Bonds are said to be "retired" when their term has expired and the face value has been repaid.

In public project financing, bonds are often issued for very long terms, perhaps 10 or 20 years. Over this period, interest rates may fluctuate widely. During the bond period, the bond rate may therefore become more or less attractive relative to prevailing interest rates. Investors may freely buy and sell bonds (subject to bond conditions) during the bond period, but the actual price paid for a bond will rise or fall depending on prevailing economic conditions and the term remaining on the bond.

8.7.2 Equity Capital

Common Stock   Private corporations have the capability of issuing common stock—shares in the company—as a way of raising equity capital. The price of common stock is set, essentially, by what the market will bear. It can be affected by prevailing economic conditions, the history of the company, the nature and potential of the product or service the company offers, and similar factors.

Preferred Stock   Preferred stock can also be issued by a private corporation. Preferred stockholders are guaranteed a dividend before one is paid to common stockholders. If the company is dissolved, obligations to preferred stockholders must be met before those to common stockholders. Preferred stockholders may also have special privileges—for instance, the right to elect senior officers of the corporation—which are not available to common stockholders. The dividends payable to preferred stockholders are usually specified as a fixed percentage of the stock par value. Preferred stock can therefore be a more secure investment than common stock and may be less susceptible to market fluctuations.

Retained Earnings   Profits made by a company may either be paid as dividends to stockholders or kept for reinvestment in the company. If kept, these profits are referred to as "retained earnings" and are available as a source of equity capital for future projects. It is seldom the case that all of a company's profits are kept as retained earnings, simply because stockholders usually demand a share of the profits.

Depreciation Funds   Depreciation funds are those set aside from profits as a source of capital to replace aging equipment. Until such time as they are needed for that purpose, they are available for investment and thus constitute a source of equity capital for the company. Care must be taken to ensure that equipment-replacement demands are protected, however, so that future capacity is not limited by present capital demands.

Leasing   Traditionally, leasing has been thought to offer an income tax advantage over debt financing, because the full amount of lease payments is usually tax deductible, while only the depreciation of purchased assets can be claimed as an income tax deduction. Under recent income tax amendments, however, the advantage of leasing and debt financing has diminished. Some leases incorporate maintenance costs and may therefore offer savings, so that saved funds can be reinvested elsewhere. The greatest advantage of leasing may, however, be that it allows a company to keep abreast of fast-changing technology without tying up capital funds in obsolete equipment.

8.7.3   Allocation of Capital

In a company or government with many projects under way simultaneously, capital must be allocated among projects in a systematic fashion. The following is a typical sequence of capital allocation (De Garmo et al. 1997):

1. Preliminary project planning and screening
2. Determination of the cost of enterprise capital
3. Review of capital expenditure policies and evaluation procedures
4. Preparation of an annual capital expenditure budget
5. Project implementation and review
6. Communication of results

Various techniques are used to decide which projects should, and should not, be allocated capital. Among these are the methods described in Section 8.1 (i.e., present-worth method, future-worth method, annual-worth method, internal rate of return, benefit-cost analysis, etc.). If several projects are under consideration simultaneously, optimization techniques, including linear programming methods, may be helpful in determining the best combination of investments. Linear programming was used, for instance, in a recent study of point and nonpoint source control measures in the Bay of Quinte, Ontario, to determine which combination of agricultural and sewage treatment plant controls would result in the most cost-effective reduction in total phosphorus loads to the Bay of Quinte (D.W. Draper and Associates Ltd. et al. 1997).

REFERENCES

DeGarmo, E. Paul, William G. Sullivan, James A. Bontadelli, and Elin A. Wicks. 1997. *Engineering Economy*. 10th ed. Upper Saddle River, N.J.: Prentice-Hall.

D. W. Draper and Associates Ltd., M. Fortin, Bos Engineering and Environmental Services, and I. W. Heathcote. 1997. *Phosphorus Trading Program Evaluation and Design: Final Report*. Prepared for the Ontario Ministry of Environment and Energy, Environment Canada, the Lower Trent Region Conservation Authority, the Trent-Severn Waterway, and the Bay of Quinte RAP Implementation Advisory Committee, Kingston, Ontario.

Goodman, Alvin S. 1984. *Principles of Water Resources Planning*. Englewood Cliffs, N.J.: Prentice-Hall.

Loughlin, James C. 1977. The efficiency and equity of cost allocation methods for multipurpose water projects. *Water Resources Res.* 13(1).

MISA Advisory Committee. 1991. *Water Conservation in Ontario Municipalities: Implementing the User Pay System to Finance a Cleaner Environment*. Technical Report. Toronto: Ontario Ministry of the Environment.

Tobias, Daniel, and R. Mendelsohn. 1991. Valuing ecotourism in a tropical rain-forest reserve. *Ambio* 20(2): 91–93.

U.S. Water Resources Council. 1983. *Economic and Environmental Principles and Guidelines for Water and Related Land Resources Implementation Studies*. March 10. Washington, D.C.: U.S. Water Resources Council.

9

Legal, Institutional, and Administrative Concerns

Humans are social animals and live together in social groups. In every society, no matter how advanced or how primitive, conflicts arise and must be resolved if the society is to continue to exist peacefully and sustainably. The framework for this resolution usually consists of a system of written or unwritten rules that state the boundaries of acceptable behavior within the society and specify sanctions, the penalties that will be imposed on those who violate the rules that prevail in their society.

These rules are not cast in stone. As internal and external forces shape a society, laws may come to be perceived as unjust or immoral. Gradually, members of the society may begin to disregard a law or to actively protest it. As opposition becomes widespread, initiatives are begun to revise or replace the law. In democratic societies, for instance, the legal system itself often contains provisions to amend or abolish outdated rules. A society's rules—its legal system—therefore reflects a sort of social consensus as to what that society currently judges to be morally correct and just.

Current legal and administrative frameworks in English-speaking nations reflect such a consensus—and an ongoing debate about how much government protection of the environment is "enough." Our laws also reflect the development of a thousand years of rule making in a variety of societies, as well as the social and economic pressures that shaped those rules. The following discussion provides an overview of that development from prehistoric times to the present. The author believes this history to be important in understanding the development of current views on water management, water allocation, and water pollution control. Readers who are interested primarily in current legislative and administrative systems are directed to Section 9.2 and following sections.

9.1   THE EVOLUTION OF MODERN ENVIRONMENTAL PROTECTION LEGISLATION

9.1.1   Behavioral Codes in Primitive Societies

Behavioral codes have probably existed from the earliest beginnings of human society, and they are found everywhere. Typically, these systems are based either on compensation—the precept of "an eye for an eye"—or on punishment.

Under compensation-based systems, only adequate compensation will satisfy the injured party and put an end to a dispute. These systems take many forms, depending on the values of the society that developed them. They are generally established by custom and handed down orally from one generation to the next. Bohannan (1967) offers the following examples of compensation-based behavioral codes, sanctions, and retribution among various peoples of the world. The Yukon Indians of Northern California, for instance, believed that every invasion of privilege or property must be exactly compensated, with compensation taking the form of wealth, including dentalium shells, woodpecker scalps, obsidian blades, and deerskins. In this system, an individual accused of killing a man of social standing would be required to pay 15 strings of dentalium, with perhaps a red obsidian and a woodpecker scalp headband, and would have to hand over a daughter to the family of the slain man. (Killing a common man would carry a penalty of only 10 strings of dentalium.)

In punishment-based systems, sanctions imposed against the offender are generally weighted in proportion to the misdeed. In modern aboriginal societies in Australia, societal codes of behavior are unwritten but appear to be based on the principle of equivalent retaliation and compensation. In general, a man pays less to members of his own clan than he pays to outsiders as compensation for injuries, because as a clan member he is entitled to share in all compensation. Some groups use older members of the tribe as the agents of public opinion in enforcing sanctions against offenders, requiring the parties to meet, and directing the proceedings.

Compensation and punishment rituals may be complex, involving the facing of opposing parties, the singing of songs by the accused and his family, and the exchange of ritualized feints and blows. Bohannan notes that the purpose of these complex rituals seems to be as much to bring an end to the dispute as to bring compensation to the injured party.

These elements of conflict resolution, compensation, and punishment are present in modern, written legal systems as well. Indeed, they are the foundation of "common law," which in turn underlies much of the statutory law—and common practice—in English-speaking countries around the world. Even the use of elders as arbiters in a dispute has its parallels in the Supreme Court system and in modern rabbinical practice.

9.1.2 The Emergence of Written Law Codes

The history of law is discussed in most basic law and administration texts, such as Fitzgerald (1977), Kernaghan and Siegel (1991), Stephenson (1975), and Jennings and Zuber (1991). The following discussion draws from a variety of sources.

Written law codes appeared first in the Middle Eastern kingdom of Babylonia in the twenty-second century B.C. The Hammurabic code, assembled in about 1700 B.C., established a long list of rules governing specific offenses, such as adultery, theft, and faulty workmanship by a house builder. In a sense, these codes were formalized forms of earlier unwritten customs. Their written form, however, made them more straightforward to enforce and less flexible to apply.

By about 500 B.C. written codes had emerged in the civilizations of India and China, and the complex social and religious laws of the Israelites had been assembled into a written code. This so-called Mosaic code reflected the teachings of Moses, the great Israelite leader of the 1200s B.C., and, like the Indian and Chinese codes, stressed moral principles and the obligations of a citizen under the law. The Mosaic code, which includes the Ten Commandments traditionally considered to have been given to Moses by God, was later incorporated into the Hebrew religious writings and the first books of the Christian Bible. Rabbinical interpretation of this code continues to form the basis of rabbinical courts and parallels the English system of common law and the power of the courts to make law.

The first Greek law code was drafted by the Athenian politician Draco in 621 B.C. Draco's code was, again, a list of specific offenses and the (usually harsh) penalties attached to them. The Draconian code was replaced with a less severe code in the 590s B.C. At the same time, the Athenian assembly was made more representative and given increased lawmaking powers. With time, these elected assemblies of Athenian citizens gained power in decision making and are considered by many historians to represent the founding of democratic government.

The Greek system of law was, however, hampered in its development by an overabundance of lay administrators and by large and cumbersome decision-making bodies. Roman law was able to evolve further through the efforts of individuals (jurisconsults) who specialized in interpreting the law. Roman law was founded on a list of customary rules called the Laws of the Twelve Tables, written about 450 B.C. This simple system was frequently amended and enlarged over the next four hundred years, to the point where an average citizen was not capable either of understanding or of applying it. Specialists in law, like the jurisconsults and the rabbinical schools (and like the common-law judges discussed later), used traditions of wisdom, reason, and interpretation of general principles to develop concrete guidance for problem solving.

In the sixth century A.D., the Roman Emperor Justinian introduced his *Corpus Juris Civilis* (body of civil law), a code that covered the whole field of law. The Justinian code persists to this day as the basis of much canon (church) law in the West and, ultimately, as the foundation of the law in most civil-law countries.

9.1.3 The Emergence of Feudalism: Ties to the Land

In the late 400s A.D., the West Roman Empire, which had its capital in Rome, fell to invading Germanic tribes, who brought with them strong clan loyalty and primitive law codes consisting mostly of fines for specific offenses. By the 800s, most of the legal and cultural institutions developed by the Romans had been replaced by a system of allegiance to individual lords rather than to a central government. The legal and economic systems that developed through the Middle Ages developed from this landlord-serf relationship and centered on a system of large "manors" or estates owned by the wealthy and worked by peasants. Under this system, peasants were bound to the land they worked—essentially a part of the property and inseparable from it if the property were to be sold. As the feudal/manorial system grew, cities and towns declined and a network of large rural estates burgeoned. Remnants of this system persist throughout Europe and Great Britain in large ancestral land holdings.

Feudal law remained the fundamental legal system in Western Europe until about 1300. As trade grew and human activities became more complex, however, peasants began to leave their manors to seek work in urban areas and cities once again expanded. With this changing social structure came a revival of an economic system based on payment with money for goods and services and the need for a legal system sufficient (as the feudal system was not) to cope with complex property and commercial disputes. In Europe, scholars at the University of Bologna began to train law students from across Europe in the principles of Justinian's *Corpus Juris Civilis*. Interest in the code soon spread to other European universities and, with time, Roman law began to replace feudal law throughout mainland Europe.

In England the legal system took a different direction. England already had a strong legal tradition based on the rights of judges to make decisions suited to local customs and conditions. By the early 1100s, this system had grown chaotic and there thus began an effort on the part of the monarchy to establish a system of royal courts to apply similar rulings in similar cases. This court system soon established a body of "common law," law that would apply equally anywhere in England. As English common law developed over the years, it established many precedents that limited the powers of government and the monarchy and protected the rights of the people.

9.1.4 The New World and Free Enterprise

Europe in the eighteenth century therefore exhibited three major legal traditions: common law, Roman law, and feudalism. In England, the common-law system was, and continues to be, well established and powerful. Laws need not be written to have force: what is necessary is that a court upholds long-standing traditions of property rights and personal justice. Roman law has formed the basis of most Western European law codes, including those of France and Spain, from the fourteenth century. Even feudalism, a relic of the Middle Ages, persists in some forms. For example, in early nineteenth-century Norway, servants constituted the largest single category of the farm population (Starr and Collier 1989). In 1835 there were 124,600 servants, 103,000 freehold farmers, and 100,000 cotters. The relations between the farmer and the cotter, and the conditions of the servant, were based on contracts. Two decrees, in 1750 and 1752, stipulated that if the cotter himself had cleared the plot allocated to him, he was entitled to a lifelong contract and the right of his widow to remain on the land. The cotter's rights and duties were not specific in these statutes, and the contracts differed greatly between various districts and regions, depending on local custom.

The French and American Revolutions, and the opening of the New World to settlement, proved to be a turning point in the history of property law. In Europe many factors, including feudalism and a scarcity of arable land, had contributed to the evolution of strict controls on land transfer. This framework was ill suited to an environment where rapid development was to be encouraged. Early North American law codes therefore tended to abandon restrictions on land transfers and instead focus on the rights and obligations involved in buying and selling land. Contract law was similarly important in the New World, where free enterprise was highly valued. In the growing free enterprise system, business people wanted to have the freedom to regulate their dealings largely by contract without the interference of government.

9.1.5 Settlement and Law in the New World

Early European settlers in North America faced daunting challenges in clearing forests and crossing natural obstacles such as great rivers and deep ravines. As a result, early development proceeded slowly, hampered by reliance on hand tools and animal-drawn equipment. By about 1800, for instance, Upper Canada's (now Ontario) rural population numbered about 15,000, scattered in tiny settlements through the dense forest and often remote from developing urban centers (Careless, 1984). By contrast, the town of York in that province had fewer than 700 residents.

Governments and private interests saw great potential in the vast natural resources of these new territories and sought to exploit them as quickly as possible. By 1850, the United States government owned 80% of the land in the country, most of it appropriated from Native American tribes. By 1900 more than half of this land had been sold or given away to private interests. Ostler et al. (1996) note that by artificially lowering prices, these land transfers encouraged wastefulness in resource exploitation, the rate of which increased tremendously following the coming of the railway in the middle of the century. From the 1850s rail lines spread rapidly to major centers in the United States and Canada, linking water traffic (including that from the Erie Canal, opened in 1825) with upland routes. For urban centers, the railway brought improved communications and accelerated economic growth. It also opened access to rural areas, allowing the rapid establishment of farm communities with cities as their economic core.

The railway had two important impacts on the rate and pattern of settlement. First, railway access greatly increased the rate at which trees could be brought out of logging areas and transported to urban centers for shipment throughout the United States and elsewhere. It thus vastly accelerated the rate at which watersheds were stripped of trees and erosion could accelerate. And second, the establishment of the railway network in turn encouraged the development of industrial centers like Chicago and Toronto, which were close to rail and water routes, for the efficient distribution of goods and services.

The new railway cities had busy port and manufacturing areas and increasingly dense urban development. Their harbors were filled with all manner of shipping vessels—and the assorted wastes and debris from fast-growing populations. The intense pace of industrial activity took precedence over maintenance and sanitation, and in many cities decaying structures and debris were allowed to accumulate in harbor waters, sometimes obstructing ship passage so that ships had to be towed to their berths.

Municipal water supply and sewage services lagged well behind industrial development. Most homes did not have piped water supplies until the beginning of the twentieth century, but continued to draw water directly from streams or from private wells. As the population increased, so did the number of backyard privies and leaking cesspools, and thus the contamination of wells and water supplies. Untreated sewage flowed from homes to lakes and rivers in a haphazard system of pipes originally designed to drain roadways. Urban growth also meant rapid population growth and an increase in the typhus and cholera that would plague North American cities until the last quarter of the nineteenth century.

There were several direct consequences of this industrial growth and wasteful resource extraction. Ostler et al. (1996) describe the warnings published by contemporary authors like Ralph Waldo Emerson, Horace Greeley, and Henry David Thoreau. The state of urban lakes, rivers, and harbors grew so atrocious that it could no longer be ignored, and citizens became more vocal in questioning the quality of their drinking water. In 1882 , an editorial in the Toronto (Canada) *Globe* called the city's water supply "drinkable sewage."

By the mid-1880s public health officials had become more aware of the causes of infection and better able to combat them. Dedicated public health agencies began to emerge, and with them regular inspections of unsanitary conditions in homes and businesses, particularly leaking privies and cesspools, refuse dumps and unscrupulous food suppliers.

These two concerns, about excessive resource wastage and the protection of public health as a responsibility of government, were driving forces in the development of modern environmental and water management legislation. They also prompted the formation of nongovernment organizations, like the Sierra Club in 1892 and the Audubon Society in 1905, which remain important in watershed management today.

Legislation such as the U.S. River and Harbors Act (later the Refuse Act) of 1899 and the Ontario Public Health Act of 1884 arose from concerns about the maintenance of public water supplies, navigation, and sewage disposal. From these statutes grew much of modern environmental legislation. These statutes were not originally intended as environmental conservation instruments, but were rather concerned with public or private nuisance (a common law concept). Nevertheless, they reflected a growing realization on the part of the public that resources are finite and that the government (and the law) has, after all, a legitimate role in resource conservation and public health.

9.1.6 The Beginnings of the U.S. Legal System

When the American colonists declared their independence from England in 1776, they based their claims partly on ancient Greek and Roman ideas of natural law. These ideas, developed in detail by various French philosophers of the 1700s, promoted the idea that the natural law gives all people equal rights. The U.S. Declaration of Independence echoed this idea in the phrase "... all men are created equal [and] are endowed by their Creator with certain unalienable Rights."

English common law was, however, at the heart of the American claims for independence. Many American leaders of the time were lawyers who had been trained in the common law. Common-law principles (e.g., the rights of the people) also influenced the development of the Declaration of Independence, the U.S. Constitution, and the Bill of Rights.

Although American courts in theory had the same power to make laws that English courts had, a series of U.S. Supreme Court decisions in the early 1800s strengthened this power. The Court's decision in 1803 in the case of *Marbury v. Madison* was especially important. In this decision, the Court declared a federal law unconstitutional for the first time. The principle of judicial review was thus firmly established, enabling U.S. courts to overturn laws they judged unconstitutional.

Despite the fact that English common law had provided the foundation for the Declaration of Independence and the Constitution, parts of the common law, particularly those related to property transfers, were impractical for the new, rapidly expanding nation of the United States. Land was scarce in England, and so common law narrowly restricted the transfer of land from one owner to another. But much of the land in the United States was unsettled, and the nation was constantly expanding its frontiers. To ensure the nation's growth, people had to be free to buy and sell land. American property law therefore began to stress the rights and obligations involved in land transfers, and the English laws that restricted such transfers were discarded.

By the early 1800s, Americans had begun to develop a flourishing economy based almost entirely on free enterprise. The rapid growth of the U.S. economy in the 1800s brought an enormous increase in contract law, which is used in a free enterprise system to regulate commerce.

9.1.7   The Beginnings of the Canadian Legal System

Canada's legal history dates from the legal system established by the first French settlers in the 1600s. The French set up a civil-law system in the areas they colonized, including what is now the province of Quebec. They based their system on one of the major local law codes in France, a code known as the Custom of Paris.

In 1763, Great Britain gained control of France's Canadian possessions and introduced a common-law system. But French Canadians objected to giving up their legal traditions. In 1774, the British Parliament passed the Quebec Act, which allowed French Canadians to follow their traditional system in private-law matters. Common law, however, remained the basis of all other law in Canada. In 1866, Quebec adopted a private-law code based on the Code Napoleon.

In 1867, the British North America Act, passed by the British Parliament, created the Dominion of Canada and gave that country limited self-government and a constitutional framework. Under the British North America Act (now called the Constitution Act 1982), jurisdiction (law-making power) was split between the federal and provincial governments, based on a common-law framework at the federal level and on prevailing legal systems in each province. Each province except Quebec elected to base its legal framework on common law; Quebec retained its civil-law system in matters of private law.

9.1.8   The Development of Modern Legal Frameworks

In the nineteenth century the emphasis on free enterprise and the rights of the individual encouraged a spirit of entrepreneurialism in Canada and the United States and indirectly set the stage for the development of public health and resource management legislation. Three interrelated factors contributed to this development: the rate of uncontrolled resource exploitation, the advent of the railway, and the rapid growth of urban centers.

The main civil- and common-law systems in the United States and Canada remained largely unchanged throughout the nineteenth century. By 1900, U.S. private law dealt mainly with the protection of property rights and businesses, with freedom of contract a central doctrine. Complete freedom of contract had served the needs of America's rapidly expanding economy during the 1800s, but by 1900 many businesses in the United States were using this freedom to increase their profits at the expense of their employees, stockholders, and customers. For example, factory owners claimed that efforts to protect the rights of workers interfered with the owners' rights to contract freely with their employees. Employees often had to accept unfavorable contracts or lose their jobs.

During the 1800s, most Americans rejected the notion that the law should be able to interfere in private business matters, but as concerns about social welfare increased through the end of the nineteenth century and the beginning of the twentieth, public attitudes toward the law began to change. Today the prevailing societal view is that the private interests of some members of society should not deprive other members of their rights. Legislation and court decisions during the 1900s have reflected this belief, especially by stressing the social aspects of contract law. For example, Congress and the state legislatures have passed many laws to help ensure the fairness of employment contracts. Some of these laws regulate working conditions and workers' wages and hours. Other laws guarantee the right of workers to organize and to strike.

Legislation and court decisions also changed many features of property and tort law during the last century. The social obligations of property owners have been enforced by zoning laws and by laws prohibiting environmental pollution. During the 1800s tort law held that a person could collect for an injury only if another person could be proved at fault. But the development of private and public insurance programs during the early years of this century helped to establish that a person should be paid for accidental injuries regardless of who was at fault. This "no fault" principle has made it unnecessary to sue for damages in certain cases.

Today, the United States, Canada, and other former British colonies continue to base their national legal systems on the common law, although to a large extent statutes have been written in an attempt to clarify and, in some cases, alter interpretations of (unwritten) common law. These systems are in contrast with those in civil-law countries, whose legal systems are based exclusively on the Justinian Code or the Napoleonic Code—systems of written statutes. In civil-law countries, and in the U.S. state of Louisiana and the Canadian province of Quebec, which were colonized by France and whose legal systems are patterned after the French civil-law system, every legal decision must be based on an existing statute, and not on precedent. These countries organize their written laws (statutes) into codes, which provide the final answer in any question of law. In common-law countries the judges in effect make the laws, and old common-law traditions of personal rights are embedded in constitutional law.

There are several important points to be drawn from this discussion:

- Canada and the United States (with the exceptions of Louisiana and Quebec) share a tradition of common law and value highly the rights of the individual.

- In both countries laws are made in the courts through the decisions of judges, based on legal precedents.
- Some aspects of common law have been extracted and written down, either to clarify them or to alter permanently their interpretation; these written laws, called statutes, include many environmental and water laws.
- Notwithstanding the existence of statutory law, common-law rights and responsibilities continue to bind people in the United States and Canada.

9.2 COMMON-LAW CAUSES OF ACTION

As discussed earlier, common-law rights and responsibilities continue to bind people even when statutory law also appears to exist in the same domain.* For example, a citizen has the right to charge a polluter under statutory law such as the U.S. Clean Water Act or the Canadian Environmental Protection Act, but that individual may also have the right to bring a private action against the offender in nuisance, under common law.

There are several main differences between common law and statute law. One—the fact that common law is unwritten—has been described. A second major difference lies in the relief that is available under each type of law. Under common law, the age-old principle of compensation—an "eye for an eye"—can apply (cf. Section 9.1.1). Under common law, compensation is often termed "damages." Alternatively, a plaintiff can seek an injunction to stop a defendant from continuing an offensive action. This injunction may either be temporary or permanent, depending on the circumstances of the case. It is a long-held tradition in common law that an injunction will never be granted if damages can give an adequate remedy in the case. Under statute law, fines are paid to the government, not to individuals (but the cost of prosecution is also borne by government. Statutes can, like common law, require the cessation of an offensive activity.

9.2.1 Common-Law Causes of Action in Water and Environmental Management

Under common law, a "cause of action" is a right to sue an individual or group—that is, to bring a complaint (suit) before a judge and ask for compensation or an injunction to stop the offense. More than one cause of action can be cited in a single complaint.

There are numerous common-law causes of action. In terms of water and environmental management, the most common of these are nuisance, riparian

---

*Common law is sometimes called "private law," because it is the law of individuals. Statute law is therefore "public law" in that it is the law of major public bodies acting on behalf of the citizenry.

rights, strict liability, trespass, and negligence, which are briefly discussed in the following paragraphs. The interested reader is referred to Estrin and Swaigen (1993) for a more detailed discussion of common-law rights and remedies; much of the following discussion is drawn from that source.

Nuisance   Nuisance is a very old concept under common law. It embodies the societal precept that one person should not interfere with another's peaceful enjoyment of property. Suits in nuisance are useful remedies for air and water pollution, noise, vibration, smells, soil contamination, flooding, and similar intrusions into peaceful enjoyment of property (Estrin and Swaigen 1993). Nuisance may be either "private" (unreasonable interference with another person's use or enjoyment of owned or occupied land) or "public" (unreasonable interference with the use of public lands and water; affecting more than one or two people).

Traditionally, nuisance is measured by the degree of impact on "peaceful enjoyment." Liability is deemed to exist if the activities of the defendant have caused real harm to another person's health or property. The plaintiff must be able to prove that the defendant could have foreseen the consequences of his or her action but did not take steps to prevent the nuisance from occurring. Examples of actions in nuisance have included suits against farmers who spread manure over saturated land, thus allowing contaminated runoff to flow into receiving waters, and against companies who sprayed nearby land with pesticides, allowing the spray to drift over the plaintiff's land. Nuisance can, therefore, be the result of either direct or indirect action; the test of nuisance is that harm must be proved to have occurred. Property may be either owned or occupied (for instance, as a tenant) by the plaintiff; in other words, the plaintiff need not be the owner of the land.

Public nuisance actions are often brought by public agencies (in Ontario, for instance, by the attorney general) rather than by an individual. They may therefore be subject to political influences and, indeed, may not come forward at all because of partisan concerns. Generally speaking, public nuisance actions must show that a public nuisance is also a private nuisance and, therefore, that at least one person's property or health has been significantly affected by the offense.

Traditional defenses against nuisance include the following:

- *Statutory authority.* The law required that the defendant take the action that caused the effect; the effect was therefore unavoidable.
- *Prescription.* The activity had been conducted for a long time in the same place without complaint and should be allowed to continue.
- *Acquiescence by plaintiff.* The plaintiff knew of the situation and failed to make a complaint within a reasonable period of time.

Defenses that are not usually successful in nuisance actions include:

- That the land was used for a reasonable purpose
- That the "best available technology" was used to avoid the effect, but the effect occurred despite the use of this technology
- That the activity existed before the plaintiff arrived on the scene

Under this common-law precept, therefore, individuals have a responsibility to their neighbors (society at large) to conduct themselves in such a way that they do not interfere with others' "peaceful enjoyment" of water or property. In court, the test of nuisance is whether harm actually occurred, and whether the defendant tried hard enough to avoid that harm. This latter concept, sometimes called "due diligence," runs throughout common law and underlies most public and private obligations under statutory law as well.

Riparian Rights   Like nuisance, riparian rights is a very old concept in common law. In essence, riparian rights are the rights of a property owner (or tenant) to use the water (stream, river, or lake) flowing through or past that person's land. Under riparian rights, the landowner (or tenant) has the right to enjoy the continued flow of the water in its natural quantity and quality, undiminished and unpolluted. Typically, riparian rights refer to visible flows of water such as those in a defined channel. They are not, therefore, usually taken to encompass groundwater flows. In some areas of the United States and elsewhere, the doctrine of riparian rights has been supplanted by prior appropriation as a way of allocating water rights. This distinction is discussed in more detail in Section 9.4.4.

Strict Liability   Strict liability simply says that the owner of dangerous goods or materials is responsible for seeing that those goods or materials do not cause harm to neighboring people or properties. Strict liability as a cause of action implies that some abnormal use is being made of the land, otherwise dangerous materials would not be held there. The classic judgment in a strict liability action is usually taken to be the eighteenth-century decision in *Rylands vs. Fletcher* in England. In this case an individual dammed a large quantity of water (the "dangerous substance") on his land. The dam broke, allowing the water to escape and flood a neighbor's mine. The owner of the flooded mine therefore had a legitimate action in strict liability against his neighbor.

Because of the implication of "dangerous materials" and unusual use of land, many pollution episodes can be taken as instances of strict liability. Examples of successful actions in strict liability include those involving the release of noxious gases, mercury, radioactivity, and lead into the environment.

Trespass   Trespass is a familiar concept, and its common usage is consistent with its meaning under common law: individuals should not enter others' property, nor should they deposit materials there, without the express permission of the property owner or tenant. Legal precedents in trespass actions have demonstrated several important elements in this cause of action. First, there must be physical intrusion by the defendant onto or over the land owned or occupied by the plaintiff. (There is some question as to how far above the land this doctrine applies; for instance, are jets flying at 35,000 feet over the land surface "trespassing"?) The physical intrusion can take the form of people entering the land or of objects or materials deposited on the land. To have an effective action in trespass, there must also be intent on the part of the defendant to enter the property; in other words, the intrusion must have been intentional and negligent. Finally, unlike an action in nuisance in which indirect or consequential harm is a legitimate cause of action, in trespass there must be direct harm resulting from the defendant's intrusion onto the property. Intrusion alone is not a sufficient cause of action. To have a successful defense, the defendant must prove that the entry was unintentional; usually it is argued that the property was not marked "private" in such a way that the defendant could be expected to realize he or she was entering private lands. This is why many landowners post "no trespassing" signs at regular intervals on fenced property boundaries. And the landowner must demonstrate that real and direct harm resulted from this intrusion. So, for instance, entry onto private lands for the purposes of bird-watching might not constitute trespass unless the bird-watcher damaged plants or structures, or left behind litter, during the intrusion.

Negligence   Negligence has been defined as "conduct that falls below the standard regarded as normal or reasonable in a given community" (Estrin and Swaigen 1993). Clearly, this standard will vary from community to community and from business to business. For this reason, the court will usually judge the standard of conduct expected of a business operator against usual conduct in other businesses of the same type. For example, if a judge was trying to determine whether a pulp and paper mill had been negligent in allowing the release of partially treated effluent into a receiving water, that judge would compare the precautions taken by the mill against typical operating procedures in place at other pulp and paper mills. In cases where materials that are particularly hazardous are in use, or the receiving environment is particularly sensitive, higher standards may reasonably be expected to apply. Once again, this is the idea of due diligence: did the individual, or company, try hard enough to ensure that the pollution did not occur?

There are several difficulties with assessing due diligence in real cases. For one thing, business operations differ significantly, and no two can be expected to be exactly alike. Second, pollution control systems can be highly complex and tailor-made for the particular site or receiving water. Precautions that would be reasonable, and indeed necessary, at one location may therefore be unreasonable or even unnecessary at another. Third, given the complexity of natural systems (for instance, the difficulty of predicting meteorological conditions with associated rainfall and flooding), no individual can be expected to anticipate, and guard against, everything that might possibly go wrong, so it is not always clear as to what is due diligence and what is negligence.

These common-law principles seem familiar and comfortable to us. They are the rules we learn in kindergarten: you do your very best not to hurt other people, not to damage other people's property, and not to annoy people unnecessarily. If you break these rules, society will require you to compensate the offended party or to halt your activity altogether.

These same principles underlie all of the environmental statutes discussed in Section 9.3. In large part, these are requirements that are implicit in common law. Writing them down as statute law allows the regulatory agency to impose very strict and specific interpretations on common law. It also allows governments to impose sanctions on offenders that benefit the public purse, punish the offender, or both, rather than sanctions that benefit only the plaintiff or cause the offense to cease.

9.3   THE MAKING OF LAWS

9.3.1   The Legislative Process

A law is a binding rule, whether unwritten but widely endorsed (as in the case of common law) or written and formally endorsed through a public affirmation process (as in the case of statute law). Statutes are, in effect, modifications or clarifications of common law that are written down, approved by elected governments, and enforced by public agencies. In both the United States and Canada, proposed statutes are scrutinized by all political parties and undergo a multistep approval process. As a result, they cannot easily be modified without a second multistep, multiparty approval process.

Generally speaking, there are two types of statutes. Criminal laws prohibit the kinds of actions that our society deems fundamentally wrong and unacceptable—acts like theft, murder, and rape. Public welfare laws address more "everyday" matters, regulating acceptable conduct (for instance, on the part of engineers), setting out the powers of major public bodies (for instance, universities), granting decision-making power, and so on.

New statutes or regulations of any kind (including amendments to existing laws) are made through the legislative process. The multistep process in each country is intended to provide checks and balances on lawmaking—in other words, to ensure that no single voice is able to create new laws that will bind the wider society.

The legislative process differs between Canada and the United States, so the two systems must be described separately. In each country, each of the three branches of government—legislative, executive, and judicial—has a role, although this role differs somewhat in the two nations. The legislative branch in the United States is Congress; in Canada it is the House of Commons and Senate. State and provincial legislatures are also considered part of the legislative (lawmaking) branch of government, as are municipal councils. The executive branch of government is the group of elected representatives and their staffs, including the president of the United States, the prime minister of Canada, and their senior advisors. The executive branch also includes the mayors, executive committees, or boards of control of municipalities. The executive branch of government is usually made up of people who are chosen from among a group of elected representatives, but may also include individuals who are appointed rather than elected. The judicial branch of government consists of the courts and judges. As shown under common law, the judicial branch is central in law-making in that it protects the rights of the individual citizen, is (in theory) independent and impartial, and interprets the law and legal precedents in judgments about new cases.

The following paragraphs describe the first of these branches, the legislative branch, in the United States and Canada.

United States   The legislative process in the United States involves Congress (comprising the Senate and the House of Representatives) and the executive branch (primarily the president). Of these, Congress has the most central role. It was granted power under the Constitution to "make all laws which shall be necessary and proper" to carry out the various powers delegated to it under the Constitution. These powers include the "power of the purse" (power over government taxation and spending) and the power to coin money, regulate trade, and wage war.

Congress consists of two bodies, the Senate (100 members) and the House of Representatives (435 members). Every state has at least one seat in the House. Both major political parties (Republicans and Democrats) are represented among this membership. All members of these bodies are elected. In addition to making laws, Congress approves the president's choices for senior government appointments and Supreme Court justices and approves (or rejects) treaties with other countries.

Congress uses a system of committees to draft, review, and revise proposed legislation ("bills") so that the actual time spent in congressional debate is used efficiently. Bills may be either public, affecting the public good, or private, affecting a single individual. If a bill is favorably received by a congressional committee, it is referred to the Rules Committee, which decides which bills will be debated on the floor and what conditions may apply in that debate. The Rules Committee has the power to block floor debate on a proposed bill and thus, indirectly, can block a bill from passing into law. If a bill is allowed into floor debate, it may be approved, amended, or rejected, all by majority vote or unanimous consent (the preferred mechanism). It can also be sent back to committee for further discussion and revision.

The main steps in the process are as follows:

1. A law is proposed, by any member of Congress, by anyone on a member's staff, by executive officers of government, by the president, by a political party at a national convention, or by any other interested party.

2. The proponent of the new law must seek a sponsor, a member of Congress who will formally introduce the bill. Several members of Congress may cosponsor a single bill.

3. The new bill is assigned a number with a prefix indicating its origin: S for the Senate and H.R. for the House. Bills are also often nicknamed after their sponsors or topics.

4. The bill is formally introduced in Congress.

5. The bill passes to an appropriate congressional committee for debate. There are currently 16 standing committees of the Senate and 22 standing committees of the House of Representatives. Special ad hoc committees may also be appointed for special tasks. Standing committees cover areas such as agriculture, finance, foreign relations, labor and human resources, and so on.

6. The congressional committee either "reports" the bill (that is, passes it on to the next step) or "tables" it (that is, shelves it, intending no further action). The committee may hold public hearings to determine what action to take on a bill. The task of the committee is to screen out bills that are unlikely to have broad public support and will therefore fail in congressional debate. Thus, most bills that reach the Senate or the House are likely to succeed and pass into law.

7. If the bill is reported, it passes to the Rules Committee for decision about calendaring (scheduling for formal debate on the floor). The Senate has only one calendar for public and private bills. The House has five calendars: for bills that raise or spend money, for other public bills, for private bills, for noncontroversial bills, and for motions to remove a bill from committee.

8. The bill is passed to the House or Senate (as appropriate for its place of origin) and debated. It may be passed, amended, rejected, or passed back to committee for further discussion.

9. After the bill has passed one house of Congress, it is referred to the other house. Approval must be obtained from both houses for the bill to become law. The second house often approves a bill without change.

10. After approval has been obtained from the second house, the bill passes to the president for approval. The president has 10 days to sign or veto the bill. If the president fails to sign or veto the bill within this period (which excludes Sundays) and Congress is in session, the bill is passed into law. The president may choose this course if the executive does not wish to be a party to a particular law but sees that there is widespread public support for it. If, on the other hand, Congress has adjourned, the bill does not become law; this is sometimes called a "pocket veto." Presidential vetoes are rare, reflecting strong feeling on the part of the president that the bill is somehow unconstitutional, too expensive, or will be too difficult to enforce. Vetoes can be overridden in the legislature.

This lengthy debate ensures that all members of Congress have the opportunity to scrutinize proposed legislation, request changes, and satisfy themselves that the bill will serve the needs of their constituencies. In this way, the process ensures that the public, through their elected representatives, has a say in determining which bills become law and which are rejected. State governments tend to parallel the federal system but are not necessarily identical in terms of powers and processes.

Although government departments are created by Congress, once created they become part of the executive branch. As such, their interests and objectives may sometimes differ from those of elected representatives. Administrative arrangements are discussed in more detail in Section 9.5.

The strength and centrality in lawmaking granted to Congress under the Constitution ensures that lawmaking power in the United States will clearly reside at the federal level, although powers can be, and have been, formally delegated to the states by Congress. As a result, most U.S. legislation of importance to environmental and water management interests is federal legislation. Although additional state laws may also apply in a given situation, federal law provides a clear and unequivocal foundation for all water management activities in the country.

In Canada, this is far from the case, because of the division of powers under the British North America Act of 1867, now the Constitution Act 1982. These differences are discussed in the following section.

Canada The Canadian legislative system, like the U.S. system, is bicameral—that is, consisting of two houses. It, like the parliaments of other former British colonies like Australia, India, and South Africa, is modeled on the parliamentary system of Great Britain. In Canada, Parliament has two houses, the Senate and the House of Commons. Senators are appointed by the government in power and generally have less power than their counterparts in the House of Commons. Members of Parliament are the elected representatives that serve in the House. As in the U.S. Congress, Members of Parliament (often termed MPs) represent a constituency of the public. Every Canadian province has a number of seats in the House of Commons, and all major political parties (Conservative, Liberal, New Democratic, Reform, Bloc Québecois) are represented there.

As in the U.S. process, bills may be proposed by any member of the public, by elected officials, by bureaucrats, or by any other person, but they must be brought forward to the House by a sponsor. A bill must pass through three "readings" in the House of Commons and must be approved in each reading. As part of this process, a law is usually "gazetted"—that is, distributed for wider scrutiny among members of the legal profession. If approved, the bill passes to the Senate for endorsement and then to the governor general for "royal assent." Royal assent in fact means approval by the queen; the governor general, as the queen's representative in Canada, is delegated the power to approve laws on behalf of the monarch. The main steps in the Canadian legislative process are as follows:

1. A law is proposed, by any MP, by anyone on a member's staff, by executive officers of government, by a member of the government bureaucracy, by a member of the public, by a political party at a national convention, or by any other interested party.

2. The wording of the bill is then drafted and reviewed by the political party who will endorse the bill. Private members' bills are also possible and need only be drafted and approved by the sponsoring member.

3. Most often, bills are introduced by the political party currently in power. When this occurs, the bill undergoes extensive scrutiny in Cabinet, the committee of senior ministers of the party in power.

4. Cabinet may refer the bill to committee for further discussion. The interests of the various members of Cabinet may vary widely. For instance, the federal minister of industry may support a bill that would encourage aggressive development in Northern Canada, whereas a minister of the environment may be opposed to such a bill. Normally, these differences are resolved in Cabinet so that the party in power can present a united position in the House. (Failure to secure this support can result in split voting and failure of the bill in the first or subsequent readings.)

5. When approved by Cabinet, the bill passes to the legislature for approval. The bill must receive three readings in the legislature. The first reading is essentially a mechanism for introducing the bill to the elected representatives.

6. After the second reading, the MPs may debate the contents of the bill, propose amendments, or reject all or part of it. As in the U.S. system, there is provision in this process for public hearings in the event that a bill is particularly far-reaching or controversial.

7. The third reading presents the bill, with any revisions that were incorporated after the second reading, to the House for a vote. No further discussion occurs after the third reading, and it is rare for changes to occur at this stage.

8. The Senate reviews and usually ratifies the decision of the legislature regarding the proposed bill.

9. The bill is passed to the governor general in the federal process, or (in the case of provincial legislation) to the provincial lieutenant governor, to receive approval from that individual as the queen's representative in the process (royal assent).

10. Having received royal assent, the bill becomes law. It may not, however, be binding until it is "proclaimed in force." Proclamation in force normally occurs upon granting of royal assent, but may also be delayed if Cabinet believes that the bill may be onerous or costly to implement. Some bills have taken up to six years after being passed in the legislature to be proclaimed in force.

Like the U.S. process, the Canadian legislative process is designed with a system of checks and balances. All parties have an opportunity to review proposed legislation and request changes in it. Thus, the resulting statute, in theory, has the support of a majority of citizens, as represented by their elected representatives in the House of Commons.

9.3.2 Subordinate Legislation

Statute law is intended to be carefully thought out, robust, and difficult to change. As a result, it usually contains wording that is clear and simple—and that has broad public support. Such a law may not include the details of, for instance, allowable effluent quality from an individual discharger, or the ways in which certain types of waste must be managed. These details are often appended to the law in regulations or other so-called subordinate legislation. In Canada, for example, the Canadian Environmental Protection Act simply states that the governor general "in council" has the power to make laws regarding, for instance, the control of pollution from industrial sources. The statute itself does not specify which industries, which types of pollution, or which types of controls may be appropriate; these details (where available at all) are contained in regulations appended to the Act. Regulations may be extremely detailed, to the point where individual manufacturing facilities and even individual effluent discharge pipes may be listed, with allowable discharge levels for a long and detailed list of contaminants.

Subordinate legislation, which includes documents such as discharge permits as well as regulations and "Codes of Practice," is not, unlike statute law, drafted with the cooperation of all political parties. Subordinate legislation is prepared by bureaucrats, often without external scrutiny. It may receive careful attention by members of the party in power, but may never be seen by those outside government. For these reasons, while statute law is difficult to change, subordinate legislation—permits, regulations, and so on—is relatively easy to change and perhaps more subject to political forces than statute law. In many cases, subordinate legislation reflects only the views of the party currently in power, rather than a wider societal consensus.

9.3.3 The Division of Powers

As described earlier, the United States Constitution vested considerable law-making power in the U.S. Congress, and it is clear even two hundred years later that the body remains the central and dominant legislative force. While Congress can, and does, delegate power to the states, there is still a strong federal oversight role. If a given state is not sufficiently assiduous in enforcing its own and federal laws, federal agencies often step in and see that the enforcement is carried out.

The situation is considerably different in Canada, where the division of powers is much less clear. The Canadian Constitution was drafted in 1867, at a time when people were more concerned with resource exploitation and resource ownership than they were about environmental protection or human health. As a result, the Constitution is silent on many issues that are important today, such as water quality and water diversions. On the other hand, resource ownership and oversight of business practices are clearly spelled out. In terms of water management, the result is that federal and provincial jurisdiction (the power to make laws) over the environment and water resources sometimes, even often, overlap.

In general, the Constitution grants jurisdiction over matters of national significance, and which cross provincial boundaries, to the federal level of government. These areas include the power to:

- Regulate international/interprovincial trade and commerce
- Regulate navigation and shipping
- Regulate seacoast and inland fisheries
- Impose taxes and spend money raised by taxation as it chooses, as long as it does not interfere with the rights and responsibilities of the provinces
- Make criminal laws (interpreted also as protection of public health)
- Regulate works that are interprovincial or international in nature: shipping, railways, telegraphs, interprovincial pipelines
- Make laws for the "peace, order and good government of Canada" (generally interpreted to mean laws of national significance—for instance, control of nuclear power)

The provinces were granted jurisdiction over matters of business and property and were allowed to retain ownership of their natural resources. Provincial powers include:

- Control of natural resources.
- Management and sale of public lands belonging to the province and the timber and wood on those lands.
- Establishment and control of municipal institutions (in other words, the power to set up municipalities and delegate responsibility to them).
- Power over property and civil rights (power to make contracts and carry on businesses).
- Power over all matters of a local or private nature in the province.

Over the past quarter-century, water management responsibility has been split between the federal and provincial levels of government. The provinces (with their municipalities) traditionally have had primary responsibility for water supply and pollution control through constitutional rights over their own natural resources. In recent years, however, the federal government has taken an increasingly active role in regulation writing and enforcement, but this role may again be altered under proposed "harmonization" and "regulatory efficiency" initiatives, begun in December 1994 and intended to make federal and provincial laws mutually compatible.

9.4 EXISTING LEGAL FRAMEWORKS FOR WATER AND ENVIRONMENTAL MANAGEMENT

Most authors date the rise of the modern environmental movement, and public awareness about environmental degradation, to the publication of Rachel Carson's book *Silent Spring* in 1962. Over the next decade, several fundamental pieces of environmental legislation were enacted in Canada and the United States, including the U.S. National Environmental Protection Act (1970) and the Canadian (federal) Fisheries Act (1970). The 1960s also saw rapid growth in the field of ecology, as well as widespread data collection efforts to map ecological systems in lakes, rivers, and oceans. Older agencies with narrow mandates (for instance, sanitary engineering) were transformed into environmental protection agencies. In 1970, the US Environmental Protection Agency was established by President Richard Nixon. In Canada, federal and provincial departments of the environment were set up.

A detailed discussion of current environmental legislation is beyond the scope of this book, and in any case would be quickly outdated as statutes and subordinate legislation are continually changing. The following is intended merely as an overview of major legislation pertinent to water management at the time of writing.

9.4.1 Major United States Federal Statutes

A number of United States federal, state, and local agencies have some level of jurisdiction over water and water-related activities. At the federal level, for example, the U.S. Environmental Protection Agency's Office of Water, Office of Pollution Prevention and Toxic Substances, Office of Prevention, Pesticides, and Toxic Substances, and the Pollution Prevention Council all have jurisdiction regarding water use and water pollution. The National Oceanic and Atmospheric Administration within the U.S. Department of Commerce has jurisdiction over certain water issues. The Natural Resources Conservation Service, and various agricultural agencies (including the Agricultural Extension Service and the Forest Service, both of the U.S. Department of Agriculture) have varying degrees of authority over agricultural activities and, thus, over nonpoint-source pollution.

At the state level, various state departments, including those interested in natural resources, environmental quality, and agriculture, are involved in the management of water resources. At the local level, planning commissions and zoning boards can affect water bodies and point and nonpoint sources of pollution through ordinances and building regulations. County governments are also involved in regulation—for example, in floodplain and shore land management and rural zoning—and often in implementation of animal waste laws. Many agencies therefore share the responsibility for water management. As is the case in Canada, these various agencies have over the years developed ways of separating their mandates to minimize overlap and unnecessary resource expenditure. The current major U.S. water-related legislation is discussed in following paragraphs.

The Clean Water Act   The Federal Water Pollution Control Act (commonly referred to as the Clean Water Act, or CWA) was first enacted in 1972 in an attempt to consolidate several water pollution control statutes, including the River and Harbors Act of 1899. The Clean Water Act, which was reauthorized in 1987, aims to improve water quality for the "protection and propagation of fish, shellfish, and wildlife" and for "recreation in and on the water." Originally, the Act intended to eliminate the discharge of all water pollutants by 1985, but in fact contained provisions only for the control of point-source pollution. The significant contributions of nonpoint sources and groundwater flows were largely overlooked when the Act was drafted.

Since 1977, the Clean Water Act has contained provisions for civil litigation against people who violate pretreatment standards for wastewater discharges. The following discussion about this and subsequent Acts is largely drawn from EcoLogic International (1995).

The Act contains seven major provisions for the control of water pollution:

1. A requirement that point-source dischargers of wastewaters hold valid discharge permits (the National Pollution Discharge Elimination System, or NPDES, permit system).
2. Requirements for the control and prevention of accidental or intentional spills of oils and toxic substances.
3. Regulations pertaining to dredged and fill materials and dredging and filling activities, including those on wetland areas.
4. Grant and subsidy programs for sewage treatment plants ("publicly owned treatment works," or POTWs).
5. A requirement that states establish a planning process for point and nonpoint sources of pollution.
6. A provision that states must share the responsibility for regulating land use practices in order to pursue national water quality objectives outlined in the CWA. (States are, however, allowed discretion in choosing a process for carrying out this requirement, since much of the planning is intended to be regionally oriented.).
7. A framework for the states concerning the structure of nonpoint-source management programs (although no requirement that states adopt such a program: a management program is required only if a state wants to be eligible for Section 319(b) funds).

Normal agricultural activities, such as plowing, seeding, cultivation, minor drainage, harvesting, farm pond construction, and irrigation ditch maintenance, are exempt from the permitting process required under the Clean Water Act. These broad protections for agriculture also exist in Canada.

The Great Lakes Critical Programs Act (CPA) amends Section 118 of the Clean Water Act and requires the US EPA to develop water guidance for the Great Lakes system. This guidance is to consist of not only numerical limits on pollutants in the Great Lakes aimed at protecting human health, aquatic life, and wildlife, but also direction to the Great Lakes states on the minimum water quality standards, antidegradation policies, and implementation procedures necessary to achieve the aforementioned numerical limits.

The result of the CPA's mandate is the Great Lakes Water Quality Guidance (also termed the Great Lakes Initiative), originally published in the Federal Register on April 16, 1993. This guidance focuses primarily on point-source loadings to the Great Lakes. For this reason, the US EPA and the Great Lakes states have agreed to establish a multimedia approach to identify and address gaps in effectively preventing, controlling, or eliminating toxic loadings to the Great Lakes Basin from non-NPDES sources. This effort, called the Great Lakes Toxics Reduction Effort, addresses five pathways omitted from the Great Lakes Water Quality Guidance, including contaminated sediments; atmospheric deposition; storage, handling, and transport (spills); stormwater, combined sewer overflows, and urban runoff; and waste sites. At the time of writing, the Great Lakes Initiative does not address agricultural nonpoint-source pollution.

Coastal Zone Act Reauthorization Amendments of 1990 Section 6217 of the Coastal Zone Act Reauthorization Amendments of 1990 (CZARA) requires coastal states with approved coastal zone management programs to address nonpoint-source pollution sources that impact or threaten coastal waters. The legislation specifies that states must submit Coastal Nonpoint Pollution Control Programs for approval to both the US EPA and the National Oceanic and Atmospheric Administration (NOAA). Failure to submit an approvable program will result in a state's losing a portion of its federal funding under Section 306 of the CZMA and Section 319 of the CWA.

Section 6217(G) of the legislation requires that the US EPA, in consultation with NOAA and other federal agencies, publish guidance specifying management measures to restore and protect coastal waters from specific categories of nonpoint-source pollution, including agriculture. In January 1993, the US EPA published its *Guidance Specifying Management Measures for Sources of Nonpoint Pollution in Coastal Waters.*

The Safe Drinking Water Act The Safe Drinking Water Act (SDWA) was first passed in 1974 and most recently amended in 1986. It establishes mandatory and nationwide drinking water standards, which are administered by state and local authorities. The 1986 amendments significantly increased EPA's requirements for disinfection, filtration, contaminant monitoring, and groundwater protection. (The Lead Contamination Act of 1988, which is not described here, further amends the SDWA to provide additional safeguards against contamination of drinking water with lead.) The Act also limits deep-well disposal of liquid hazardous wastes where those wastes may impact on drinking water supplies and establishes a Wellhead Protection Program for public water supplies drawn from groundwater.

Federal Insecticide, Fungicide, and Rodenticide Act (FIFRA) The 1947 Federal Insecticide, Fungicide, and Rodenticide Act (FIFRA), as amended in 1972, authorizes the US EPA to control pesticides that may threaten groundwater and surface waters. FIFRA mandates registration and approval by the US EPA prior to use of any pesticide, sets labeling and applicator licensing requirements, and makes improper use of pesticides a misdemeanor. The US EPA uses a step-by-step approach to determine the appropriate regulatory approach for specific pesticides: (1) the Agency first determines the pesticide's potential for leaching into ground and surface waters; (2) if such a potential exists, the US EPA considers whether establishing national labeling restrictions, enforceable under FIFRA, would adequately address these leaching concerns; (3) if restrictions would not adequately address these concerns, the US EPA must determine whether providing states with the opportunity to develop Pesticide Management Plans for the chemical will address the risk from pesticide contamination; (4) in the event that state Pesticide Management Plans would not address the problem, the US EPA is authorized to suspend or cancel the registration of these pesticides.

Necessary components of state plans include state philosophy and goals, state roles and responsibilities, legal authority, resources, assessment and planning, monitoring, prevention, response, enforcement, public awareness and participation, information dissemination, and records and reporting.

Application of FIFRA has resulted in the suspension or cancellation of pesticides containing some of the most toxic chemicals, such as DDT, dieldrin, aldrin, chlordane, heptachlor, mercury, kepone, chlorobenzilate, endrin, DBCP, and 2,4,5-T silvex. To ban a substance, the US EPA must prove that a particular chemical causes "unreasonable adverse effects on the environment." This process requires extensive study on a compound-by-compound basis and thus can result in extensive delays in the decision-making process.

Resource Conservation and Recovery Act (RCRA) The Resource Conservation and Recovery Act of 1976, as amended in 1984 and 1986, is the primary piece of federal legislation addressing solid waste disposal. Hazardous waste disposal is mandated under this Act, although some categories of solid waste that are not considered hazardous generally fall under state, rather than federal, solid waste management plans. State management plans required under RCRA must (1) identify general strategies the state has chosen to protect its citizens against adverse effects of solid waste disposal; (2) describe how the state will provide adequate sanitary landfill capacity, and (3) show that there are

adequate institutional arrangements for implementing these strategies. Although some states, such as California, require localities to prepare solid waste management plans that address agricultural production and processing wastes, including manure, prunings, and crop residues, other states, including most of the Great Lakes states, exclude agricultural wastes from their solid waste act requirements.

The Comprehensive Environmental Response, Compensation, and Liability Act (CERCLA or "Superfund")

CERCLA, which was drafted in 1980 and amended in 1982, was intended to respond to the environmental disasters of the 1960s and 1970s, like those caused by leaking hazardous waste disposal sites at Love Canal, Times Beach, and Stringfellow, and to perceived weaknesses in RCRA in respect to problems of this kind. Particular problems had been identified with abandoned sites where the costs of cleanup might be enormous. CERCLA granted the US EPA the power to carry out cleanup in situations where public health and welfare was urgently in need of protection. Public trust funds are used in these situations, and where possible the costs of cleanup are recovered from those responsible for the wastes. CERCLA does not provide for compensation to individuals, either for property damage or for direct or indirect health effects.

The Marine Protection, Research, and Sanctuaries Act (Ocean Dumping Act)

The Ocean Dumping Act, enacted in 1972, provides guidelines for the dumping of wastes and other materials into the oceans. It also imposes limits on who can do that dumping. Under the Act, certain sites are designated as approved locations for ocean dumping, and other sites as sanctuaries, to be patrolled by the Coast Guard. The Act prohibits the disposal of radiological, chemical, or biological warfare agents in the oceans, restricts the ways in which U.S. vessels can discharge wastes into the oceans, and prohibits unauthorized transportation from U.S. territory with the purpose of ocean dumping. This Act also provides oversight of the National Estuary Program, which uses existing legislation as a framework for the protection of 17 estuarine sites in the United States.

The National Environmental Policy Act

The National Environmental Policy Act (NEPA) of 1969 has the intention to create a unified government approach to environmental protection. It requires Environmental Impact Statements (EIS) by federal agencies before they undertake actions with potential to impact the environment, including direct, indirect, and cumulative impacts. The NEPA also establishes a Council on Environmental Quality, with specific regulations covering EIS, and provides for the preparation of annual environmental quality reports to the president and Congress on the status of the U.S. environment and the development of policies for the improvement of environmental quality.

The Pollution Prevention Act The Pollution Prevention Act of 1990 reflects a growing societal concern that, where possible, pollution should be prevented before it is created, not just managed after the fact. The Act establishes national targets for recycling and establishes an Office of Pollution Prevention as an information clearing house. It establishes an emphasis, under the law, on the pollution prevention hierarchy: prevention (first), recycling, treatment, and disposal (last resort).

The Toxic Substances Control Act (TSCA, often pronounced "Tosca") The Toxic Substances Control Act, passed in 1976 and last amended in 1986, establishes several programs relating to the introduction of new and potentially hazardous substances into commerce. It lays out requirements for premanufacture testing and notification and empowers EPA to require manufacturers of certain chemicals to undertake extensive testing of those chemicals if there is a high risk of exposure or if existing data are inadequate. Through TSCA provisions, EPA has the power to limit the manufacture and release of chemicals until it is satisfied with the results of industrial testing. EPA also has the power, under TSCA, to ban chemicals of known risks and has done so in the case of PCBs, chlorofluorocarbons, and asbestos. TSCA does not have authority over products covered by FIFRA or over tobacco, nuclear materials, food or drugs covered by the Food, Drug and Cosmetics Act, or certain other materials.

The Soil Conservation Act The Soil Conservation Act of 1935 exists as one of the major pieces of federal legislation governing soil conservation. The close relationship between soil erosion and transport of various pollutants, including nutrients, heavy metals, and pesticides, therefore makes this an important mechanism for the control of nonpoint-source pollution. The intent of this Act is to foster, develop, and implement practices that will positively impact conservation of soils. The Act originally focused on land treatment activities, but subsequent amendments have authorized small watershed management programs, rural development activities, and so on. The Act is now the central basis for provision of technical assistance to landowners through cost-share programs and education for soil and water conservation purposes. Under this legislation, farmers with highly erodible land are required to plan and apply soil conservation systems in order to remain eligible for farm program benefits.

The Clean Air Act Atmospheric sources comprise a large portion of the total loadings of some pollutants, including dioxins and furans, mercury, and lead. Section 112(m) of the 1990 amendments to the Clean Air Act requires the US EPA to identify and assess the extent of atmospheric deposition of hazardous air pollutants to the Great Lakes and other designated "Great Waters" such as Lake Champlain and the Chesapeake Bay. The Great Waters program is intended to examine sources and deposition rates of air pollutants, the relative contribution of atmospheric pollution to total pollutant loadings, and potential adverse health effects resulting from the volatilization and reemission of chemicals from agricultural lands and subsequent deposition to these waters.

Food Security Act of 1985 (also referred to as the Farm Bill) The 1985 Food Security Act (FSA) contained provisions which, though not specifically water quality-oriented, have the potential to reduce agricultural nonpoint-source pollutant loadings. The FSA was enacted to establish greater consistency between farm production and soil and water conservation programs. Essentially, the Act provides for more forceful obligations for soil and water conservation in the form of cross-compliance for those farmers dependent on government commodity supports and other subsidy or support programs. A few pilot projects were conducted under this legislation, but no nationwide program was ever developed. Wetlands reserve conservation easements are not currently funded under the Act. The Act was amended in 1990 and again in 1995 as the 1995 Farm Bill.

Great Lakes Water Quality Agreement Although not a statute, the U.S.-Canadian Great Lakes Water Quality Agreement (GLWQA), as amended in 1987, is a treaty-based obligation with some weight under the law. The Agreement, which arises from the Boundary Waters Treaty of 1909 between the United States and Canada, sets out guidelines for reducing point and nonpoint sources of pollution into the "boundary waters" shared by the two nations, including binational channels such as the St. Croix (Maine and New Brunswick) and the entire Great Lakes Basin. The GLWQA specifies water quality targets for surface waters, sets out a research agenda, and binds the United States and Canada to a range of control measures for point- and nonpoint-source pollution, including sewage treatment measures, reduction in phosphorus loadings, control of toxic substances, and similar programs. Annex 13 requires the two countries to identify land-based activities contributing to water quality problems described in Remedial Action Plans (RAPs) or Lakewide Management Plans (LaMPs) and to develop and implement watershed management plans, consistent with the RAPs and LaMPs, to reduce pollutant inputs.

Clearly, U.S. federal legislation contains many provisions for the control of point and nonpoint sources of water pollution. Generally speaking, these provisions are consistent in tone and intent, facilitating cross-compliance and program consistency. At the federal level, the Clean Water Act provides broad authority to maintain and preserve the integrity of the nation's waters. Collaboration between the various federal, state, and local agencies is clearly vital to the implementation of effective water management programs. For example, the planning and implementation of controls for nonpoint-source pollution requires collaboration between US EPA, USDA (U.S. Department of Agriculture), and NOAA at the federal level, including the provision of joint technical and financial assistance mechanisms. In addition, these agencies must engage in regular communication regarding agricultural nonpoint-source pollution control so as to avoid the promulgation of overlapping or contradictory regulations. Most important, the federal government must provide the necessary financial support to the states to ensure that they are able to implement any nonpoint-source management programs that may potentially be required under federal law. The same interagency, interjurisdictional collaboration is necessary in most other water management situations.

9.4.2 Major Canadian Federal Statutes

In Canada, as discussed in Section 9.3, regulation of water management involves all levels of government: federal, provincial, and, to a lesser extent, municipal. As a result, many departments and agencies have authority over similar water-related activities. At the federal level, Environment Canada, Fisheries and Oceans Canada, Natural Resources Canada, Agriculture Canada, and Health and Welfare Canada all have varying degrees of control over water and water-based activities.

de Loë (1991) identifies five agencies at the Canadian federal level, and at least six at the provincial level, with clear mandates in the management of Ontario water resources. He describes these institutional arrangements as developing in an ad hoc fashion and lacking integration. While he notes that this pattern is typical of water management in many jurisdictions, it is clear that the lack of an overall guiding strategy, delivered by a single agency, complicates the development of integrated water management policy in that province.

In addition to federal agencies, most provinces have agencies with authority over water management. As in the United States, these include ministries of environment (or environmental protection), natural resources, agriculture, and energy. At the municipal level, planning procedures, zoning, and bylaws can also have an effect on water resources management activities. With at least three levels of government (four, if regional municipalities are included), and multiple departments having control over the same activities, there is significant potential for the legislation governing these activities to overlap and become fragmented. Numerous authors have attested to this fragmentation (e.g., Heathcote 1993, de Loë 1991, Estrin and Swaigen 1993).

The federal level of Canadian legislation has jurisdiction over nonpoint-source pollution from agricultural sources mainly through the departments of Environment Canada, Health and Welfare Canada, Agriculture Canada, and Fisheries and Oceans Canada. The Canada Fisheries Act and the Environmental Protection Act are the most significant federal statutes in regard to water pollution control at this level. Although neither Act directly mentions nonpoint sources or groundwater pollution, the general prohibitions outlined in the statutes are thought to encompass these issues.

The Canadian Environmental Protection Act   The Canadian Environmental Protection Act (CEPA) of 1988, administered through both Environment Canada and Health and Welfare Canada, has the potential to impact point- and nonpoint-source pollution in that "the purpose of this Act is to provide for the protection and conservation of Canada's natural environment." CEPA provides for the protection of the environment through the application of preventative and remedial measures and the establishment of levels of environmental quality that function to protect the environment from the release of toxic substances. The Act purports to adopt an ecosystem approach to regulations, in that it attempts to control the release of toxic contaminants throughout the environment rather than dealing separately with air, land, and water issues.

CEPA creates a framework within which nationwide environmental standards controlling toxic substances in Canada can be developed. Section 8 of the Act outlines the authority of the minister to formulate nonregulatory instruments, such as release guidelines and codes of practice, that will provide environmental guidance to industries and regulators. The Act also provides a framework for the federal environment minister to suggest standards and practices to fellow cabinet members to govern pollution from activities within other federal departments.

The Fisheries Act   The Fisheries Act, administered by Fisheries and Oceans Canada in conjunction with provincial agencies, is intended to protect fish and other marine animals and their respective habitats in all internal and coastal waters of Canada. The Act was first passed in 1868 for the protection of fisheries resources. Its modern form was first enacted in 1970 and amended in 1991 to increase the penalties for violations under the Act.

Section 35 prohibits the "carrying on of any work or undertaking that results in the harmful alteration, disruption or destruction of fish habitat." Section 36(3) of the Act further states, "No person shall deposit or permit the deposit of a deleterious substance of any type in water frequented by fish, or in any place under any conditions where the deleterious substance may enter any such water." A deleterious substance, as outlined in Subsection 34(1), is defined as "any substance that if added to any water would degrade or alter or form part of a process of degradation or alteration of the quality of water so that it is rendered or is likely to be rendered deleterious to fish or fish habitat or to the use by man of fish that frequent that water." The term *deposit* has also been defined by Subsection 34(1) to mean any discharging, spraying, releasing, spilling, leaking, seeping, pouring, emitting, throwing, dumping, or placing. It should also be noted that under Subsection 40(5) a "deposit" as defined in Subsection 34(1) takes place regardless of whether or not the act or omission resulting in the deposit is intentional. Under the Act there is also a duty to report the deposit of a deleterious substance. Subsection 38(5) requires prevention (where possible) and mitigation of adverse effects. The Act further empowers the federal Minister of Fisheries and Oceans to request plans, procedures, schedules, analyses, samples, and other information relating to any work or undertaking that results in or is likely to result in the alteration, disruption, or destruction of a fish habitat or the deposition of a deleterious substance in waters frequented by fish. With this information, an evaluation of the potential impact can be made and the measures, if any, that would prevent the deposit or mitigate the effects thereof can be examined.

Under the Fisheries Act, the governor-general in council is empowered to make regulations for the proper management and control of the seacoast and inland fisheries, the conservation and protection of fish, the obstruction and pollution of any waters frequented by fish, and the conservation and protection of spawning grounds.

The Fisheries Act remains one of Canada's most powerful environmental statutes. Its power derives, in part, from its simple wording and from the fact that it is usually easy to demonstrate that fish were, or could be, present in a given water body and that the materials discharged to that water body are in some way deleterious to those fish. There is no need to prove that fish, in fact, died or that contaminants were discharged in toxic amounts. As a result, the Fisheries Act remains a cornerstone of Canadian water management.

The International Rivers Improvements Act The International Rivers Improvement Act regulates the construction, operation, and maintenance of dams and diversions in international rivers. Essentially, the Act requires that such "improvements" proceed only with the formal approval of the federal Minister of Transport. Like the Navigable Waters Protection Act, it has become important not so much because of its specific provisions, but because it can trigger the federal environmental assessment process and thus interfere with provincial dam-building or diversion activities.

Navigable Water Protection Act The objective of the Navigable Water Protection Act, administered by Transport Canada, is to prevent work that affects navigable waters from being undertaken unless first approved by the Minister of Transport. This Act's relevance to water management relates to a general prohibition, to the effect that no person shall deposit any material or rubbish into navigable water that is liable to interfere with navigation.

Fertilizers Act The Fertilizers Act is administered by Agriculture Canada and is intended to prevent the sale or import of any fertilizer or supplement that has not been registered, packaged, or labeled as prescribed in the Act. Section 9 of the Act prevents the sale of any fertilizer or supplement product that is harmful to plant growth. Section 9 states, "No person shall sell any fertilizer or supplement that contains destructive ingredients or properties harmful to plant growth when used according to directions accompanying the fertilizer or supplement or appearing on the label of the package in which the fertilizer or supplement is contained." Although this Act prohibits the use of fertilizers that are harmful to plant growth, it does not contain any prohibitions against the use of fertilizers that are harmful to the environment in general.

The Canadian Environmental Assessment Act The Canadian Environmental Assessment Act (CEAA) was passed in 1995, in response to nationwide calls for reform in environmental assessment procedures. It replaced less formal policies that had previously existed and that had been administered by Environment Canada. The CEAA established the Canadian Environmental Assessment Agency and specified four types of assessment: screening, comprehensive study, review by a mediator, or review by a panel. All projects receive an appropriate level of assessment based on their complexity. The federal government undertakes annually about 6,600 environmental assessments of all types. More than 99% are screenings. Since 1973 the federal government has referred more than 50 environmental assessments to a panel for public review.

The Canada Water Act The Canada Water Act of 1970 provided for research and planning of water management programs by the federal government. Part III of the Act was incorporated into the Canadian Environmental Protection Act in 1988; the remaining portions of the Act have been used only to set up federal-provincial or interprovincial water management agreements.

Canada Wildlife Act The Canada Wildlife Act, administered by Environment Canada, imposes controls on point- and nonpoint-source pollution in areas designated for the protection of wildlife. In such protected areas, this Act outlines prohibitions on activities deemed harmful to wildlife. These activities may include the use of vehicles, the dumping of wastes, and the disturbance of soil or vegetation. Numerous scientific projects, such as environmental impact studies and research into the effects of pollutants on wildlife, are performed by the Canadian Wildlife Service under the Canadian Wildlife Act.

The Pest Control Products Act The Pest Control Products Act, administered by Agriculture Canada, is the principle statute governing pesticides in Canada. The intent of the Act is to prohibit any person from manufacturing, storing, displaying, distributing, or using any pest control product under unsafe conditions. Under the Act, a pest control product is defined in Section 2 as "any product, device, organism, substance, or thing that is manufactured, represented, sold, or used as a means of directly or indirectly controlling, preventing, destroying, mitigating, attracting, or repelling, any pest and includes any compound or substance that enhances or modifies or is intended to enhance or modify the physical or chemical characteristics of a control product to which it is added and any active ingredient used for the manufacture of a control product."

The Pest Control Products Act regulates the import, sale, and use of pest control products in Canada through the use of a registration system. All pest control products in Canada must be registered. Section 4 of the Act prohibits any person from importing or selling any pest control product unless it has been registered, packaged, and labeled according to prescribed conditions.

The Minister of Agriculture has the responsibility of registering pesticides when he or she is of the opinion that the control product has merit and will not pose an "unacceptable risk of harm" to public health, plants, or the environment. During the registration process the onus is on the pesticide manufacturer to demonstrate product safety and merit through numerous scientific tests on the control product's effectiveness, toxicity, environmental hazard, residue persistence, bioaccumulation, carcinogenicity, and other related matters. The scientific data are then reviewed by a board consisting of representatives from Health and Welfare Canada, Environment Canada, Fisheries and Oceans Canada, and Forestry Canada. Although information and data on the control product are reviewed by a number of departments, Agriculture Canada has the final decision concerning registration. Public involvement in the registration process has so far been prevented. The Act has been criticized for being dated, in comparison with other Canadian statutes, and indeed has not been significantly amended since 1969. Numerous problems and concerns have been identified in regard to the Act and its control product registration process. Such problems include the lack of public involvement in the registration process, the lack of public access to information on health and safety data concerning control products, and the use of questionable testing techniques by chemical manufacturers in the registration process. These concerns are currently under review by a government panel charged with recommending revisions to the Act.

In summary, Canadian federal statutes governing water-related activities provide a comprehensive, if fragmented, framework for water management. In general, they underemphasize or entirely overlook contributions from nonpoint sources of pollution, as well as to and from groundwater. The emphasis in most statutes is clearly on management of point sources to surface waters. Indeed, a major weakness in this framework is the specific exemptions and other forms of relief that have been granted to agricultural operations. Almost without exception, exemptions for agriculture (e.g., exemption from permitting requirements, exemption from normal waste disposal requirements) are granted in all primary federal and provincial environmental statutes. These exemptions keep agricultural operations free from the environmental regulations and constraints imposed on other sectors of Canadian industry.

The split responsibility for water management in Canada demands close and continued interaction between the various federal, provincial, regional, and local agencies charged with water management. In a binational channel like the Niagara River, for example, effective management demands, at minimum, liaison with U.S. federal and state agencies. Environment Canada, the Ontario provincial Ministry of Environment and Energy, the regional municipality of Niagara (an "upper tier" or regional municipal body), local municipalities (individual cities like Fort Erie, Welland, and Niagara Falls), and a variety of other agricultural, natural resources, tourism, and industrial agencies at several levels of government. In general, this collaboration works well, and multiagency planning initiatives are the rule rather than the exception in Canada. This is a situation that may be unfamiliar to the novice, however, and one that requires careful management in the establishment of planning structures and processes.

9.4.3 State and Provincial Legislation

The first point that must be noted in regard to state and provincial legislation is the vast variability from one jurisdiction to another. In part, this variability arises from the differing social and economic histories of the various regions and the social priorities that have arisen as a result of those forces. For instance, in water-poor areas, considerable emphasis may be placed on allocation of water rights, while in water-rich areas the law may be silent on this issue.

Most states and provinces have a general "environmental protection" statute that is designed to protect air, water, and terrestrial ecosystems from pollution or mismanagement. Most also have environmental assessment legislation (see Chapter 10), requiring environmental assessment of major public projects. In some cases environmental assessment requirements may be built into general environmental statutes. Some jurisdictions have specific legislation dealing with the management of public utilities (water supply, sewerage, electric power generation, and so on). Finally, occupational health and safety legislation is found in most regions and, although not expressly directed at environmental concerns, often contains provisions for the control of pollutants in the indoor environment and in the bodies of workers.

It is beyond the scope of this book to inventory the huge number of state and provincial statutes, regulations, and policies that may bear on water resources management. It can, however, be stated that optimal water management almost always requires action, and therefore legislation, at the local or, at least, regional level. Often, however, these local ordinances (for instance, land use planning restrictions) are poorly enforced because local government and soil conservation district officials are hesitant to restrict landowners' property rights (Holloway and Guy 1990) and may even be subject to legal challenge by landowners.

As was observed at the federal level, state and provincial protections for groundwater, and control of nonpoint sources of pollution, sometimes lack the force of those for surface water protection and point sources of pollution. In general, however, state and provincial water pollution control laws, wetlands protection laws, soil and water conservation laws, and zoning laws provide the statutory authority necessary to create and implement sound water management programs.

9.4.4 Allocation of Surface and Ground Water Rights

Riparian Rights and Reasonable Use in Surface Water Allocation
Systems of water allocation have, in large part, developed in response to local water disputes. There are two fundamental systems of allocating water rights. The oldest of these is riparian rights, discussed in Section 9.2. Under riparian rights, a landowner is entitled to use the water that flows through his or her property, unimpeded in quantity and unimpaired in quality. The term *riparian* refers to the banks of a lake or stream (usually the latter).

English settlers brought the riparian rights doctrine with them and applied it in early settlements in eastern North America. Over time, however, the restrictions of riparian rights—a landowner could not adversely affect the quality or quantity of water flowing through his land—became too burdensome for the entrepreneurial American spirit. In the eastern United States, riparian rights were gradually replaced with the doctrine of "reasonable use," which permitted landowners to modify water quality and flows as long as the use was deemed to be reasonable for the purpose to which the land was put.

Although "reasonable use" in theory permits withdrawals consistent with land use, some jurisdictions have chosen to modify the doctrine by imposing a permitting system on water extractions. In this system, the landowner must secure a permit from the appropriate state agency before withdrawals can be made. In some jurisdictions, permits need only be obtained if extractions are expected to exceed some threshold level. Permits may restrict the amount of water that can be extracted, the place where extractions can be made, the time period over which water is extracted, or any combination of these factors. Permits may specify the use to which the water is to be put and may prohibit changes in that use without prior approval from the permitting agency. Finally, permits may specify water use priorities in the event of water shortages; in other words, they may require that under certain conditions the permit holder may or may not extract water, depending on the proposed water use.

Today, riparian rights (reasonable use) water allocation is usually found in areas that are relatively rich in water—for instance, in eastern Canada and in much of the eastern United States. In these areas, when water shortages occur, landowners are expected to reduce their water use equally, not in order of seniority. There is usually no restriction on the ways that water can be used in a riparian rights system.

Appropriation Rights in Surface Water Allocation   In water-poor areas of the United States, particularly the West and Southwest, state economies were built on water-intensive activities like mining and agriculture (Viessmann and Welty 1985). In some of these areas, water sources were located far from the place where water was needed, requiring diversion and pumping over land that might be held by another party—and affecting that landowner's riparian rights. The settlers felt that their interests were not well served by riparian rights in these cases and, instead, developed a doctrine called "appropriation rights." Under this doctrine, appropriators are assigned rights based on their priority of application for those rights. The phrase "first in time, first in right" is sometimes used to describe appropriation rights.

Under an appropriation rights system, a would-be water user must apply to the appropriate state agency, stating the beneficial use to which the water will be put. If the application is successful, it is assigned a date that determines its priority. Applications are usually successful if they do not conflict with existing water uses. Conflicts arise in times of water shortage, when there simply may not be enough water to go around. At these times, the most recent appropriator may not be allowed any water at all. Appropriators must use their water regularly or they will risk losing their water rights through nonuse. Appropriation rights may be sold or transferred as long as the purported use of the water does not change. If it does, the appropriator may be required to apply for a new permit.

There are several difficulties with appropriation rights. First, they place heavy emphasis on "beneficial uses," which historically have included uses with economic value, such as mining, agriculture, and municipal extractions. They have not traditionally included maintenance of minimum flows for the purpose of preservation of aquatic organisms or habitat (although some western states are now incorporating such definitions). And finally, the "use it or lose it" emphasis in appropriation rights creates an incentive to use water rather than to conserve it, even though appropriation-rights areas are often those that would benefit most from water conservation efforts.

Allocation of Groundwater Rights   Groundwater is less visible, less easily mapped, and more variably described than surface water. As a result, various jurisdictions define it differently and allocate its use using different principles. In general, these principles differ somewhat from those used for the allocation of surface water rights.

Under common law, groundwater is not encompassed by riparian rights, and traditionally landowners were able to withdraw unlimited quantities of water for any purpose without fear of retribution. This cavalier attitude may have arisen from a time when technology allowed only small extractions and there were few users sharing any given aquifer. In recent centuries, however, population densities have increased and technology has improved to the point that very large water extractions can be made easily, and many users may share a groundwater source.

Three groundwater-allocation doctrines are commonly used in the United States. The first is reasonable use, which, as in surface water extractions, implies sensible, conservative use, with all users having equal rights. The second doctrine is correlative rights, whereby rights are allocated in proportion to the extent of ownership of overlying land. Under this doctrine, the more land you own, the more groundwater you may extract. Finally, prior appropriation is used for groundwater allocation as it is for surface water allocation. The user must apply for a groundwater extraction permit from the appropriate agency and, in being granted approval, is also given seniority in water rights.

Certain areas such as Native American reservations, national parks, national forests, and other federal lands have rights to the water necessary to carry out their functions. Native American water rights are separate and distinct from those allocated under the various aforementioned doctrines. Aboriginal water rights are normally granted first-in-time status for the water body serving the needs of the reservation.

Table 9.1 lists the current surface and groundwater allocation systems in the 48 contiguous United States. In Canada, the pattern is much the same, with eastern provinces employing a reasonable use approach, with or without permitting, and western provinces relying on appropriation rights.

Table 9.1   Current Surface and Groundwater Allocation Systems in the 48 Contiguous United States, Excluding the District of Columbia

State	Surface Water Allocation	Groundwater Allocation
1. Alabama	Riparian (reasonable use)	Reasonable use
2. Arizona	Appropriation (permit)	Reasonable use/Appropriation/ Permit
3. Arkansas	Riparian (reasonable use)	Correlative rights/reasonable use
4. California	Appropriation/riparian (permit)	Correlative rights/appropriation/ conjunctive Management.
5. Colorado	Appropriation/permit	Appropriation/permit
6. Connecticut	Riparian (reasonable use)	Absolute ownership
7. Delaware	Riparian (reasonable use)/permit	Reasonable use/permit
8. Florida	Riparian (reasonable use)/permit	Correlative rights/permit
9. Georgia	Riparian (reasonable use)/permit	Absolute ownership/reasonable use/permit
10. Illinois	Riparian (reasonable use)	Absolute ownership
11. Idaho	Appropriate/permit	Appropriation/permit
12. Indiana	Riparian (reasonable use)	Absolute ownership/reasonable use
13. Iowa	Riparian (reasonable use)/permit	Reasonable use/permit
14. Kansas	Appropriation/Riparian (Permit)	Appropriation/permit
15. Kentucky	Riparian (reasonable use)/permit	Absolute ownership/reasonable use
16. Louisiana	Riparian (reasonable use)	Law of capture
17. Maine	Riparian (reasonable use)	Absolute ownership/reasonable use
18. Maryland	Riparian (reasonable use)/permit	Reasonable use/permit
19. Massachusetts	Riparian (reasonable use)	Absolute ownership/reasonable use
20. Michigan	Riparian (reasonable use)	Reasonable use
21. Minnesota	Riparian (reasonable use)/permit	Reasonable use/permit
22. Mississippi	Appropriation/permit/Preexisting riparian rights confirmed	Absolute ownership/reasonable use/permit in some areas
23. Missouri	Riparian (reasonable use)	Reasonable use
24. Montana	Appropriation/permit	Appropriation/permit
25. Nebraska	Appropriation/permit	Reasonable use/permit
26. Nevada	Appropriation/permit	Appropriation/permit
27. New Hampshire	Riparian (reasonable use)	Reasonable use
28. New Jersey	Riparian (reasonable use)/permit	Reasonable use/permit
29. New Mexico	Appropriation/permit	Appropriation/permit
30. New York	Riparian (reasonable use)	Reasonable use
31. North Carolina	Riparian (reasonable use)/permit	Reasonable use/permit
32. North Dakota	Appropriation/permit/preexisting riparian rights confirmed	Appropriation/permit
33. Ohio	Riparian (reasonable use)	Reasonable use
34. Oklahoma	Appropriation (reasonable use)/permit	Reasonable use/permit
35. Oregon	Appropriation/permit/preexisting riparian rights confirmed	Appropriation/permit

Source: Viessman and Welty 1985.

Table 9.1 (*Continued*)

State	Surface Water Allocation	Groundwater Allocation
36. Pennsylvania	Riparian (reasonable use and natural flow)	Reasonable use
37. Rhode Island	Riparian (reasonable use)	Absolute ownership/reasonable use
38. South Carolina	Riparian (reasonable use)	Absolute ownership/reasonable use/some permits
39. South Dakota	Appropriation/permit (some riparian rights persist)	Appropriation/permit
40. Tennessee	Riparian (reasonable use)/permit	Reasonable use/permit
41. Texas	Appropriation/permit/preexisting riparian rights confirmed	Absolute ownership
42. Utah	Appropriation/permit	Appropriation/permit
43. Vermont	Riparian (reasonable use)	Absolute ownership/reasonable use
44. Virginia	Riparian (reasonable use)	Reasonable use/permit
45. Washington	Appropriation/permit/preexisting riparian rights confirmed	Appropriation/permit
46. West Virginia	Riparian (reasonable use)	Reasonable use
47. Wisconsin	Riparian (reasonable use)/permit	Reasonable use/permit
48. Wyoming	Appropriation/permit	Appropriation/permit

9.5 ADMINISTRATIVE AND INSTITUTIONAL SYSTEMS

Administrative law consists chiefly of (1) the legal powers that are granted to administrative agencies by the legislature and (2) the rules that the agencies make to carry out their powers. Administrative law also includes court rulings in cases between the agencies and private citizens. Basically, under their respective Constitutions, nations like the United States and Canada have granted powers to various levels of government. Through those powers, they have further allocated responsibility ("jurisdiction") among government departments and agencies.

Although administrative arrangements may appear to be trivial forces in comparison with environmental legislation, they can in fact be very important in promoting or blocking the flow of information and in addressing national, regional, and local concerns.

The United States and Canada are countries that cover large geographic areas. It is not, therefore, surprising that their administrative systems reflect a decentralization of responsibility. Both countries have "head offices" located in their respective capitals of Washington, D.C., and Ottawa, Ontario. Each also has a wide network of satellite offices located throughout the country, and those regional offices work closely with local citizens and industries on issues of regional importance.

The following discussion illustrates current administrative structures within US EPA and Environment Canada as representative of other structures at the federal and state/provincial levels in those countries.

US EPA The United States Environmental Protection Agency (EPA) is an independent agency, or department, of the executive branch of government (Ostler et al. 1996). It is headed by one administrator, a deputy administrator, and nine assistant administrators, all of whom are appointed by the president with confirmation by Congress. EPA has 10 regional offices, each directed by a regional administrator.

EPA's mandate is to administer and enforce a range of major environmental statutes and, in certain circumstances, to administer and enforce state and local environmental laws. The laws administered by EPA include the Clean Air Act, the Clean Water Act, the Safe Drinking Water Act, the Ocean Dumping Act, the Pollution Prevention Act, TSCA, FIFRA, CERCLA, and RCRA. Where existing state and local enforcement programs are deemed to be ineffective, EPA has the authority to step in and enforce them in support of EPA's mandate. Enforcement usually consists of notification of violation, followed by informal negotiations if the violation is not corrected, followed by administrative hearings and, finally, by legal action in U.S. district court. Penalties for violations of statutes enforced by EPA can include heavy fines, criminal conviction, and even imprisonment. EPA can revoke permits and licenses for programs under its mandate and can impose cleanup orders to correct environmental damage.

EPA also administers national effluent-discharge permit programs for industries, publicly owned treatment works, stormwater runoff, and a variety of other programs such as those related to wetland protection and spill prevention.

Information about EPA and its activities can be found in EPA's World Wide Web page: www.epa.gov.

Environment Canada Environment Canada, like US EPA, is an independent agency, or department, of the executive branch of the federal government. It is administered by a minister, who is a senior (elected) member of Parliament appointed to the position by the prime minister. This appointment does not receive formal ratification like that provided by the U.S. Congress for senior administrators in that government. The minister is supported by a staff which he or she personally appoints, and which is expected to resign when the minister steps down. Members of the minister's staff are not elected officials but may be present or former bureaucrats or other individuals. Reporting to the minister is the deputy minister, the most senior bureaucrat in the agency. The deputy minister is not an elected official and usually remains in the position even after the minister has stepped down. Under the deputy minister are a number of assistant deputy ministers, usually with responsibility for specific program areas such as Finance and Management, Human Resources, and Corporate Policy. Also reporting to the deputy minister, and not necessarily through an assistant deputy minister, are several directors general. The number of these changes with departmental reorganization, but there are typically four to seven in total. Each director general is responsible for a program area or geographic area; for example, there is a regional director general for each regional office of Environment Canada.

Environment Canada administers a number of environmental agencies with specific mandates, including Action 21 (Canada's response to Agenda 21), the Atmospheric Environment Service, the Canadian Environmental Assessment Agency, the Canadian Ice Service, the Canadian Meteorological Centre, the Climate and Water Information Branch, the Ecological Monitoring and Assessment Network, the Environmental Effects Monitoring Program, and the National Water Research Institute.

The federal statutes administered by Environment Canada include the Canada Water Act, the Canada Wildlife Act, the Canadian Environmental Assessment Act, the Canadian Environmental Protection Act, the International River Improvements Act, and the Interprovincial Trade Act. Environment Canada also assists in the administration of the Arctic Waters Pollution Prevention Act, the Export and Import Permits Act, the Fisheries Act, the Pest Control Products Act, and the Transportation of Dangerous Goods Act.

Traditionally, Environment Canada has deferred to the provinces in matters of environmental control and enforcement, preferring to take an arm's-length, advisory role instead. Over the past 10 years, however, the agency has increased its regulatory presence through the Canadian Environmental Protection Act and the Canadian Environmental Assessment Act, both of which are duplicated in most provinces' environmental legislation. Accompanying this change has been an increase in enforcement activities, which tend to follow the same sequence as those in US EPA: notification of violation, informal negotiations, formal hearings, and, finally, legal action.

Information about Environment Canada and its activities can be found in Environment Canada's World Wide Web page, the "Green Lane": www.doe.ca/envhome.

9.6   TRANSBOUNDARY WATER ISSUES AND FREE TRADE

Relationships between the United States and Canada, and the United States and Mexico, under the North American Free Trade Agreement (NAFTA) are still evolving. Various authors (e.g., Eaton 1996, Heathcote 1996, and Holm 1988) have expressed concern about the potential for impacts on state and provincial laws, on water diversions, on water as a salable good, and similar issues under NAFTA. Water-poor areas see the trade agreement as a mechanism for obtaining water for food production and other essential services. Water-rich areas are worried that the agreement will make water diversions easier and drain resources away from those who currently own them.

Although Canadian federal analysts have argued that water is not covered by NAFTA or the environmental side agreement, private analysts (e.g., Linton and Holm 1993) conclude that the agreements cover "all natural water other than sea water" and not merely bottled water, as has sometimes been suggested. While certain items are specifically exempted, water is not among them: "ice, snow and potable water not containing sugar or sweetener" are considered tariff items in the agreement.

Authors such as Makuch and Sinclair (1993) note that the agreement does not bind the states or provinces and thus cannot be used to enforce state or provincial environmental laws. Furthermore, the definition of *environment* in the agreement excludes laws regarding natural resource management, so no enforcement of such laws is possible under NAFTA. Some analysts therefore fear that the agreement will encourage accelerated extraction of natural resources, including water.

Canadians' greatest fear in this regard is that NAFTA will facilitate major water diversions such as the GRAND Canal scheme, a northern diversion/hydroelectric project whose ultimate goal is to generate electricity for sale in the United States. Accelerated extraction of groundwater for bottled water sale is another major concern. Conservative Canadian analysts, including those at the Fraser Institute in British Columbia, justify water exports to the United States on the grounds that the water will be used to irrigate vegetables grown for export back to Canada. Other authors are less optimistic that Canada will reap benefits proportional to the costs of such diversions.

At present there is consensus that major water diversions are an unlikely prospect, if only because of their high cost and complex environmental impacts. Yet it appears clear that if water were to be diverted from a river and reserved for the use of Canadians only, NAFTA would allow the United States to launch a trade challenge to obtain a proportional share of the resource.

Transboundary issues, including trade agreements, are emerging as especially important in the implementation of multijurisdictional watershed management plans. Chapter 12 discusses these problems in more detail, presenting case studies from the Danube River, the Rhine River, and other multijurisdictional basins.

REFERENCES

Aubert, Vilhelm. 1989. *Continuity and Development in Law and Society*. London: Norwegian University Press.

Bohannan, Paul. 1967. *Law and Welfare*. Garden City, N.Y.: The Natural History Press.

Cairns, J. 1992. *Restoration of Aquatic Ecosystems*. Washington, D.C.: National Research Council (U.S.), National Academy Press.

Careless, J. M. S. 1984. *Toronto to 1918: An Illustrated History*. Toronto: James Lorimer and Co., Publishers, and National Museum of Man.

de Loë, R. 1991. The institutional pattern for water quality management in Ontario. *Can. Water Resources Journal* 16(1):23–43.

Eaton, David J., ed. 1996. *The Impact of Trade Agreements on State and Provincial Laws*. Proceedings of a conference sponsored by the U.S.-Mexican Policy Studies Program, University of Texas at Austin, November 10, 1995.

EcoLogic International Inc. 1995. Agricultural nonpoint source pollution regulatory framework: United States federal and state. In *An Agricultural Profile of the Great Lakes Basin: Characteristics and Trends in Production, Land-Use and Environmen-*

tal Impacts, edited by the Great Lakes Commission. Ann Arbor, Mich.: Great Lakes Commission.

Estrin, David, and John Swaigen. 1993. *Environment on Trial: A Guide to Ontario Environmental Law and Policy.* 3d ed. Toronto: Emond Montgomery Publications Ltd.

Fitzgerald, Patrick. 1977. *This Law of Ours.* Toronto: Prentice-Hall.

Government of Canada. 1991. *The State of Canada's Environment.* Toronto: Queen's Printer.

Hayman, D., and C. Merkley. 1986. *Upper Thames River Rural Beaches Strategy Program—1986.* Toronto: Ontario Ministry of the Environment.

Heathcote, Isobel W. 1993. An integrated water management strategy for Ontario: Conservation and protection for sustainable use. In *Environmental Pollution: Science, Policy and Engineering*, edited by B. Nath, L. Candela, L. Hens, and J. P. Robinson. London: European Centre for Pollution Research, University of London.

Heathcote, Isobel W. 1996. Canadian Water Resources Management. In *Environment and Canadian Society*, edited by T. Fleming. Toronto: Nelson Canada.

Holm, Wendy. 1988. *Water and Free Trade: The Mulroney Government's Agenda for Canada's Most Precious Resource.* Toronto: J. Lorimer.

Holloway, James E., and Donald C. Guy. 1990. Emerging regulatory emphasis on coordinating land use, soil management, and environmental policies to promote farmland preservation, soil conservation, and water quality. *Zoning and Planning Law Report* 13:49.

Jennings, Winfield Holmes, and Thomas G. Zuber. 1991. *Canadian Law.* 5th ed. Toronto: McGraw-Hill Ryerson.

Kernaghan, Kenneth, and David Siegel. 1991. *Public Administration in Canada: A Text.* Scarborough, Ont.: Nelson Canada.

Linton, J., and W. Holm. 1993. *NAFTA and Water Exports.* Canadian Environmental Law Association special report. Toronto: CELA.

Makuch, Z., and S. Sinclair. 1993. *The NAFTA Environmental Side Agreement: Implications for Canadian Environmental Management.* Canadian Environmental Law Association special report. Toronto: CELA.

Ostler, Neal K. and John T. Nielsen (eds.). 1996. *Environmental Regulations Overview.* Volume 2, Prentice Hall's Environmental Technology Series. Englewood Cliffs, N.J.: Prentice Hall.

Saxe, Dianne. 1994. *Ontario Environmental Protection Act, Annotated.* Aurora, Ont.: Canada Law Book Inc.

Starr, June, and Jane F. Collier. 1989. *History and Power in the Study of Law.* Ithaca and London: Cornell University Press.

Stephenson, Ian S. 1975. *The Law Relating to Agriculture.* Westmead, England: Saxon House.

Swaigen, John. 1990. The "Right-to-Farm Movement" and environmental protection. *Canadian Environmental Law Reports* (New Series) 4(2):121–128.

Viessman, Warren, Jr., and Claire Welty. 1985. *Water Management: Technology and Institutions.* New York: Harper & Row.

10

Environmental and Social Impact Assessment

Environmental assessment (EA) has been called one of the more successful policy innovations of the twentieth century (Sadler 1996). Begun as a practical response to a persistent administrative problem, EA is now applied in more than 100 countries around the world to help decision makers make informed judgments about projects that have potential to impact the environment.

10.1 THE HISTORY OF ENVIRONMENTAL ASSESSMENT POLICY

The roots of EA lie in congressional concerns in the late 1960s. At that time many government agencies were frustrated by the waste of time and money that resulted when a major public project that was well underway encountered public opposition. Often, the concerns expressed by the public revealed that project planners had not considered important indirect or cumulative effects of the project. These deficiencies were difficult to rectify once the project was under construction. Congress wanted to establish a legislative requirement that federal agencies would carry out a complete and balanced assessment of impacts *before* construction was allowed to proceed. Congress was also concerned about the social and economic well-being of individuals and communities in addition to, and as linked with, biophysical impact. Of particular concern were cumulative, secondary, and "off-site" environmental effects (Gibson and Savan 1987). Although reactive legislation (and common-law remedies) were available to punish those who adversely affected the environment, Congress sought to prevent those impacts by careful data collection and informed decision making.

To this end, Congress drafted two action-forcing clauses, which, at the time, must have seemed modest if not downright unimpressive, but which have since had far-reaching consequences around the world. In 1969, Congress passed the National Environmental Policy Act, which includes an "action-forcing" provision (Section 102(2)(b)) to the effect that:

> All agencies of the Federal Government shall ... identify and develop methods and procedures ... which will ensure that presently unquantified and environmental amenities and values may be given appropriate consideration in decision making, along with economic and technical considerations.

Section 102(2)(c) of the Act contains the basis of most modern environmental assessment legislation:

> All agencies of the Federal Government shall ... include in every recommendation or report on proposals for legislation and other major Federal actions significantly affecting the quality of the human environment, a detailed statement ... on (i) the environmental impact ... (ii) adverse environmental effects which cannot be avoided ... (iii) the alternatives to the proposed action ... (iv) the relationship between local short-term uses ... and ... long-term productivity, and (v) any irreversible and irretrievable commitments of resources.

Congress intended these provisions to ensure that proposed projects were evaluated on the basis of clearly stated purposes, adequate and balanced consideration of alternatives, and careful evaluation of potential impacts on the social, economic, cultural, and biophysical environment. In the past projects had not been exposed to this rigorous evaluation, with the result that problems had often cropped up during or after construction, after large quantities of time and money had already been expended. Congress also hoped that by forcing comprehensive and early data collection, more cumulative, secondary, and off-site environmental effects could be foreseen before costly projects were actually built.

Many countries have now adopted EA as a central component of their environmental policy framework. Although the form and process of EA vary from country to country, the intent of the legislation is virtually the same all over the world. So fundamental was the National Environmental Policy Act in proposing this approach, in fact, that the U.S. Council on Environmental Quality (1993) has called it the "Magna Carta" of the field.

In both Canada and the United States the environmental assessment process is open to public scrutiny at all stages. This means that the scope of the assessment, and the way that each element is analyzed, must be well justified and competently performed. If the analyst is careless about what is included, or how impacts are assessed, the government has the right to halt progress on the project until an adequate assessment has been compiled. As a result, the EA process sometimes functions as a mechanism for building social consensus about major public activities—what Baber (1988) has called "procedural democracy."

In essence, EA forces the proponent of a major project to evaluate the proposed undertaking in light of:

- A clearly stated purpose
- Adequate consideration of alternatives
- Comprehensive evaluation of the potential impacts on the social, economic, cultural, and biophysical environment of the undertaking and of alternatives to the undertaking
- Preventive or mitigative measures that would be required to reduce or eliminate anticipated impacts

In addition to the many countries that have adopted EA as part of their policy framework, international lending agencies like the World Bank and international aid organizations employ EA in the evaluation of projects in client countries. In 1985, the European Union (EU) established minimum provisions for environmental impact assessment (EIA) in a directive to member states (these provisions were not binding until 1988). Canada, New Zealand, and Australia were among the first countries to follow the lead of the United States in implementing EA. Originally, Canada opted for an administrative process (Australia chose a legislated approach), which was replaced in 1992 with the Canadian Environmental Assessment Act. Now all 10 provinces and both territories have EA processes that bind provincial (territorial) activities. Indeed, Ontario's definition of *environment* under the Ontario Environmental Assessment Act remains the most powerful in that province's environmental legislation and may be the most comprehensive in Canada. In Ontario's Environmental Assessment Act, *environment* is defined as:

- Physical features (air, land, water) and
- Buildings, structures, machines, or other devices or things made by man;
- Any solid, liquid, gas, odor, heat, sound, vibration, or radiation resulting directly or indirectly from the activities of man;
- Biological subjects (plant and animal life, including man) and
- Human and ecological systems (the social, economic, and cultural conditions that influence the life of man or a community); and
- Any part or combination of these elements

10.2 OVERVIEW OF THE EA PROCESS

There are many forms of EA. Vanclay and Bronstein (1995) report at least a dozen categories and subcategories, including environmental impact assessment, social impact assessment, technology assessment, policy assessment, economic assessment, health impact assessment, environmental auditing, and sustainability assessment. Sadler (1996) defines four key terms in EA:

1. *Environmental assessment* (EA) is a systematic process of evaluating and documenting information on the potentials, capacities, and functions of natural systems and resources in order to facilitate sustainable development planning and decision making in general, and to anticipate and manage the adverse effects and consequences of proposed undertakings in particular.

2. *Environmental impact assessment* (EIA) is a process of identifying, predicting, evaluating, and mitigating the biophysical, social, and other relevant effects of proposed projects and physical activities prior to major decisions and commitments being made.

3. *Social impact assessment* (SIA) is a process of estimating the social consequences that are likely to follow from specific policy and government proposals, particularly in the context of national EA requirements.

4. *Strategic environmental assessment* (SEA) is a process of prior examination and appraisal of policies, plans, and programs and other higher-level or preproject initiatives.

Environmental assessment often encompasses three stages: preliminary assessment, detailed assessment, and follow-up. Preliminary assessment is used to determine whether a project is covered by EA legislation or policy, whether an EIS is required, the necessary nature and extent of the EA process, and scoping. Scoping, in terms of EA, generally includes the same tasks as described in Chapter 3: determining the boundaries of the plan, and identifying the issues, processes, and components that are most important in the analysis. Detailed assessment includes analysis of the impacts and mitigation necessary for the undertaking and a range of alternatives to the undertaking, usually including the "do nothing" option. Follow-up includes monitoring and audit functions to determine the actual impacts of the project and to ensure that necessary mitigation measures are in place. A typical EA process is illustrated in Figure 10.1.

Although this process appears to be straightforward, it can be very difficult for the practitioner to decide what to include in the assessment, what to exclude, what to emphasize, and so on. Figure 10.2 illustrates a hierarchy of principles that can assist in the successful application of EA.

The difficulty in applying a theoretical hierarchy such as that shown in Figure 10.2 lies in determining *whose* values should be incorporated in the EA process. Ideally, of course, there should be social consensus on values. Who, after all, could argue with "integrity," which Sadler (1996) defines as conforming "with accepted standards and principles of good practice"? Yet in environmental assessment processes such as the Canadian federal EA process in regard to the vast proposed Voisey's Bay nickel mine in coastal Labrador, aboriginal interests and values may not at all align with Western values. "Accepted standards" of behavior in the affected Innu and Inuit societies differ markedly from those in typical English-speaking groups.

Consider the following simple example of this problem. A white, English-

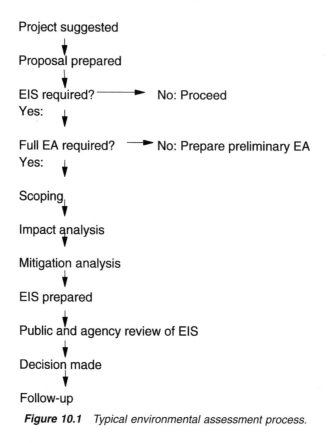

Figure 10.1   *Typical environmental assessment process.*

speaking hydrologist may make an inventory of water resources and conclude that there are primary and secondary streams, small lakes, wetland systems, estuaries, and oceans. By contrast, an Innu elder may identify 20 different categories of "small lakes": grassy lakes, shallow lakes, muskeg lakes, swampy lakes, chains-of-lakes, small deep lakes with no apparent source, small lakes, very small lakes, and so on (Clément 1995). "Good practice" to the elder may include a category-by-category analysis of impacts on each type of lake. The appearance of the water, and the plants and animals it supports, would be an important part of this analysis. (The Innu are traditional hunter-gatherers who rely heavily on the food resources of the back country.) To the white hydrologist, good practice may consist simply of assessing probable changes in lake levels or streamflows, with a statistical assessment of expected high and low values. To the Innu, deep lakes with no apparent source are particularly special, because they are thought to have spiritual or even magical properties. Damage to these lakes (for instance, through tailings disposal practices) might be viewed by the Innu as far more important than, say, damage to a muskeg lake. This distinction would be lost on the average English-speaking hydrologist.[*]

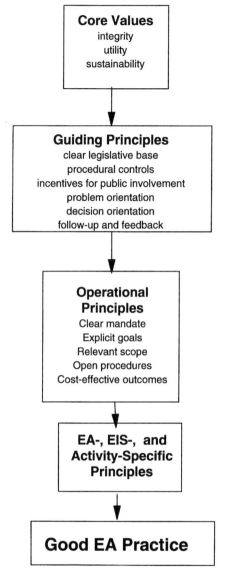

Figure 10.2 *Hierarchy of environmental assessment principles (adapted from Sadler 1996).*

*As a white, English-speaking scientist, I recently had a lesson in values when I planned to attend an Innu environmental assessment meeting wearing a brooch shaped (very realistically) like a jumping frog. I think of frogs as representing clean water and a healthy environment. This particular brooch, in my view, combined aesthetic appeal with a positive environmental message. But the Innu had a different view. I was surprised to learn from a colleague that the Innu have a category of animals they believe to be disgusting or repulsive, among which is the frog. I removed the brooch.

Values also become critical when conflicts arise and a choice must be made. For instance, a mining engineer may select technology A over technology B because A is lower in cost but still does a reasonable job of pollution control. But if technology A is much noisier than technology B (and would thus affect caribou migration, for example), or if technology A will affect those deep, sourceless lakes more than technology B, the Innu elder might opt for B instead. Who is "right"?

Many EA texts implicitly adopt Western values without even acknowledging the fact that consensus on values rarely exists. There is a large literature on "traditional ecological knowledge" (TEK), some of which addresses this issue. The interested reader is referred to Inglis (1993) and Freeman and Carbyn (1988). Clément (1990, 1995) provides detailed reviews of Innu ethnobotany and ethnozoology.

Finally, business interests sometimes object to the time and cost involved in full-scale EA. Yet over the long term, EA can provide important guidance as to what businesses should expect to encounter—and to pay for. The World Business Council on Sustainable Development (1995) has said:

> Running a business encompasses a broad range of activities, not least of which is the management of potential risks associated with failure to handle environmental matters adequately. Industry does this by managing its environmental affairs with due care and by being increasingly eco-efficient, taking account of scientific, technical, and economic factors, and the requirements of environmental legislation as a starting point.
>
> At the heart of sound environmental management is the assessment of effects, real or potential, on the environment as a consequence of business activities and the planning and implementation of measures to avoid or mitigate that damage. Environmental assessments can assist companies in their quest for continuous improvement by identifying ways of maximizing profits through reducing waste and liabilities, raising productivity and demonstrating a company's sense of duty towards its customers and neighbors.

Section 10.3 describes some of the most common methodologies for EA. Sadler (1996) provides a discussion of the broader context of EA and an evaluation of the effectiveness of current practice.

10.3   CHALLENGES IN ENVIRONMENTAL ASSESSMENT

10.3.1   Scoping

The problem of scoping was discussed at length in Chapter 3, and many of the concepts presented there also apply to EA. Scoping sets the boundaries for EA, it asks the important questions, and it identifies the information needed for successful decision making. Sadler's (1996) review shows that all jurisdictions have difficulty with scoping, especially in prioritizing issues and identifying the impacts of real concern. In part, this may be because underlying values and expectations differ among the participants in an EA process. If these fundamental values and expectations are not clarified and the debate on scoping drags on, a number of problems are likely to occur:

- EIS (Environmental Impact Statement) documents become large and poorly focused in an attempt to satisfy the range of interests.
- Important issues are identified late, forcing major revisions to the EIS and thus significantly increasing expenditures of time and cost.
- Impact analysis becomes a "laundry list" or catalog rather than a focused assessment.

Environmental assessment is now required for most major public projects. Guidelines developed by the U.S. Council on Environmental Quality (CEQ) (1981) provide advice on all aspects of EA. In terms of scoping, the CEQ guidelines advise:

1. Start scoping when you have enough information to present an initial list of alternatives and environmental issues.
2. Prepare an information package so that participants can make an informed contribution to the process.
3. Design the scoping process.
4. Give the public adequate notice of opportunities for participation in the scoping process.
5. Hold a public meeting to hear public opinions.
6. Review comments received from the public, to identify major issues and potential impacts.
7. Develop work schedules and budgets that will allow you to address the major issues and impacts in a timely and efficient manner.
8. Where issues are contentious, try alternative process formats (e.g., workshops, review committees, independent moderators, etc.) to resolve conflicts.

In general, EA is aimed at determining the direction, magnitude, and significance of impacts that may arise from the development of a project. "Direction" means positive or negative impacts, both of which should be included in the analysis. For example, a dam project may have positive impacts on (would reduce) flood losses, but negative impacts on fish habitat (through inundation and excessive sedimentation). "Magnitude" means how big the effect will be—for instance, the areal extent of an impact, or the difference between pre-project conditions and postproject conditions. "Significance" means the importance of the effect to the community. For instance, a major flood may have a large magnitude (cover a large area, result in much higher than normal streamflows) but low significance if the inundated land is not inhabited, contains no valuable structures or machinery, and the inundation will not affect resources contained in that land. On the other hand, a flood of the same magnitude, but in a major city, may have enormous significance. Most jurisdictions also evaluate "residual" impacts—the effects that are expected to remain after mitigative measures have been put in place.

Methods of assessing impact are by no means perfectly precise or accurate. These issues have been addressed elsewhere in this book, particularly in Chapters 3, 6, and 7. Chapter 8 presented methods of incorporating risk and uncertainty into planning that may also have application in EA. Typically, an EIS contains the following elements:

- A description of the proposed action or project
- A statement of the purpose of, and need for, the project
- A description of the environment likely to be affected by the project
- A discussion of prevailing land use plans, policies, controls, and applicable legislation in the affected area
- A list of alternatives to the project (for instance, "do nothing")
- A list of alternative ways of carrying out the project
- A discussion of the anticipated positive and negative effects on the environment of the preferred project and each feasible alternative to it
- A discussion of the steps that could be taken to mitigate those effects, for the preferred project and each feasible alternative
- A discussion of the relationship between short-term uses of the environment and long-term productivity
- Any irreversible or irretrievable commitments of resources that would be involved in the proposed action if it were implemented

Table 10.1 shows a suggested outline for preliminary or detailed EIS documents, based on CEQ guidelines and other sources:

Applicable guidelines for EA vary considerably from jurisdiction to jurisdiction. Those engaged in the preparation of EIS documents or in planning EA processes should consult the appropriate federal, state, or provincial agencies for advice on the guidelines for their region and type of project.

10.3.2   Evaluation of Impact Significance

Many basic EA texts discuss the problem of evaluating the significance of impacts. Possible approaches include those described in Chapters 3, 5, 6, and 7, for instance, identification of acceptable target values and evaluation of the likelihood of postproject effects resulting in values in excess of those targets. "Best professional judgment" and other qualitative approaches are also commonly used. Some authors (e.g., Hilden 1996) recommend using quantitative

Table 10.1   Suggested Outline for EIS Documents (After Goodman 1984)

1.0 Project Description
    1.1 Purpose of action
    1.2 Description of action
        1.2.1 Name
        1.2.2 Summary of activities
    1.3 Environmental setting
        1.3.1 Environment prior to proposed action
        1.3.2 Related federal activities
2.0 Land Use Relationships
    2.1 Conformity or conflict with other land use plans, policies, and controls
        2.1.1 Federal, state, and local
        2.2.2 Clean Air Act and Federal Water Pollution Control Act
    2.2 Conflicts and/or inconsistent land use plans
        2.2.1 Extent of reconciliation
        2.2.2 Reasons for proceeding with action
3.0 Probable Impact of the Proposed Action on the Environment
    3.1 Positive and negative effects
        3.1.1 National and international environments
        3.1.2 Environmental factors
        3.1.3 Impact of proposed action
    3.2 Direct and indirect consequences
        3.2.1 Primary effects
        3.2.2 Secondary effects
4.0 Alternatives to the Proposed Action
    4.1 Reasonable alternative actions
        4.1.1 Those that might enhance environmental quality
        4.1.2 Those that avoid some or all adverse effects
    4.2 Analysis of alternatives
        4.2.1 Benefits
        4.2.2 Costs
        4.2.3 Risks
5.0 Probable Adverse Environmental Effects That Cannot Be Avoided
    5.1 Adverse and unavoidable impacts
    5.2 How avoidable adverse impacts will be mitigated
6.0 Relationship Between Local Short-Term Uses of the Environment and the
    Maintenance and Enhancement of Long-Term Productivity
    6.1 Trade-off between short-term environmental gains at expense of
        long-term losses
    6.2 Trade-off between long-term environmental gains at expense of
        short-term losses
    6.3 Extent to which proposed action forecloses future options
7.0 Irreversible and Irretrievable Commitments of Resources
    7.1 Unavoidable impacts irreversibly curtailing the range of potential uses of the
        environment
        7.2.1 Labor
        7.2.2 Materials
        7.2.3 Natural
        7.2.4 Cultural
8.0 Other Interests and Considerations of Federal Policy That Offset the Adverse
    Environmental Effects of the Proposed Action
    8.1 Countervailing benefits of proposed action
    8.2 Countervailing benefits of alternatives

approaches when effects can be predicted with some degree of accuracy (or perhaps when the degree of accuracy is known) and negotiation with stakeholders when available information is sparse, uncertain, or contradictory. The approaches described in Chapters 3, 5, 6, and 7 of this book work well in the former case; the techniques described in Chapter 4 may be useful in the latter.

The Canadian Environmental Assessment Agency (1995) has proposed the following tests for environmental impacts.

An effect is considered "adverse" if it causes one or more of the following:

- Effects on biota health
- Effects on rare or endangered species
- Reductions in species diversity
- Habitat loss
- Transformation of natural landscapes
- Effects on human health
- Effects on current use of lands and resources for traditional purposes by aboriginal persons
- Foreclosure of future resource use or production

An effect is considered "significant" if it falls into one or more of the following categories:

- Large
- Geographically widespread
- Of long duration or high frequency
- Irreversible
- In a sensitive ecological context

A significant effect may be considered "likely" if it has either:

- A high probability of occurrence, or
- Significant scientific consensus about its occurrence.

Sadler (1996) recommends four steps in the evaluation of significance:

1. Assessment of the nature and extent of impacts (e.g., type, duration)

    Impacts are more likely to be significant if they are extensive over space or time.

    Impacts are more likely to be significant if they are intensive in concentration relative to assimilative capacity, or if they exceed allowable environmental standards or thresholds.

2. Assessment of the likely effects on the receiving environment (e.g., sensitive area, land use, community traditions)

   Impacts are more likely to be significant if they adversely and seriously affect ecologically sensitive areas, heritage resources, other land uses, community life-style, and/or indigenous peoples' traditions and values.

3. Assessment of the magnitude of impacts (e.g., low, moderate, high)
4. Identification of options for impact mitigation (e.g., reduction, avoidance)

Sadler also recommends incorporation of tests of significance at various stages in the EA process, with focus on the following questions:

1. How adverse are the predicted effects (e.g., change, loss, foreclosure)?
2. How do these vary in scope and intensity (e.g., in their effect on ecological and resource values)?
3. How significant or serious are the impacts (e.g., irreversible versus inconsequential)?
4. How probable is it that they will occur (e.g., high risk versus low risk)?

In a sense, therefore, environmental assessment is a risk assessment of a project before it is undertaken. The term *risk assessment* is used in the sense of formal risk analysis, in which both the effect (for instance, toxicity) and probability (risk) are taken into account in decision making. Environmental assessment seeks to identify the possible significant impacts of a project in the context of their likelihood. Section 10.4 discusses a range of methodologies commonly used for the assessment of impacts.

10.4 ENVIRONMENTAL ASSESSMENT METHODOLOGIES

Numerous authors (e.g., Jain and Urban 1975; Henderson 1982; Goodman 1984) have described methodologies available for EA. The following discussion provides an overview of the main methods currently in use, with comments about their suitability (or lack of suitability) in certain situations.

10.4.1 Matrix Analysis

Matrix methods may be the most familiar and easily understood of all EA methodologies. One of the earliest, and certainly the best known, of these is that proposed by Leopold et al. (1971) for the U.S Geological Survey. The matrix developed by these authors lists 88 environmental characteristics on the vertical axis and 100 actions with potential to affect the environment on the horizontal axis. The cells of the matrix are then filled in to indicate where actions may affect environmental characteristics.

Leopold and his colleagues proposed a method of scoring that employs a diagonal line drawn across the cell, from the lower left to the upper right corner. In the top half of the cell, a number from 1 to 10 is entered, indicating the "magnitude" of the impact. In the lower half of the cell, another score, also from 1 to 10, shows the "importance" of the impact. Higher numbers indicate greater magnitude or importance, as the case may be. A plus sign (+) can be used to indicate a positive impact. Attention can be drawn to cells with particularly high scores by circling or highlighting them so they stand out from the matrix. Figure 10.3 shows a portion of such a matrix, adapted from the matrix developed by Leopold et al. (1971).

This matrix is intended to be more or less universal. Its large size makes it impractical to use in toto, however, so its authors suggest that users develop a reduced matrix incorporating only rows and columns applicable to the project under study. The authors do not suggest summation of scores, such as might be used in other multiattribute decision-making methods (see, for example, Section 5.5.4). Indeed, they point out that cells cannot be compared directly; in other words, a score of 10 in importance in terms of "airports" may not equate directly to an importance value of 10 for railroads or canals. The values in individual cells do, however, provide a basis of comparison between various project alternatives. That is, it is possible to compare the magnitude and importance scores for a cell formed by the intersection of "erosion" and "canals" among preproject conditions, preferred project conditions, and the conditions that would obtain under various alternative approaches to the project.

Many variations of the Leopold matrix are possible. Although broadly applicable, the matrix can, for example, be made more specific for a certain kind of activity, such as a landfill site or an irrigation project. In these projects, it may be possible to itemize individual actions and to separate similar actions in different places. For example, while this matrix simply cites "paving" as one of the land transformation actions, a more detailed and specific matrix could be created by listing "paving of roadways," "paving of administration complex parking lot," and "paving of airstrip." Alternatively, the matrix could specify steps in the paving process: "laying of gravel sublayer," "compaction of gravel layer," "application of asphalt layer," and so on. Similarly, the environmental characteristic "water quality" could be broken down into individual water quality parameters, such as the indicators described in Chapters 3 and 5. These may include, for instance, fecal coliform bacteria, total phosphorus, total suspended solids, and total lead as indicators of a range of water quality concerns. Environmental components or characteristics can be added as appropriate to the climate and culture.

In short, the Leopold matrix provides an excellent starting place for EA, and one that is both easily understood and easily customized for the purposes of the project at hand. It serves as a visual summary of analyses that appear in narrative form in the EIS document. Figure 10.4 illustrates the use of a Leopold-type matrix in the Canadian federal-provincial (Nova Scotia) EA of the proposed Halifax-Dartmouth Metropolitan Sewage Treatment Facility.

Instructions:

1. Identify all actions (horizontal axis) that are part of the proposed project.

2. Under each of the actions, place a slash across each cell where an impact is possible.

3. Place a number from 1 to 10 in the upper left corner of each cell to indicate the magnitude of impact (10 is highest), using a plus sign (+) to indicate that the impact is beneficial.

4. In the lower right corner of each cell place a number from 1 to 10 to indicate the importance of the impact (e.g., regional vs. local; 10 is highest).

5. The text that accompanies the matrix should include a discussion of the significant impacts (columns and rows with many boxes marked and individual cells with high numbers).

		Modification of Regime									Land Transformation and Construction								
Physical and Chemical Characteristics	Proposed Actions	Exotic species	Habitat modification	Alteration of flows	Alteration of drainage	Irrigation	Canalization	Burning	Surface or paving	Noise and vibration	Urbanization	Airports	Roads, highways	Railroads	Transmission lines	Barriers, fencing	Dams, impoundments	Cut and fill	Blasting, drilling
Earth	Mineral resources																		
	Soils																		
Water	Surface																		
	Groundwater																		
	Water Quality																		
Air	Temperature																		
	Quality																		
Process	Flooding																		
	Erosion																		

Figure 10.3 Example portion of information matrix (after Leopold et al. 1971).

Figure 10.4 Portion of environmental effects matrix used in the Halifax-Dartmouth Metropolitan Sewage Treatment System EA (adapted from Jacques Whitford Environment Ltd. 1992).

Valued Ecosystem Components	Dredging	Infilling	Clearing, Site Preparation	Excavation, Tunneling, Blasting	Vehicle Operations	Diffuser Effluent	Combined Sewer Effluents	Air Emissions	Presence of Artificial Island
Terrestrial habitat	○	○	○	○	○				
Osprey	○	○		○	○				○
Great blue heron	○	○		○	○				
Air quality	—	○	—	○	○			○	
Marine benthic community	○	⊖		—	○	+	+		
Marine water quality	○	○		—	—	+	+		
Marine sediment quality	○	○		—	○	+	+		
Coastal physiography	—	○				⊖	⊖		○

— no impact
○ insignificant negative impact

+ significant negative impact
⊖ positive effect

10.4.2 Overlays and Geographic Information Systems

Overlays and transparencies have been used for decades to evaluate the intersection of activities and environmental components. This technique has much to recommend it in terms of simplicity and ease of use. The basic approach is to take a base map, usually a topographic map, of some appropriate scale, and to place over that base one or more transparencies on which are drawn information about environmental characteristics and/or the expected impacts of the proposed action.

For example, a base map of a subwatershed area could be overlaid with a transparency showing the location of bathing beaches, another showing historical and archeological structures, a third showing sensitive habitat, and so on. These three overlays alone will indicate the conjunction of uses of different types (heavy traffic at bathing beaches could, for instance, have adverse effects on nearby sensitive habitat). Once sensitive areas have been mapped in this way, new overlays can be prepared showing the anticipated effects of the proposed project. For instance, one transparency might show the expected range of noise effects during construction. Overlaying this on the beach/habitat maps would quickly reveal whether those uses are likely to be impacted by noise during construction.

Color can be a useful adjunct in overlay mapping. Single-color overlays can, for instance, employ a range of color intensity to indicate degree of concern (or degree of protection needed). A single-color overlay of sensitive habitat could show areas of local significance in 10% shading and highly-valued areas of national significance (perhaps endangered species habitat) in 30% or 50% shading. Similarly, impact overlays could employ graded color, so areas of minor noise impact would be shown with 10% shading, moderate impact by 30% shading, and intense impact by 50% shading. When overlaid, the maps of sensitive habitat and noise impacts will show intense colors in areas where the beneficial use is highly valued *and* the impact is likely to be intense: "significant" impact.

Table 10.2 lists a range of attributes that have been mapped using overlay techniques.

The advent of Geographic Information Systems (GIS) in the past decade has vastly increased the power of the overlay method in EA. GIS systems are essentially computerized maps containing georeferenced data of almost any kind. The term *georeferenced* implies that each piece of information—for instance, soil type—has map coordinates associated with it. Most GIS systems (and there are many different ones available) incorporate basic mapping capability, a relational database, and a suite of interpretive tools, such as statistical analyses. Using a GIS, it is possible to store a wide range of information that would formerly have been held only in map form, and then to extract that information in different combinations as necessary for a given analytical purpose.

Many, perhaps most, public agencies are now using GIS to store and display data. The technique lends itself well to visualization, and some agencies

Table 10.2   Environmental Characteristics and Impacts Suitable for Overlay Mapping

Environmental Components and Characteristics	Environmental Effects
Land ownership (public, private, reserve, etc.)	Dust
Land use	Noise
Slope	Increased traffic
Soil type	Odor
Major traffic corridors	Soil erosion
Bathing beaches	Increased flooding
National, state, and local parks	Burning
Major public structures (e.g., hospitals, universities, armed forces bases, airports, landfills, disposal sites, etc.)	Paving
	Compaction
	Blasting and drilling
Major private interests (e.g., industrial installations, agricultural lands, golf courses, commercial centers, etc.)	Removal of vegetation
	Bulldozing and leveling
	Altered slope stability
Areas of major aesthetic value	Altered vegetative cover (altered species composition)
Sites of major cultural or historical interest	Radiation
Sites of major religious or spiritual value	Impaired aesthetics
Sites of aboriginal significance	Lowering of water table
Species diversity	Increased air pollution (by parameter if desired)
Sensitive habitat (e.g., endangered species)	
Wetlands and estuaries	Increased water pollution (by parameter if desired)
Water quality (zones or individual parameter ratings)	
Floodplains	
Drinking water aquifers	

are now taking advantage of this feature to increase public access to data and collect public feedback on different planning options. A computer display terminal may, for instance, be set up in a public library or shopping center, linked to the main agency database. A member of the public can view regional and local maps on the computer, "point and click" to select an area of particular interest, zoom the display to obtain a close-up view, and then request different types of display: which of the area bathing beaches were closed to swimming in the past calendar year? which agricultural areas contain soils with a high risk of erosion? which of these are close to streams?—and so on.

GIS techniques formed the basis of a recent major study of agricultural land use and environmental impacts in the Great Lakes Basin (Great Lakes Commission 1996). The study aimed to develop an overview of agricultural practice and the impacts of agriculture on the environment of the Great Lakes Basin. A GIS was chosen as the framework for data storage and analysis, in part because GIS systems were already in use in the U.S. Department of Agriculture and Agriculture Canada, the two federal agencies that would be central in the study.

In practice, the GIS proved difficult to use. This difficulty stemmed from differences in data collection practices and GIS technology used in the two agencies. As noted earlier, there are many GIS systems available. Although the data can usually be "exported" from one and "imported" into another, these data transfers are seldom straightforward or easy. More difficult in the Great Lakes Commission's study was the problem of scale of data collection. In the United States, most data were collected at a county scale. In Canada, the watershed was the more common unit. As a result, data were difficult to overlay accurately and large areas lacked data altogether.

Finally, where one data set did not exactly match another in areal extent, orphan polygons were created in the GIS maps. Whereas in a manual overlay system, the analyst's eye may make a judgment that this little piece could be lumped in with this or that area (perhaps because the analyst has direct knowledge of the area involved), in a GIS this is much more difficult. A GIS employs computer-based computation to analyze data (for instance, to compile summary statistics for an area) and cannot automatically distinguish between an intended map unit (e.g., a county) and an orphan polygon created by default when two overlays do not match exactly. If the matching of one GIS layer to another is generally poor, many such orphan polygons will be created and the necessary computations can increase exponentially, with associated increases in computing time and costs. Techniques are available to deal with orphan polygons—for instance, by using a system of rules to merge orphan polygons with adjacent larger map units. To date, however, these techniques have been cumbersome to use. In the Great Lakes Commission work, the most detailed data was available for the U.S. Great Lakes states, and in some cases lack of data prevented analysis of the Canadian side of the basin altogether.

The technique is, nevertheless, a powerful one, primarily because of its flexibility in data display. Figures 10.5 and 10.6 show examples of GIS results from the Great Lakes Commission study.

Increasingly, GIS systems are being linked to predictive water quality models, so that predictions about in-stream impacts can be continually updated as better land use, water quality, soils, and other data become available. GIS-based agency databases therefore serve two purposes: as a data storage system and as input to predictive computer simulation models. It should be emphasized that the linkage between GIS and computer simulation models remains awkward in most cases, often requiring several intermediate "translation" steps. The technology is continually improving, however, and both GIS systems and computer models are now being developed with the goal of linkage in mind.

10.4.3 Checklist Methods

A checklist approach can be used to itemize impacts with and without a given project. Sometimes these methods are informal, and sometimes they involve complex scoring and weighting schemes. In the simplest form, a list of envi-

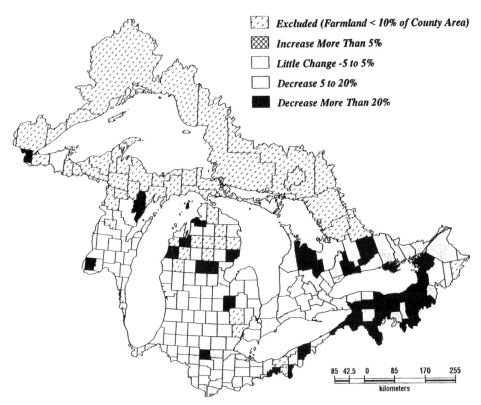

Figure 10.5   *Population density change in the Great Lakes Basin: Change in individuals per sq km of county area 1981/1982 to 1991/1992 (based on 1991 Canadian Census of Agriculture, 1992 United States Census) (Great Lakes Commission 1996).*

ronmental components is made on the vertical axis, while the horizontal axis contains the various project options. Table 10.3 illustrates such an approach.

The list of environmental components can be tailored to suit the particular environment of the project. Similarly, the "projects" along the horizontal axis can be broken down into specific activities. Separate checklists can be used for categories of activities. For instance, the impacts of accidental events can be evaluated with a checklist that itemizes environmental components down the vertical axis and individual types of accident across the horizontal axis. Table 10.4 illustrates the type of checklist that was used in the federal-provincial EA for the Halifax-Dartmouth Metropolitan Sewage Treatment Facility.

The checklist method can be adapted to include some information about the magnitude and direction of impact—for instance, by extending the range of symbols used, by replacing symbols with positive or negative numbers, or by the use of footnotes, bold type, or flags to indicate highly significant effects.

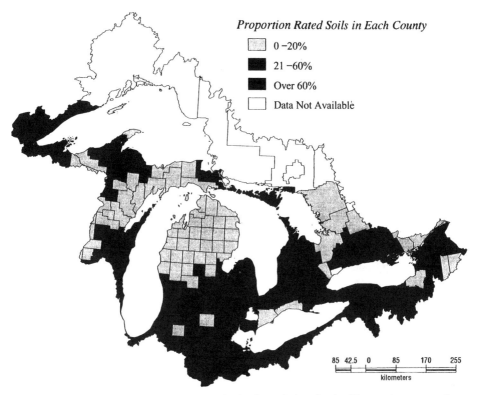

Figure 10.6 *Change in herbicide usage in the Great Lakes Basin: Change in area receiving herbicide 1981/1982 to 1991/1992 (based on 1991/1981 Canadian Census of Agriculture, 1992/1982 United States Census) (Great Lakes Commission 1996).*

10.4.4 Combination Methods

There are probably as many methods for EA as there have been environmental assessments. In part, this is because analysts discover shortcomings in existing methods and modify them to suit their own needs. The main challenge in EA methodology is not so much in data collection (although that too has its difficulties) as in data presentation. The goal is to present data on the proposed undertaking, as well as a range of alternatives to it, and to list for each the impacts on a range of environmental components and the magnitude, direction, and significance of those impacts. Ideally, the presentation should be simple so that it is accessible by any member of the community. This is a tall order indeed.

As one example of a "combination" method loosely based on the Leopold matrix and checklist methods, Jacques Whitford Environment Ltd. (1992) developed a scheme for cataloging the impacts of a proposed regional sewage treatment facility on the busy Halifax (Nova Scotia) Harbor. The harbor has a thriving commercial fishery and complex hydrodynamics that would likely

Table 10.3 Checklist Method for Environmental Assessment

Environmental Component	Impact Without Project	Impact with Project A	Impact with Project B	Impact with Project C
Terrestrial habitat		•	•	•
Aquatic habitat		•	•	•
Groundwater flows		•		
Surface water flows	•	•		
Groundwater quality		•		
Surface water quality		•		
Sediment quality	•	•	•	•
Endangered species		•	•	•
Air quality			•	
Noise			•	
Odor			•	
Dust			•	
Human health		•	•	•
Historical and cultural artifacts				
Commercial interests			•	•
Industrial interests				
Aboriginal communities		•	•	•
Residential communities			•	

• indicates that impact is expected.

be altered by discharges from a large-capacity sewage treatment plant. Jacques Whitford's scheme employs several elements:

- A symbol to indicate direction and magnitude of impact
- A numerical value for geographical extent of the impact
- A numerical value for the duration of the impact
- A numerical value for the frequency of impact
- A number indicating the basis of the ratings (i.e., professional judgment only, available data and professional judgment, abundant data supported by professional judgment)

An example of the Jacques Whitford scheme appears as Figure 10.7. The method is not easily understood without frequent reference to a legend, but it does capture several different considerations in a single matrix cell and thus extends the Leopold matrix approach significantly. In an effort to reduce the complexity of impact-assessment matrices, however, Jacques Whitford has simplified the list of environmental components to the major elements listed in Table 10.4. This approach may underestimate the impacts on specific environmental components (for example, a particular water quality parameter) while overemphasizing the importance of "valued environmental components" such as the great blue heron and osprey that were selected for inclusion.

Table 10.4 Checklist of Environmental Impacts (After Jacques Whitford Environment Ltd. 1992)

Valued Ecosystem Component	Fire or Explosion	Petrochemical or Fuel Spills	Spills of Other Hazardous Materials	Failure of Emissions Treatment	Breaks in Collector System	Breakage of Diffuser	Interruption of the Oil-from-Sludge Process	Electrical Outage
Terrestrial habitat	●	●						
Osprey	●	●						
Great blue heron	●	●						
Air quality	●	●		●				
Marine benthic community	●	●			●	●		
Marine water quality	●	●			●	●		
Marine sediment quality					●	●		
Coastal physiography	●	●						

● indicates that impact is expected.

Note: This combination methodology uses a rectangle containing a symbol indicating probable impact (see Figure 10.4) and four numbers providing additional information on the geographic extent, duration, and frequency of the impact, and the basis for these ratings.

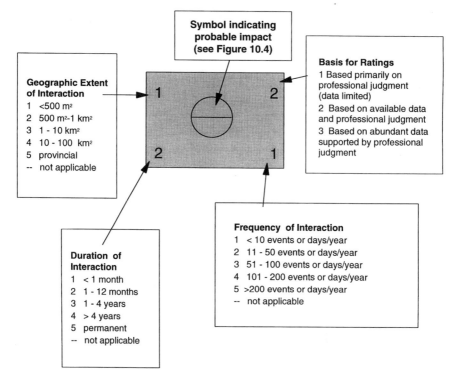

Figure 10.7 *Example "combination" EA rating methodology (after Jacques Whitford Environment Ltd. 1992).*

Goodman (1984) offers a helpful review of various EA applications, showing how basic methodologies have been modified to suit the needs of particular EA processes.

10.5 SOCIAL IMPACT ASSESSMENT

Social impact assessment is now an integral part of most water resources planning exercises and, indeed, is required by law for many. Such an assessment provides a way for government agencies and other proponents of major projects to anticipate impact on people and communities, especially when the benefits and costs of a project are not equally distributed. It is often the case that major multiobjective public projects have benefits distributed over a wide geographic area, but impacts that are concentrated in the local community. The same is true of many major private projects such as mine developments, pulp and paper mills, and similar undertakings. While these projects may have enormous economic, and therefore social, benefits, those benefits seldom accrue entirely to the community directly affected by the project. Social impact assessment attempts to describe the effect on the community of such projects.

As described earlier in this chapter, impact assessment is based on values: what is a significant impact to one group may be considered a negligible impact, or even ignored, by another. Many groups bring values to the planning table. The affected community usually has at least one group, and may have several different groups, whose perspectives are informed by their fundamental views of what is right and wrong. Agency staff and other public officials have perspectives that differ again, perhaps because they incorporate consideration of a wider geographic region or an emphasis on service provision over other considerations. Corporate, commercial, and institutional interests can all be defined in some processes. Each of these groups may see an effect as "good" or "bad," depending on its perspective.

Chapter 4 described a range of techniques suited to public involvement and public consultation in water resources planning. Many of these also have value in formal social impact assessment. Goodman (1984) suggests that the 12 social science tools most frequently used in this context are:

1. Institutional analysis
2. Policy profiling
3. Value mapping
4. Social profiling
5. Content analysis
6. Small-group process techniques
7. Human-cost accounting
8. Community impact assessment
9. Ethnographic field analysis
10. Questionnaire and nonparametric statistical analysis
11. Population projections
12. Trend and cross-impact analysis

The first three of these methods were discussed in Chapter 4 and represent (along with (6) and (10)) methods of determining community response to, and concerns about, a proposed project. Chapter 3 discussed methods for compiling population projections.

Social profiling and content analysis attempt to map the range and depth of public concerns in a systematic fashion. (Use of the public involvement techniques described in Chapter 4 can produce a vast quantity of information that is difficult to interpret and understand without a content-analysis framework.)

Social profiling is discussed in more detail by Flynn and Schmidt (1977); Dunning (1982) provides a review of content analysis techniques.

Human-cost accounting attempts to measure the psychological and behavioral impacts of a project. Affected individuals are positioned on a trauma scale, which is then related to standard benefits scales paid by agencies like the U.S. Veterans Administration for similar disabilities. Other human health impacts can be translated into costs using hospital-bed-days and estimates of the costs of economic disruption. Community impact assessment examines demographic changes in the community (for instance, worker influx) as a result of project activities. As described in Chapter 3, worker influxes (as one example) can be translated into demand for goods and services of various kinds, and thus can be used to forecast other kinds of community impacts.

Ethnographic field analysis, a method of on-site observation of community structure and culture, usually requires the assistance of an expert.

Some of these techniques (e.g., social profiling, small-group processes, questionnaires) are well suited to defining problems. Others (e.g., small-group processes, policy profiling, and values mapping) are useful in generating alternatives (and see Chapter 5 of this book). Most of the techniques are useful in compiling an inventory of the society that will be affected by the project. The last nine of the list have particular value in analysis and comparison of plans and in selection of a preferred alternative.

As with environmental impact analysis, impacts that may be considered include both "direct" (immediate, obvious) impacts, such as those related to population relocation, major demographic changes, and economic trends, and "indirect" (delayed, less obvious) impacts, such as impacts on the health care and education systems, on community attitudes and values, and on "quality of life." Both kinds of impacts may be assessed over few years (short term) or many (long term).

Table 10.5 lists the social science variables commonly considered in social impact assessment.

Social impact occurs at the level of the individual, the family, the neighborhood (or community group), the community, and the region. At the *individual or family level*, Maslow (1954) suggests a hierarchy of needs, beginning with basic survival and physiological needs, followed by security and self-sufficiency needs, then by social interaction needs, then by esteem, and, finally, growth or self-actualization. At the *community level*, concerns center on community cohesion and the structures and values that promote that cohesion. These may include considerations such as community stability, opportunities for public participation in the community, and the presence or absence of community structures to enhance cohesion. At the *regional level*, analysis generally focuses on patterns of economic growth, employment, infrastructure, and demographic changes.

Any of the basic methodologies described in Section 10.3 can be used in social impact assessment, with the "environmental characteristics" or "components" replaced with social characteristics or components—for instance, those

Table 10.5 Typical Social Impact Components and Indicator Variables Used in Social Impact Analysis (After Goodman 1984)

Component	Indicators
A. Urban and Community Impacts	
1. Income distribution General Low-income households	General income, average income levels, income distribution
2. Employment distribution General By ethnic group	Employment levels, unemployment levels, employment of different ethnic groups, availability of suitable employment
3. Population Geographic distribution Composition	General, number, trends, demographic characteristics, mobility and migration, housing, future projections
4. Fiscal condition of government State Regional Local	Taxes, community finances, degree of indebtedness
5. Quality of community life Attitudes Infrastructure Facilities Organization/disorganization	Attitudes toward community, perceptions of effects, general lifestyle patterns in the area, crime, displacement, poverty, mass communication, transportation facilities, formal and informal community structures, health care, education, sanitation, public utilities, welfare, recreational opportunities, cultural opportunities, aesthetics, real or perceived impact on particular subpopulations, economic resources of the area (timber, fish, mineral, etc.)
B. Life, Health, and Safety	
Risk of flood, drought, or other disaster	Attitudes and perceptions about risk and impacts, safety services of police, fire protection, health and sanitation
C. Displacement	
1. Of people	Attitudes and perceptions about risk and impacts, community disorganization, and infrastructure
2. Of businesses	Attitudes and perceptions about risk and impacts, community disorganization, and infrastructure
3. Of farms	Attitudes and perceptions about risk and impacts, community disorganization, and infrastructure

in Table 10.5. The applicability of matrix and checklist methods will be easily seen, but overlay and GIS-assisted analysis can also be used, with overlays showing average income level, population density, land use, and other features.

In summary, social impact assessment seeks to define the social systems that currently exist in a watershed or other geographic area and to determine how management actions will alter those social systems. In some jurisdictions (for instance, Ontario) social and commercial systems are included in the definition of "environment" under environmental assessment legislation. In other jurisdictions, separate legislation governing social impact assessment may apply.

10.6 STRATEGIC ENVIRONMENTAL ASSESSMENT

Strategic environmental assessment (SEA) is an emerging area with significant potential to affect water resources management. As discussed in Section 10.2, SEA is a process of prior examination and appraisal of policies, plans, and programs and other higher-level or preproject initiatives. According to Sadler (1996), there is still considerable debate as to the scope of SEA and its potential application to environmental policy, as compared with plan or program levels. Sadler makes the case that SEA of policies, plans, and programs is necessary and important because it:

- Builds environmental considerations into all levels of decision making, not just project approvals
- Helps to determine the need for the feasibility of government initiatives and proposals
- Avoids the foreclosures of options and opportunities that arise when assessment occurs only at the project stage
- Addresses environmental issues and impacts that are best dealt with or can be considered only at the policy or program level (for example, those that cannot readily be divided into projects)
- Establishes an appropriate context for project EIA, including pre-identification of issues and impacts that are likely to warrant detailed examination

In Canada, the federal government established a process of policy and program assessment as part of a Cabinet directive of 1990; a parallel review process is not in place at the provincial level. This federal process requires all agencies submitting policy and program proposals to document the environmental impacts of those policies or programs. The actual level of compliance with this policy is difficult to determine, but it is clear that not all federal agencies carry out the required policy EA to the level intended in the original Cabinet directive.

In the European Union, internal policies call for EA of new legislation and actions likely to affect the environment. Member states are increasingly incorporating SEA-type reviews of policies, plans, and programs.

In the United Kingdom, there are no formal policy requirements for SEA, but the Department of the Environment has issued "good practice" guidelines that require EA of policies drafted by government agencies. Policy assessment incorporates benefit-cost analysis, scoping, identification of issues, and development of a policy impact matrix.

In the United States, the National Environmental Policy Act of 1969 (NEPA) in theory applies to all legislation and other major federal actions, but in practice only program-level EIS reports (PEISs) are prepared. These cover a wide range of resource management, waste disposal, and other types of environment-related programs and are increasingly used in long-term planning, dealing with cumulative effects, tiering actions requiring project EIA, and evading costly litigation (Sadler 1996). Sadler also notes that, to date, NEPA requirements have not been applied to broad government policies. Other than in California, SEA is not actively practiced at the state level.

SEA is hampered in many jurisdictions by a lack of political will and a limited societal support base. There is, however, considerable support for the notion among EA professionals. A typical SEA process contains the following elements:

1. Accountability by agencies who initiate policies, plans, or programs for assessing the environmental effects of those initiatives
2. Assessment of effects early in policy, plan, or program design
3. Scope of the SEA sufficient to include all potential impacts or consequences of the policy, plan, or program
4. Clear objectives and terms of reference for the SEA
5. Inclusion of a full range of effects, including biophysical, economic, and social, as appropriate for the policy, plan, or program
6. Provision for public involvement
7. Public reporting of the SEA and decisions
8. Explicit inclusion of environmental factors in policy-making
9. Tiering to bring SEA to the level of project EIA where possible
10. Monitoring and follow-up of measures
11. Independent oversight of process implementation, agency compliance, and government-wide performance

Sadler (1996) suggests the following methods for SEA:

1. *Literature search*—to determine the state of knowledge and the probable linkages between policy actions and environmental effects. State-of-the-environment reports and environmental policy plans may be helpful adjuncts.
2. *Expert judgment*—using Delphi techniques, workshops, or other small-group techniques (see Chapter 4).
3. *Analytical techniques*—including scenario development, checklists and matrices, GIS, benefit-cost analysis, multicriteria analysis, or indicator variables to determine probable areas of impact.

4. *Consultative tools*—such as interviews, workshops, key informant interviews, sectoral discussions, multistakeholder roundtables, and similar techniques.

10.7 MONITORING AND FOLLOW-UP

No matter how careful or comprehensive the EA, it can only be guesswork until the project is actually in place. Postproject monitoring and follow-up is an essential component of the overall EA, both to determine compliance with legal requirements imposed as a result of the EA and to identify any new or unforeseen effects. Postproject monitoring is also important to determine the effectiveness of mitigation measures that may have been simulated but not actually tested in the field.

Ideally, EA is an iterative process through which impacts are assessed, and later reassessed when water management changes are again contemplated. Postproject EA is essential to this iteration. Without it, an EIS becomes a costly one-time effort, not the basis for ongoing guidance for future planning initiatives.

Several types of monitoring may be required:

1. Compliance with legal obligations, permits, policies, and similar requirements
2. Monitoring of rare or endangered species habitat, sensitive or highly valued habitat (e.g., wetlands, estuaries), wilderness areas, historical sites, and sites of particular spiritual value, to determine whether impacts are occurring at or in excess of anticipated levels
3. Monitoring of impacts that were and continue to be highly controversial, to ensure that adequate mitigation is in place and to provide additional data to inform the scientific or public debate
4. Monitoring to confirm the effectiveness of mitigation measures
5. Remapping of altered environmental conditions following project implementation; i.e., as the basis for a future EA or to inform the design of future monitoring or mitigation programs

Normally, requirements for monitoring and follow-up are specified at the conclusion of the EA process and are designed around the particular issues, controversies, and biophysical, social, and economic environments affected by the project. Where issues have remained unresolved, or where particularly sensitive species, habitats, or human environments are affected, monitoring and follow-up programs may be especially detailed in space, time, and range of indicators. A typical follow-up program includes several components:

- Surveillance and inspection to ensure that EA terms and conditions are complied with

- Routine, repetitive monitoring of the affected environment to measure changes in selected indicator variables
- Compliance monitoring, to ensure that legal requirements (permits and approvals) are met
- Response to unacceptably altered environmental conditions
- Third-party audit of surveillance and monitoring results

In a sense, then, post-EA monitoring and follow-up provides a quality control function for the EA process and generates an ongoing source of information that provides both the basis for response and mitigation and a foundation for future planning initiatives.

10.8 THE SHORTCOMINGS OF ENVIRONMENTAL ASSESSMENT

A Lack of Long-Term Data   One of the most serious weaknesses of EA is that it often relies on small data sets. For many assessments, intensive data collection begins only when the EA process is initiated, ending when the assessment ends. This means that, at most, one to two years of data are used as the foundation for assessing "impacts" on systems that may be highly complex. This weakness is particularly significant in terms of water resources impacts, which may be seriously underestimated if long-term flow and quality patterns are not well understood.

Failure to Capture Long-Term Environmental Effects   The two weaknesses of limited databases and simple assessment tools mean that EA cannot accurately predict long-term environmental effects, even though the general direction and likelihood of effects may be known. The difficulty arises in determining whether long-term effects are likely to be "significant" or "negligible." Indeed, it is this point that often forms the focus of public opposition to proposed public activities: an agency charged with responsibility for EA has decided that long-term effects will be minimal and the project should go ahead, while members of the community believe that long-term effects will be serious and irreversible, so the project should be rejected. EA methodology does not currently provide an objective way to prove one outcome over the other.

Failure to Capture Subtle Environmental Effects   Environmental assessment is, necessarily, a rather crude tool, typically based on a modest data-collection exercise (because of resource limitations) and employing largely guesswork about the future impacts of proposed actions. The further into the future these impacts are assessed, the less accurate the predictions become.

Even gross effects, such as alterations in sedimentation regime, may be difficult to predict with accuracy. Subtle effects, such as the altered flavor of a traditional meat, may be impossible to capture with typical environmental assessment methods such as overlays and matrices.

Failure to Identify and Incorporate Indirect Effects   The vast complexity of natural ecosystems cannot be understood in a year or two of data collection and analysis. Consequently, it is simply impossible for the analyst to understand and predict all possible consequences of a proposed action. Direct effects—for instance, those related to excavation, blasting, heavy vehicle traffic, infilling, inundation, and similar activities—may be understood in a general sense. But indirect effects—for instance, the impacts on human health of long-term exposure to dust and noise—may be difficult, if not impossible, to anticipate, even though they may eventually be devastating in the community.

An Emphasis on Economic Development, Sometimes at the Expense of Ecological Integrity   As Tollan (1993) points out, an EIA seldom questions economic development as such or considers the long-term goals for healthy ecosystems. This has led to a widespread emphasis on "mitigation" of impacts and, thus, a presumption that impacts are necessary if economic development is to occur. Tollan emphasizes the need to use the "precautionary principle" in EA. The precautionary principle appears as Principle 15 of the Rio Declaration on Environment and Development, signed in Rio de Janeiro in June, 1992, at the United Nations Conference on Environment and Development:

> The precautionary principle states that where there are threats of serious or irreversible damage, lack of full scientific certainty should not be used as a reason for postponing measures to prevent environmental degradation. The precautionary principle is seen as a basis for sustainable development.

Too often, EA processes instead rely on postproject monitoring to detect problems, and on reactive measures to correct them. The precautionary principle would instead mean that where serious or irreversible damage may occur, the actions likely to cause that damage should not proceed. In other words, lack of scientific certainty about the outcome of an action should not be used as a reason to go ahead with the action.

Failure to Incorporate "Traditional Ecological Knowledge"   As suggested in earlier sections of this chapter, environmental assessment is founded in values. The decision as to when a "valued resource" has been "impacted" depends very much on the perspective of the analyst. As many authors (e.g., Goodman and Edwards 1992; Sadler 1996) have pointed out, modern environmental assessment methodologies are rooted in the "Western-knowledge" tradition developed in cultures that are economically advanced. Typically, this means establishing a hypothesis ("no impact will occur"), designing a study to test that hypothesis, collecting data in a systematic fashion, drawing conclusions from the results of this study, and reevaluating the original hypothesis in light of those conclusions.

This scientific-method tradition is alien to many cultures. Many cultures, for example, rely on an oral tradition through which knowledge of the environment is passed from elders to younger members of the society. "Impacts" as judged in such a society may be tangible and significant, but different from those assessed by a scientist from the Western-knowledge tradition. An example of such an impact is altered taste of caribou flesh following low-level flying in Labrador, Canada (Larry Innes, Environmental Coordinator, Innu Nation, Sheshatshiu, Labrador, May 15, 1997; personal communication). Innu elders report subtle changes in the flavor of this meat, which is central to the Innu culture (as discussed in Section 10.2, the Innu are primarily a hunter-gatherer society, heavily reliant on caribou). This subtle change would not likely be apparent to a Western scientist even if attention were drawn to it, yet it is a serious impact as judged by the Innu.

Such a failure to incorporate "traditional ecological knowledge," as it is known, is an acknowledged weakness in current environmental assessment methodologies. Traditional environmental knowledge bases frequently incorporate many decades, or even centuries, of observation of environmental phenomena, often passed down through story telling or other oral traditions. They are thus immensely valuable in extending the short-term "scientific" observations that may be made during the data-collection phase of a typical environmental assessment process. Additional research and guidance from aboriginal communities is needed to improve the ways that this rich knowledge base can complement Western-knowledge approaches.

Sadler (1996) documents the failure of contemporary EA methodology to predict impact on the environment. In his study, which was commissioned by the Canadian Environmental Assessment Agency and the International Association for Impact Assessment, Sadler conducted a survey of EA practitioners throughout the world, concluding that:

> Current [environmental impact assessment] practice is unsuccessful or only marginally successful in making verifiable predictions, in specifying the significance of residual impacts, and in providing advice to decision makers on alternatives.

Whatever its weaknesses, however, EA forces us to examine alternatives and systematically review their advantages and disadvantages. This is a fundamental change from the project evaluation methodologies in use three decades ago and goes a long way toward fulfilling the goal stated by Congress when it passed the National Environmental Policy Act in 1969.

REFERENCES

Baber, W. 1988. Impact assessment and democratic politics. *Impact Assessment Bulletin* 6:172–178.

Canadian Environmental Assessment Agency. 1995. *A Guide to the Canadian Environmental Assessment Act.* Ottawa: Queen's Printer.

Clément, Daniel. 1990. *L'ethnobotanique montagnaise de Mingan*. Report 53 (Collection Nordicana). Québec: Centre d'Etudes Nordiques Université Laval.

―――. 1995. *La zoologie des Montagnais*. Paris: Editions Peeters.

Dunning, C. M. 1982. Content analysis. In *Social Impact Assessments Training Manual*. Fort Belvoir, Va.: U.S. Army Corps of Engineers, Institute for Water Resources.

Flynn, Cynthia B., and Rosemary T. Schmidt. 1977. *Sources of Information for Social Profiling*. Fort Belvoir, Va.: U.S. Army Corps of Engineers, Institute for Water Resources.

Freeman, Milton, R., and Ludwig W. Carbyn, eds. 1988. *Traditional Knowledge and Renewable Resource Management in Northern Regions*. A joint publication of the IUCN Commission on Ecology and the Canadian Circumpolar Institute (formerly the Boreal Institute for Northern Studies). Occasional Paper No. 23. Alberta: University of Alberta.

Gibson, Robert, and Beth Savan. 1987. *Environmental Assessment in Ontario*. Toronto: Canadian Institute for Environmental Law and Policy.

Goodman, Alvin S. 1984. *Principles of Water Resources Planning*. Englewood Cliffs, N.J.: Prentice-Hall.

Goodman, A. S., and K. A. Edwards. 1992. Integrated water resources planning. *Natural Resources Forum* 16(1):65–70.

Great Lakes Commission. 1996. *An Agricultural Profile of the Great Lakes Basin: Characteristics and Trends in Production, Land-Use, and Environmental Impacts*. Ann Arbor, Mich.: Great Lakes Commission.

Henderson, James E. 1982. *Handbook of Environmental Quality Measurement and Assessment Methods and Techniques*. Publication IR-E-82-2. Vicksburg, Miss.: Environmental Laboratory, U.S. Army Engineer Waterways Experiment Station.

Hilden, M. 1996. *Evaluation of the Significance of Environmental Impacts: EIA Process Strengthening*. Canberra: Environment Protection Agency.

Inglis, Julian T., ed. 1993. *Traditional Ecological Knowledge: Concepts and Cases*. International Program on Traditional Ecological Knowledge and International Development Research Council. Ottawa: Canadian Museum of Nature.

Jacques Whitford Environment Ltd. 1992. *Environmental Assessment Report for the Halifax-Dartmouth Metropolitan Sewage Treatment Facility*. Vol. II, *Impacts and Mitigation*. Consultant report for Halifax Harbour Cleanup Inc. Dartmouth, N.S.: Jacques Whitford Environmental Ltd.

Jain, R. K., and L. V. Urban, 1975. *A Review and Analysis of Environmental Impact Assessment Methodologies*, Technical Report E-69. Champaign, Ill.: Construction Engineering Research Laboratory.

Leopold, Luna B., Frank E. Clarke, and Bruce B. Hanshaw. 1971. *A Procedure for Evaluating Environmental Impact*. U.S. Geological Survey Circular 645. Washington, D.C.: US Geological Survey.

Maslow, Abraham H. 1954. *Motivation and Personality*. New York: Harper.

Sadler, Barry. 1996. *Environmental Assessment in a Changing World: Evaluating Practice to Improve Performance: Final Report*. International Study of the Effectiveness of Environmental Assessment. Ottawa: Canadian Environmental Assessment Agency and the International Association for Impact Assessment.

Tollan, Anne. 1993. The ecosystem approach to water management. *World Meteorological Organization Bulletin*: 28–34.

U.S. Council on Environmental Quality. 1981. *NEPA's Forty Most Asked Questions.* Memorandum to Agencies. Washington, D.C.: U.S. Council on Environmental Quality, Executive Office of the President.

———. 1993. *Environmental Quality*. Twenty-third Annual Report. Washington, D.C.: CEQ.

Vanclay, F., and D. Bronstein, eds. 1995. *Environment and Social Impact Assessments.* London: John Wiley & Sons.

World Business Council on Sustainable Development. 1995. *Environmental Assessment: A Business Perspective*. Geneva: World Business Council on Sustainable Development.

11

Choosing the Best Plan

Among the many decisions that must be made in watershed management, the choice of the "best" management approach is in many ways probably the most difficult and controversial. Chapter 10 described the differences in values that may underlie discussion about water management strategies, and these differences are not easy to resolve. This chapter examines approaches to choosing a "best" plan from the alternatives that were developed (Chapter 5) and tested (Chapters 6 and 7).

11.1 EVALUATING THE EFFECTIVENESS OF OPTIONS AND STRATEGIES

Chapters 3 and 5 described the need to decide what conditions are desired in the watershed, and to select indicators of progress toward those goals. Often, but not always, there is widespread agreement on these goals: everybody wants the beaches "swimmable" again; everybody wants to be able to catch fish in the river. It is at the stage of choosing the best approach to meeting these goals that the need for good indicators becomes most apparent. What does "everybody" mean by "swimmable," for instance? Values may underlie even the most basic assumptions about the plan.

Possibly as difficult is the question of power in the decision-making process, which includes sensitive issues such as who calls and chairs meetings, which viewpoints are allowed to be heard, and which are summarily dismissed. Sometimes, regulatory agencies have the legal responsibility to make decisions about water management planning (for example, for navigational purposes) and may feel that they are weakening their mandate by allowing different views into the discussion. As Arnstein (1969) has pointed out, allowing more citizen involvement in the planning process amounts to delegation of decision-making power. Not all regulatory agencies are comfortable with this delegation. Yet successful implementation of the watershed management strategy (Chapter 12) may depend on the social consensus that underlies policy.

As discussed in Chapter 3, it is therefore essential that the wider community be involved in decisions about desired outcomes and the indicators used to track them. The International Joint Commission (IJC) has recently devoted several years to the social and technical problem of defining desired outcomes and indicators for the Great Lakes (e.g., IJC 1994a, 1994b) and continues to conduct research on how a system of indicators can best be implemented.

Assuming that the desired condition of the watershed is known or guessed, that some set of specific targets to reflect successful attainment of those conditions has been developed, and that a number of management options have been identified, the following steps can be used to reduce the number of possible management options to a workable group. (Refer to Chapters 5, 6, and 7 for more detailed discussion of the terms used in the following section.)

Step 1: Apply Planning Constraints Chapter 5 discussed the general problem of establishing planning constraints, and Chapters 6 and 7, respectively, provided advice on application of constraints in simple and detailed planning initiatives. The constraints essentially constitute a set of limits, preferably supported by the community, beyond which the plan may not go.

Application of constraints is usually a straightforward matter requiring little detailed analysis. Table 5.6 illustrates a simple example of such an analysis. Typically, application of constraints involves ruling out options that:

- Are inappropriate for the physical conditions (soils, climate, infrastructure, development, etc.) of the site
- Are too costly
- Take too long to implement
- Do not meet basic performance requirements (for instance, for maintenance of desired flow or water levels or for removal of specific water pollutants)

As important as the actual analysis of options is communication with, and involvement of, the community in decision making. If a government agency is undertaking the analysis without active participation by the public, all decisions should be clearly documented and explained in a report that is available for public and agency scrutiny and comment. Sufficient time must be allowed for public and agency review and debate. (If this requirement seems onerous, decision makers should recall that the development of a watershed management strategy is, as Bishop (1970) suggests, a process for creating social change, which must logically be endorsed by the affected society.) Public and agency commentary can provide a rich source of feedback and can identify issues or options overlooked by the decision makers, who may wish to revisit and revise the decisions made and issue an amended document to record these revisions.

Step 2: Develop a Framework for Application of Decision Criteria

The application of planning constraints should have reduced a large and confusing list of management options to a shorter list of options that are clearly feasible. The remaining analysis is directed at determining which of these best suits the needs of the community, as reflected in the decision criteria developed by the planning team.

In most cases, decision makers are attempting to make decisions that will continue to be effective for many years into the future. To do so, they must decide *what* future they want to plan for and how many intermediate "futures" should also be assessed. As discussed in Chapters 5, 6, and 7, this requires decisions about:

- An appropriate planning horizon
- An appropriate discount rate unless risk and uncertainty will be incorporated in the analysis (see Chapter 8)
- Any specific planning assumptions to be used in the analysis (for example, that "best management practices" will be assumed to have a certain level of acceptance in the community)

There is no single correct approach to these decisions. A longer planning horizon is useful for major projects and structures, but more distant futures are less certain in terms of population forecasts, prevailing interest rates, demand for goods and services, and other key planning factors. Many detailed watershed planning initiatives therefore incorporate several planning horizons, including a short-term future (2 to 5 years, highly predictable), a midrange future (5 to 10 years, less certain but still foreseeable), and a long-term future (25 to 50 years, very uncertain but desirable in terms of the expected lives of major structures). Decision makers must therefore:

- Decide whether performance will be evaluated under one or several future development conditions
- Define the characteristics of present and future development conditions, including land use planning forecasts, population densities, and location of major watershed features and structures
- Identify any other planning assumptions that will be made in evaluating these conditions—for example, public attitudes toward resource conservation or economic development

Certainly, future conditions can be chosen more or less arbitrarily: 5 years, 10 years, and 25 years, for example, with an average discount rate of 12% for

Table 11.1   Example Set of Present and Future Scenarios under which Alternative Management Strategies are Evaluated

Scenario	Plan A	Plan B	Plan C	Plan D
Present case				
5 years into future				
25 years into future— optimistic				
25 years into future— most likely				
25 years into future— pessimistic				

all. But planners may also wish to incorporate uncertainty in these analyses. For this reason, some analyses include future scenarios with low-medium-high, or best-typical-worst, projections. Table 11.1 illustrates how establishing these scenarios builds the foundation for the evaluation of management alternatives.

What values should be used to fill in the cells of this matrix? A single value would rarely suffice to represent the performance of management strategy across a range of criteria. Several approaches are possible:

1. Prepare a matrix like that in Table 11.1 for each decision criterion.
2. Prepare a matrix like that in Table 11.1, but use several columns under each "Plan" heading to display performance on various decision criteria.
3. Compile an "index" by summing performance on all criteria, using weights if appropriate.

Step 3: Develop a List of Management Scenarios to Be Modeled
Although the general future cases (Present case, 5 years into the future, etc.) are easily developed, these are not sufficient for detailed testing of scenarios. The goal of the analysis is to determine the performance of each individual management option, and combination of options ("plan" or "strategy"), on each decision criterion. And each of these cases must be examined under each of the planning horizons shown in Table 11.1. Clearly, scenario testing can rapidly evolve into a complicated and time-consuming process. Thus, it is essential that it be carried out in a systematic fashion, with careful recording of each step of the analysis (for instance, each computer simulation run).

The analyst must therefore develop a comprehensive list of scenarios that will be tested under each future condition. These should include:

- "Do nothing" cases for the present and for any separate future scenarios.
- Scenarios that examine the impact of a single management option applied to the greatest possible effect ("best case" for that option). For example,

the analyst could assume 100% adoption of conservation tillage in cropland within the watershed.

- Scenarios that examine the impact of a single option applied to a portion of the basin (e.g., a portion of the total stream length; or 50% of farms; or some similar scenario). This type of scenario might be used to model various levels of stakeholder (e.g., farmer) willingness to implement a given management practice.

Step 4: Test the Performance of Each Individual Management Option   One way to reduce the complexity of scenario testing is to review the performance of each individual management option first, and then include only very effective (and very inexpensive but moderately effective) measures in compiling plans or strategies. It is usually the case that one option will perform differently on different criteria—in other words, will not be "best" on all factors. This problem is illustrated in Table 5.8 and discussed in Section 5.5.3. The decision as to whether to include an option in subsequent plan development will depend on whether the option performs well on criteria deemed to be important (heavily weighted) in the analysis, on cost and availability of funding, and on other factors such as ease of implementation and degree of public acceptance. The analyst should eliminate from further consideration those options that

- Do not perform well on important criteria and
- Are expensive or difficult to fund, or
- Are not well supported by the community.

Testing should include both present conditions and the various future conditions of interest (see the example in Table 11.1). We could, for example, extend Table 5.8 to illustrate this concept, as shown in Table 11.2.

If many management options and future conditions are under consideration, the number of scenarios to be tested can be very large and there will be a temptation to cut corners and skip to the "obvious" conclusion. There are several reasons to guard against this tendency:

1. Each decision in the option-screening process must be clearly documented and transparent in order to ensure full accountability to the public and public agencies.
2. Cutting corners may lead the analyst to ignore less effective but low-cost and easily implemented options which may, nevertheless, have value in the development of management plans.
3. Lack of a systematic approach to option testing can result in whole categories of options being overlooked altogether.

Decisions to keep or eliminate individual options should be ratified with regulatory agencies and the community and should be fully documented and explained in the final planning report.

Table 11.2   Application of Evaluation Criteria in the Screening of Management Alternatives, Showing Inclusion of Performance under Several Possible Future Conditions

Management Alternative (Strategy)	Annualized Cost*	Suspended Sediment Reduction Efficiency			Phosphorus Reduction Efficiency		
		Present	10 years	25 years	Present	10 years	25 years
1. Do nothing (retain status quo).	$0	0%	0%	0%	0%	0%	0%
2. Construct buffer strips.	$50	55%	35%	25%	35%	25%	15%
3. Construct livestock exclusion fencing.	$35	35%	45%	55%	45%	45%	55%

*Note that consideration of annualized cost eliminates the need for comparison of cost at different points in time.

Step 5: Conduct Final Testing   Having decided which individual options are worth including in a management strategy and which can safely be eliminated from further consideration, the analyst is faced with the challenge of developing credible management strategies. To a large extent this is a subjective process, based on the interests of decision makers and the community and on considerations of plan implementation. Individual options can, for example, be grouped into strategies that focus on a particular pollution source, a particular beneficial use, or some similar rationale. For example, all agricultural tillage and crop management measures can be grouped into one scenario, incorporating the most likely levels of farmer acceptance in the watershed, and their effectiveness examined under present conditions, 5 years into the future, and 25 years into the future. The far-distant future might be examined as a worst-case estimate (e.g., maximum population growth, minimum adoption of measures by farmers) or some other scenario viewed as likely by the decision makers. Similarly, all urban stormwater management options can be grouped into a second scenario that is tested under present and future conditions, and so on. The advantage of testing like options together is to evaluate the maximum effect possible with agricultural or urban controls, respectively. With the results of this testing, it is possible to estimate, for example, the maximum reduction possible in a given pollutant if a suite of agricultural measures were implemented. This in turn allows recalculation of the relative loadings of pollutants to a receiving water under present and future management scenarios.

When this information has been obtained, "combination" plans can be compiled, incorporating a range of measures across the watershed. Here the choice is mainly subjective and will be strengthened by frequent and careful review by agency and community representatives. (The AEAM process described in Section 3.4.2 in fact uses scenario development and testing as the focus of a series of workshops. Copies of the results of computer simulation are not retained, but instead repetitive scenario testing forms the framework for a broad dialogue on issues and solutions. Although this book generally advocates a more structured approach to problem solving, the AEAM process provides many useful lessons about the process of community-agency interaction in the development and testing of management strategies.)

In the author's experience, cost can be a useful framework for plan development. Several alternative management strategies can be set up, reflecting different levels of expenditure: a minimum-cost plan, a moderate-cost plan, and a "bells and whistles" plan, for example. This approach also has value in providing a range of options for the consideration of funding agencies, because it clarifies the relationship between level of expenditure and resulting environmental improvement.

It is particularly important at this final stage of plan development and screening that the public and interested agencies have an opportunity to comment on the strategies being tested (including recommending different strategies for evaluation) and any assumptions about present and future conditions. The analyst should not regard this involvement as burdensome or interfering, but rather as an opportunity to build community support for, and consensus on, effective management strategies. This is particularly important when the strategies under consideration include those with potential to displace communities or individuals (for instance, major dam projects) but which may have benefits in proportion to or greater than their impacts. Meaningful inclusion of the public through the plan-testing process can reassure all participants about the costs and benefits of each strategy under consideration. While such inclusion may not build universal support for a contentious plan, it may reduce conflict and improve the success of implementation measures. In part, this is because a good public involvement program allows disclosure and mapping of the values that underlie conflict so that alternatives can be developed or modified to accommodate a range of values.

Step 6: Document the Results of Plan Testing and Recommend One or More Plans   At the conclusion of plan testing, it is recommended that all results be documented and the rationale for each stage of option screening and plan development be fully described. Typically, three or four plans will emerge as better than others. The difficulty, as is often the case in the testing of individual options, is that one plan may be "best" in terms of cost, while another is "best" in terms of pollutant reduction and a third is "best" in the eyes of agency officials because of its administrative simplicity.

The decision as to which of these plans is implemented is, ultimately, a political one. Water managers can facilitate the decision-making process by:

1. Presenting several effective strategies for the consideration of decision makers

2. Documenting the performance of each plan in terms of the decision criteria established at the beginning of the planning process

3. Reminding decision makers of the weights they have assigned to the various decision criteria

4. Providing supporting documentation regarding implementation considerations for each plan

The Stage 2 Remedial Action Plan report for the Metropolitan Toronto RAP area (Ontario Ministry of the Environment and Energy 1993) presented a long list of recommended management actions, each of which is assigned a priority, an estimated cost, and a recommended implementation period. The document served as the basis for negotiations with local and regional municipalities, the provincial government, and the Canadian federal government as to which agency should or would in fact fund each measure. Figure 11.1 illustrates a portion of the Toronto RAP "actions" matrix.

There is no question that funding availability and the presence of an agency "sponsor" for a measure were significant factors in defining the final characteristics of the management plan for the Toronto area watersheds. A major consideration in this debate was the implementability of different management approaches: how well each plan would be accepted by the public, how likely

Action	Lead	Partner(s)	1-5 years	5-10 years	10-20 years	
Improve ICI best management practices.	Regional municipalities	EC, LM, MOEE	■	□	□	ICI = industrial, commercial and institutional
Improve spills response and prevention.	Ministry of the Environment and Energy	EC, LM, MOT, RM	■	□	□	EC = Environment Canada LM = Local municipalities
Improve controls on agricultural practices.	OMAF	AC, FA, MTRCA, MNR, MOEE	■	□	□	RM = Regional municipalities MOEE = Ministry of Environment and Energy
Reduce sediment from construction activities.	Local and regional municipalities	MOEE	■	□		MOT = Ministry of Transport MNR = Ministry of Natural Resources
Trace and disconnect ICI cross-connections.	Local and regional municipalities	MOEE	■			MMA = Ministry of Municipal Affairs OMAF = Ministry of Agricultre and Food
Trace and disconnect residential cross-connections.	Local and regional municipalities	MMA, MOEE	■	□	□	MTRCA = Metro. Toronto and Region Conservation Authority FA = Farming Associations AC = Agriculture Canada

■ Activity initiated   □   Activity ongoing

Figure 11.1 *Portion of the Metropolitan Toronto Stage 2 Remedial Action Plan matrix of proposed actions, responsibilities, and timetables for stormwater management.*

each was to succeed over the long term, and similar considerations. Section 11.2 discusses some of these considerations. Chapter 12 describes the components of a successful implementation plan in more detail.

11.2 IMPLEMENTATION CONSIDERATIONS

Although the cost and effectiveness of a management strategy is of central interest in integrated watershed management, its "implementability" is also an essential consideration. In part, this is because a plan that is easily implemented is usually welcomed by the public and results in support for the politicians and agencies who have recommended it. In other words, a strong, popular plan can translate directly into votes. On the other hand, a plan that is highly effective but onerous to some portion of the population, very costly, or otherwise disruptive of individuals and communities will encounter resistance from the public. Resistance means delays, added costs, public acrimony, and loss of political support.

Ideally, early and meaningful public involvement will have identified issues of major public concern, will have resulted in decision criteria that reflect the reasons for those concerns, and will ultimately have prevented contentious plans from reaching the final phases of screening. In practice, however, contentious plans are sometimes retained in these final stages, either because they are highly effective or because of lack of sensitivity on the part of the analysts. This is one of the reasons to develop several preferred management strategies for presentation to decision makers.

There may also be obstacles to implementation that are not apparent either to the analysts or to the wider public. These may include, for example, statutory prohibitions on certain types of actions. While it is to be hoped that these prohibitions will be known early in the planning process, such is not always the case. A legal opinion on the viability of a given strategy may have to be delayed until the precise details of the plan are known. In Ontario's Municipal-Industrial Strategy for Abatement (Ontario's program to develop Clean Water Act-type regulations for effluent discharges), for example, it was believed that "permit-by-rule" was a viable approach for managing certain point-source dischargers. Under a permit-by-rule system, the law requires only that a specified type of technology be in place and correctly operated at a facility; no effluent standards are specified. In final legal review, however, it emerged that legislative counsel believed there to be insufficient precedent for this approach, and it was discarded in favor of a more traditional standards-based approach.

Where obstacles to implementation have been identified, it is, however, worth exploring with interested agencies and the community whether there is any way to surmount them—for instance, through financial compensation or additional structures or technologies. Some obstacles will remain no matter what mitigation is proposed; for instance, landowners are often strongly opposed to dam projects because of the potential for displacement of farms and homes. Spiritual and cultural values may create major obstacles that are not easily overcome, even if the affected group is a small minority in the community. But some obstacles (e.g., property value concerns) may be manageable if values are clearly mapped and conflict resolution processes are employed to develop outcomes agreeable to all parties.

In general, it makes sense to eliminate from detailed consideration any individual options or management strategies, however effective, that will present significant obstacles to implementation. The underlying logic here is that it is probably better to have a less effective plan that is fully implemented, rather than a highly effective plan that remains a paper exercise.

11.3   CASE STUDY: WATERSHED MANAGEMENT FOR GANDER LAKE, NEWFOUNDLAND

Gander Lake, Newfoundland (Canada), is a large freshwater lake approximately 50 km long, 2 km wide, and up to 300 m deep. Cantwell and Pinhey (1997) describe the development of a watershed management plan for the lake.

Gander Lake supplies drinking water for the town of Gander and several other communities. The rivers that supply the lake support a thriving salmon fishery, said by Cantwell and Pinhey to be among the best in North America. The area's natural beauty has recently placed it under pressure from cottage development, while forest-product companies seek to exploit the timber resources of the watershed.

In the early 1990s, the Newfoundland Department of the Environment began a water management planning initiative for Gander Lake, targeted at protection of drinking water resources. The responsible agency was the Department's water resources management division. The agency intended that this plan should not only provide for the protection of Gander Lake's water resources, but also serve as a model for other water management plans elsewhere in the province.

The process utilized a steering committee with representation from various provincial government departments, Environment Canada (the responsible federal agency), and the town of Gander. Public representatives were not included on the committee. The objectives of the initiative were to:

1. Identify the primary sources of pollutants to the lake
2. Develop a zoning plan to direct future development (especially forestry operations and cottage development)

The province placed special emphasis on scientific rigor throughout the planning process and rejected the notion of generalized planning concepts such as a uniform 50 m buffer zone all around the lake. Each decision made by the committee had to be supported by clear and specific evidence.

Gander and its neighboring towns are all relatively new, and their development is well recorded. Good records were also available for water quality in Gander Lake and from major point sources of pollution, including sewage treatment plants and industrial discharges. A review of these data soon revealed that point sources of pollution did not have a major influence on the water quality of the lake itself. Instead, nonpoint-source pollution was found to be the primary source of water pollution.

Review of available data also showed that although Gander Lake is currently oligotrophic (low in nutrient loadings and therefore not subject to nuisance algae blooms), it is sensitive to phosphorus loadings. Phosphorus thus became the focus of subsequent water management planning.

Major sources of phosphorus to the lake include direct precipitation and delivery via streams and overland runoff. Much of the phosphorus delivered to the lake was found to be attached to soil particles eroded from the land and carried by streams and runoff. Thus, erosion was also found to be an important mechanism in affecting phosphorus concentrations in the lake. Water managers compiling the plan believed that forest clearance would be the major source of erosion, and therefore phosphorus, to the lake.

The analysts for the plan (the consulting company EDM Environmental Design and Management Ltd., of Halifax, Nova Scotia) chose to employ a relatively simple computer-assisted watershed analysis, as described in Chapters 6 and 7 of this book. They employed a GIS-based predictive model based on the Universal Soil Loss Equation (USLE) using a 1 ha grid size. Digital maps of land use, development type, soil type, slope, and other factors were prepared, and a simple USLE-based predictive model ("watershed decision model") was calibrated to existing conditions.

A range of future management scenarios were then developed and tested using this model, by altering the cartographic models of suspended solids and phosphorus input to the lake under different management conditions. The model produced estimates of total annual phosphorus and suspended solids loads to the lake, both for existing conditions and for the various management strategies under consideration.

The results of scenario testing showed that planned clear-cuts in the forested portions of the watershed and additional urban development would actually have less impact than was originally expected. By contrast, cottage development and expansion of the Trans-Canada Highway (a major traffic route through the area) would be likely to have a greater impact on water quality than was originally foreseen. In part, this was because clear-cuts were proposed for areas of the watershed that were relatively resistant to erosion, while highway expansion activities would occur in more sensitive areas. The analysts explained: "In this catchment, location tended to be more critical in terms of lake water impact than actual land use."

The results of scenario testing were used to prepare a zoning plan to restrict land use within the watershed. The plan illustrates locations where development would create a negative impact on water quality. About 7% of the land in the watershed is categorized as "restricted to development" because of its sensitivity to erosion and likelihood of affecting water quality in the lake. The planners note that the finding that a majority of lake water sediment derives from only a small land area in the catchment is consistent with findings from other watersheds (cf. Wall et al. 1978). The results also allowed the identification of allowable activities in the watershed, based on the potential of those activities to disturb the soil and create erosion. Controls on the density of development were negotiated with proponents. These included limits on construction activities, stormwater management practices, and cottage development guidelines.

The Gander Lake planning initiative was relatively inexpensive but yielded important insight into watershed management practices for the community. Although the 1 ha GIS grid is relatively coarse, the model provides the basis for more detailed planning at the local level and has created a basis for dialogue with the community and with industry regarding future land use planning and water quality protection.

The output of the process was, first and foremost, a map, showing land use zones and erosion-sensitive areas. The map is easily understood by both lay people and technical experts and is an effective means of communicating the results of complex analysis.

The watershed decision model has now been accepted by the provincial government as an appropriate tool for decision making in other watersheds. Cantwell and Pinhey (1997) observe that integrating appropriate land use planning techniques and an understanding of landscape ecology will "allow planners to produce 'living plans' that respond to development demands without compromising those parts of our environment that cannot be compromised."

REFERENCES

Arnstein, Sherry. 1969. A ladder of citizen participation. *Journal of the American Institute of Planners* 35:216–224.

Bishop, Bruce. 1970. *Public Participation in Planning: A Multi-Media Course.* IWR Report 70-7. Fort Belvoir, Va.: U.S. Army Engineers Institute for Water Resources.

Cantwell, Margot D., and Jeffrey A. Pinhey. 1997. Watershed management plan for Gander Lake and its catchment. *Plan Canada* (March): 27–30.

DeGarmo, E. Paul, William G. Sullivan, James A. Bontadelli, and Elin M. Wicks. 1997. *Engineering Economy.* 10th ed. Upper Saddle River, N.J.: Prentice-Hall.

International Joint Commission. 1994a. *Bioindicators as a Measure of Success for Virtual Elimination of Persistent Toxic Substances.* Windsor, Ont.: IJC Great Lakes Regional Office.

———. 1994b. *Indicators to Evaluate Progress Under the Great Lakes Water Quality Agreement.* Report of the Indicators for Evaluation Task Force to the International Joint Commission. Ottawa and Washington, D.C.: IJC.

Lumsdaine, Edward, and Monika Lumsdaine. 1995. *Creative Problem Solving: Thinking Skills for a Changing World.* New York: McGraw-Hill.

Novotny, V., and H. Olem. 1994. *Water Quality: Prevention, Identification, and Management of Diffuse Pollution.* New York: Van Nostrand Reinhold.

Ontario Ministry of the Environment. 1987. *Technical Guidelines for Preparing a Pollution Control Plan.* Report from Urban Drainage Policy Implementation Committee, Technical Sub-Committee No. 2. Toronto: Ontario Ministry of the Environment.

Ontario Ministry of the Environment and Energy. 1993. *Metropolitan Toronto Stage II Remedial Action Plan: Clean Waters, Clear Choices.* Toronto: Ministry of the Environment and Energy, Central Region.

Wall, G. J., L. J. P. van Vliet, and W. T. Dickinson. 1978. *Contribution of Sediments to the Great Lakes from Agricultural Activities in Ontario.* Windsor, Ont.: International Joint Commission.

12

Implementing the Plan

Up to this point, this book has been concerned with developing an understanding of watershed components and processes, developing societal consensus as to the ideal condition of those watershed elements, and developing a plan to move the watershed toward the desired condition. In short, this book has so far been concerned largely with a paper exercise.

What distinguishes the successful watershed management plan from its paper counterpart is implementation: putting into action the plans and programs that planners believe will be effective in improving the state of the watershed ecosystem. Successful implementation implies a societal consensus that a plan is acceptable as a means of reaching community goals for a watershed. As Viessman (1990) observes, however:

> Water management institutions evolve from needs identified at some milestone in time. The problem is that times change, and so do needs. Unfortunately, institutions seem to march on with entrenched constituencies, and many in existence today are addressing yesterday's goals or addressing today's problems with yesterday's practices.

Figure 12.1 (reproducing Figure 1.1 for convenience) illustrates the interplay between social, economic, and biophysical forces in a watershed. It will be seen from this figure that there is a sort of dynamic tension between the watershed's water resources, its water users, the broader watershed ecosystem (including soil, air, vegetation, and animals), and the various social and economic forces at work in the watershed. Changes in social and economic forces can trigger changes in water management practices. Similarly, changes in the watershed

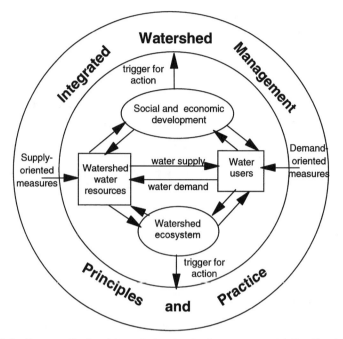

Figure 12.1 *Forces affecting integrated watershed management (after Koudstaal et al. 1992).*

ecosystem can indicate a need for altered water management practices. Supply-oriented or demand-oriented management action affect the horizontal axis of the figure: the quantity and quality of water in the watershed and the uses made of that water.

The system is dynamic because, as Viessman points out, neither the natural watershed ecosystem nor the human social and economic systems of the watershed are static. Just as watershed planning must respond to these changes, so must watershed management, including plan implementation, be dynamic rather than static.

This chapter addresses the special needs of the implementation stage, beginning with key principles of water administration and moving through the elements of a recommended implementation approach. At the end of the chapter, the administrative structures and implementation success of several case studies are examined.

12.1 PRINCIPLES OF WATER RESOURCES ADMINISTRATION

Integrated watershed planning and management has two principal goals (Goodman and Edwards 1992):

1. To plan programs and projects that are economically efficient and socially desirable, and

2. To execute projects that will be sustainable over a long period of time beyond the exodus of domestic or foreign financing, technical assistance, and the repayment of loans.

Young (1992) suggests that these goals must be fulfilled within certain constraints, namely, that:

1. Environmental quality (ecosystem functioning, soil, water, and air quality, and landscape amenity) must be maintained or improved. No economic activity should unduly disturb the regenerative capacity, the assimilative capacity, or the productivity of life-support systems.

2. All resources use must be technically and economically efficient, and resources must not be wasted.

3. Governments may fail. Planners must recognize and plan for this contingency by creating incentives for environmental improvement and plans and programs that encourage political and economic stability.

4. Future options must be maintained. This is the central challenge of sustainability: that each future generation will have the same capacity and option to solve its own problems as currently exists in the society. The creation or substitution of future options is acceptable only when they are consistent with this notion. In addition, there may be benefit in giving preference to institutional arrangements that encourage ecological, social, and economic diversity, which are more capable of responding quickly, positively, and flexibly to change.

5. Population growth should be controlled, so that future generations will have equivalent opportunities for solving their own problems and will have equal access to resources. At least primary education, and preferably secondary education, are central to this constraint.

6. Nations must conserve and enhance the value of their natural capital through controls on pricing, harvesting, and regeneration rates.

7. Where additional resources are required, nonrenewable capital (resources) should be depleted in preference to renewable resources. The logic behind this is that a sustainable future must include the life-support systems and ecological functions that underlie renewable resources. Nonrenewable resource depletion should occur at a rate that maintains the aggregate value of renewable and nonrenewable resources.

8. Wealth should be redistributed to poorer countries, because poorer countries cannot be expected to manage their resources sustainably without assistance from wealthier nations. Young suggests that this could be done by liberalizing trade arrangements to allow global markets to take greater account of environmental conditions.

Caponera (1985) reviews the work of the Institut de Droit International (IDI), the International Law Association (ILA), and the International Law Commission (ILC) of the United Nations, regarding the general principles of international water law; many of these apply also at the national and regional levels. The principles Caponera lists are not legally binding and do not impose any obligations on water management institutions. They are, however, valuable in determining appropriate implementation mechanisms. Essentially, the principles reviewed by Caponera are based on two fundamental propositions, both of which echo principles of English common law and other unwritten behavioral codes:

1. Common water resources are to be shared equitably between the states entitled to use them, and

2. States are responsible for substantial transboundary injury originating in their respective territories.

Where a watershed crosses state or national boundaries, the watershed unit can be taken as encompassing not only a biophysical system, but a "community of interests" related to water in the drainage basin. Successful implementation of watershed management plans must therefore be based on certain principles, as discussed in the following paragraphs.

12.1.1   The Principle of Reasonable and Equitable Sharing of Resources

According to the Helsinki Rules of the ILA, what is "reasonable and equitable" is to be determined in light of all the relevant factors operating in the case, including but not limited to basin geography, basin hydrology, climate, past and present water uses, economic and social needs of water users, the population dependent on the waters of the basin, the comparative costs of alternative means of satisfying the social and economic needs of each water user, the availability of other resources, the avoidance of unnecessary waste in water use, the practicability of compensation as a means of adjusting conflicts between users, and the degree to which the needs of one state (user) may be satisfied without causing substantial injury to a co-basin state (user). Water users (states) are to resolve conflict through consultation, negotiation, and agreement, all with reference to specific watershed systems.

12.1.2   The Principle of Limited Sovereignty

The principles of reasonableness and equity obviate a system whereby a state has full rights to its water resources because of territorial sovereignty. As Caponera (1985) puts it, "confronted with a factual situation, nothing prevents the independence of sovereignty from giving way to interdependence suggested by the principles of reasonableness and equity." So although a jurisdiction may claim territorial sovereignty over an area, it may nevertheless be expected to manage its water resources in a fashion consistent with the needs of downstream or adjacent water users.

12.1.3 The Duty to Cooperate in Development

If a watershed represents a community of interests, and in view of the principle of reasonableness, it seems appropriate that one jurisdiction be expected to cooperate in watershed development—for instance, by delaying new works or water uses until the needs and interests of neighboring water users can be satisfied. Caponera points out that implementing this principle is not easy. Parochial interests may prevail over "equitable utilization," especially where choices must be made among conflicting water uses and where water supplies are limited.

12.1.4 The Duty to Cooperate in Protection

According to Caponera, principles of international water law are emerging as states contend with the protection of transboundary water resources, especially in regard to pollution of water with toxic chemicals and waterborne diseases. Several elements are included here:

- Individual states are expected to have adequate legal and administrative frameworks for the protection of water within their jurisdictions.
- States are also expected to keep abreast of changing technologies for the use, recycling, and purification of waters within their jurisdictions.
- States are expected to prevent water pollution that may harm or threaten the interests of neighboring states.

Consistent with this last element is the next principle, responsibility for injury across frontiers.

12.1.5 Principle of Responsibility for Injury Across Frontiers

In general, states are responsible under general international law for acts and omissions concerning activities within their jurisdictions. In terms of water management, the relative harmfulness of pollution or water extractions may depend on the beneficial uses affected by those acts or omissions. And since interpretation of "impact" on "beneficial" use is highly context-specific, harm must be evaluated relative to each individual stream and watershed situation.

Caponera's principles of international water law are equally valid in terms of watershed cooperation and management and, indeed, underlie successful watershed management initiatives. Whether national, state, regional, or local governments are involved, integrated watershed management implies a reasonable and equitable sharing of the water resources of the basin among interested users, responsibility for harm caused, and a duty to protect the shared resources within the bounds of technical and economic feasibility.

In a sense, then, implementation means getting individuals and organizations to:

- Agree on a desired plan of action
- Agree on indicators of progress toward the shared goal
- Allocate tasks among the various water users in the basin
- Allocate the costs of water management among the various water users in the basin
- Agree on periodic review of, and revisions to, water management activities under way in the basin

This sequence of tasks means that individuals and organizations must set aside their "territorial sovereignty," with its narrow goals and potentially exploitive actions, in favor of the equitable sharing of water resources and the responsibility for water resources management. In effect, it means balancing private interests against public interests. This balancing act requires agreement among players who may have very different values and different visions of an ideal watershed condition.

There is no question that the most difficult part of implementation lies in reaching agreement among the various players. David Crombie, former mayor of Toronto and chair of the Royal Commission on the Future of the Toronto Waterfront, once said that the two biggest obstacles to resolving environmental problems are "turf and ego." There is much to support this idea in the process of implementing a watershed plan. While everyone supports the notion of a cleaner watershed and the restoration of beneficial uses, support may be likely to dwindle in the implementation phase. The main problem is usually money, although jurisdiction ("turf") can also get in the way. The difficulty is that most watershed restoration actions are costly. The Remedial Action Plan for the Metropolitan Toronto Watershed will cost in excess of $3 billion (Canadian), including capital and operating costs, if it is fully implemented (planned implementation will extend over at least 20 years). Individual agency shares of this total amount to tens of millions of dollars—a hefty price at a time when governments are trying to cut costs and shrink budgets. To avoid paying more than they need to, agencies may try to argue that another level of government has responsibility for the necessary action. At the same time, however, both agencies and politicians are eager to reap the political benefits of successful remediation ("ego"), so there may also be a struggle for political profile in the implementation process. Teclaff and Teclaff (1987) affirm this perspective:

> In multi-state basins, problems are bound to arise over the allocation of expenses among member governments. Unlike the costs of development projects, which

are often apportioned among states according to benefits ultimately received, the expenses of pollution research, monitoring, and surveillance are rarely so equitably divided. European countries have made much of the so-called "polluter-pays" principle as applied to individual polluters and polluting industries. They seem less willing to apply it as among states.

12.2 PLANNING FOR SUCCESSFUL IMPLEMENTATION

Chapter 11 discussed the development and testing of alternative management strategies, but recommended that no single plan be put forward as the preferred approach. Instead, it was recommended that two or three good strategies be offered to decision makers for consideration. The Royal Commission on the Future of the Toronto Waterfront (1992) observed that the decision as to which management approach is preferred usually lies with an elected body, such as a regional or local municipal government, or a state or provincial Cabinet. Because few plans will satisfy all interests in the community, trade-offs must be made in selecting a final plan. Ideally, these decisions should be made by elected officials who represent the larger society. Endorsement by elected officials often carries the authority necessary to expend resources of time and money and to actually "get things done." If a plan is recommended and approved only by a bureaucratic (or technocratic) group, but endorsement—and expenditure—by elected officials does not follow, the plan will likely remain on the shelf. It requires political will, encouraged by well-developed public consensus, to ensure that the plan is put into action.

Successful implementation requires several elements that have not uniformly been part of previous stages in the planning process:

1. A single lead agency to act as an advocate and facilitator for the plan with the community and with political representatives
2. Strong linkages to existing programs, including local and regional land use planning processes, water quality and flow monitoring programs, and similar programs, to optimize use of available information and minimize duplication of effort
3. Clear designation of responsibilities, timetables, and anticipated costs for project actions
4. Effective laws, regulations, and policies to provide a framework for the tasks identified in item 3
5. Ongoing tracking of the degree of implementation of management actions and of the success of those actions once implemented
6. Ongoing monitoring and reporting of progress, both to assess the effectiveness of individual actions and to sustain public and political interest in and enthusiasm for the plan

7. Ongoing public education and communication programs to consolidate and enhance the social consensus achieved in the planning process
8. Periodic review and revision of the plan
9. Adequate funding for these activities

The following sections examine each of these requirements in more detail.

12.2.1   A Single Lead Agency

Studies carried out in numerous watersheds (cf. Teclaff and Teclaff 1987) have demonstrated the importance of a simplified administrative structure for the plan. In particular, plan implementation should not require the creation of additional layers of administration, although change may be needed in existing administrative structures to accommodate the ongoing needs of the watershed plan.

For these reasons, groups like the Royal Commission on the Future of the Toronto Waterfront (now the Waterfront Regeneration Trust) have found it helpful to have a third-party sponsor/advocate for the plan. This third party should be nonpartisan and objective, yet able to work with all groups to encourage participation in elements of the plan. Often, it requires an individual who is widely respected to fulfill this need. In the case of the Royal Commission, David Crombie himself, a convert to and passionate advocate of watershed management, is that person. Speaking of his work with the Trust, Mr. Crombie says that his job is simply to get everyone sitting at the table talking to each other about who is going to do what. This, in itself, is a challenging task.

McLelland (1987) observes that the structure of the sponsoring agency may differ, depending on the level of government involved. Local agencies are primarily concerned with community water resources needs—for instance, water supply or wastewater treatment. Local governments are more likely to undertake water resources planning on a case-by-case basis as the need arises, rather than as a more forward-looking, integrated strategy. She notes the examples of local water management in China, where many rural villages have assumed responsibility for designing, constructing, and operating their own water systems and in doing so have achieved results superior to those obtained in some developing countries that have instead chosen a strongly centralized organizational structure. The success of water resources planning at the local level (in terms of watershed, we could perhaps say the subwatershed level) requires well-developed public participation. If problems are severe and the local system complex (as, for instance, in the case of Mexico City), regional or national planning may be more appropriate than local planning.

Regional planning is probably most compatible with the concept of watershed management, but the administration of water management plans at the regional level poses challenges simply because a region may encompass several or many individual jurisdictions. It is seldom clear, therefore, as to which administrative structure would best serve the needs of the water management initiative.

Regional planning was the focus of much discussion at the United Nations Mar del Plata conference of 1977 and, as discussed in Chapter 1, has been the subject of increasing interest and international effort in the 20 years since that conference. At the time of Mar del Plata, about 40% of the world's population was estimated to live within international river basins, and a large proportion of these people therefore relied on water "imported" from upstream countries (McLelland 1987). McLelland, like Caponera (1985), emphasizes the barriers to planning and management that may arise when different jurisdictions within the watershed have different interests in, and uses of, basin water resources. In France, basin authorities have been established for the management of multi-jurisdictional river basins, and the "river catchment area [is recognized] as a natural entity" (Dubosc 1992). Multijurisdictional agencies are also employed in the management of the Rhine River basin (Wilkes 1975; Kiss 1985) and the Danube River basin (Linnerooth 1990). The International Joint Commission (IJC), a Canada-United States treaty-based organization charged with oversight of the "boundary waters," provides a framework for binational management of the Great Lakes Basin and other major international waters. The U.S.-Mexico Border Environment Cooperation Commission (BECC), established in 1993, coordinates transboundary environmental infrastructure projects under the North American Free Trade Agreement (NAFTA). Other countries with long-range plans and administrative structures for the management of domestic and international watersheds include Argentina, Egypt, Mexico, and Romania (McLelland 1987).

Planning at the national level is consistent with one of the Mar del Plata Action Plan recommendations, which calls for the development of strong national programs. However, as Lee (1992) points out, centralized national planning and administration of water resources has failed, as it has failed in the administration of social and economic programs, simply because it cannot provide adequate recognition of local and regional needs and differences. McLelland (1987) echoes this sentiment, observing that although most countries in the Asia-South Pacific region have central water agencies and well-defined national water policies, only about half have been able to develop effective master plans for water management and most of these cannot implement them because of structural weaknesses in their central agencies. In less developed countries, these weaknesses typically include lack of staff, lack of financial support, and a judicial system unprepared to enforce the laws. McLelland also raises the interesting case of the Ogun-Oshem river basin in Nigeria, where two conflicting water management plans were developed, one at the federal level and the other by state authorities. Each plan responded to a different perception of need, the federal plan focusing on agricultural development and the state plan on provision of adequate water supply in urban centers. The two plans were not coordinated, so conflicts and redundancies occurred and neither plan was fully successful.

12.2.2 Strong Linkages to Existing Programs

Implementation of a comprehensive water management plan can be an expensive business; monitoring its effectiveness after implementation will add to that expense. Implementing agencies can reduce the costs of actions and of postimplementation monitoring by making strong linkages to existing programs inside and outside government agencies. Among these programs may be routine monitoring conducted by environmental and health-protection agencies, independent monitoring carried out by public interest groups or community organizations, data collection and analysis by university researchers and their students, and so on. Even elementary school students can be helpful in monitoring watershed status—for instance, in tracking the timing of manure application on agricultural fields.

Efficient utilization of these varied data sources usually requires a staff person to act as liaison with the community and other agencies. This person may also be, but is usually not, the same person charged with maintenance of a centralized database, ideally accessible by all parties. This data coordinator can be a knowledgeable volunteer but should be able to provide continuity over time and across data contributors. Each contributing agency or program would normally retain a copy of its data and file a copy with the central database.

Utilization of existing programs can also help to enhance the profile of, and community support for, the plan. Individual actions can be delegated to existing groups, either for execution or for oversight, allowing the sponsoring agency to retain an advisory role at lower cost and time commitment. Linkages to existing programs can increase media exposure of the plan, improve public awareness of issues and actions at the household level, and increase political support for ongoing programs and policies. Periodic review of the watershed plan will also be stronger and more effective in a community where previous planning initiatives have received extensive coverage and where a variety of other groups are clearly interested in, and involved with, the planning process.

Forging strong linkages to existing programs was a central feature of the Metropolitan Toronto (Ontario) Remedial Action Plan (RAP) (Ontario Ministry of the Environment and Energy 1993). A previous major planning effort, conducted in the early 1980s, had failed to produce an implementable plan despite the expenditure of millions of dollars. Gradually, private and public interest group initiatives had sprung up in the river basins to take the place of the failed government plan. These initiatives were often highly successful and enthusiastically supported in the community, but were usually local in scope and limited by slim financial resources. The Metro Toronto RAP therefore sought to extend the work of the previous government plan while building a much stronger base of public support in the community. Key to this strategy was a well-publicized effort to recognize and link up with local community projects. This was a delicate business, because the RAP team did not want to give the appearance of taking over those local projects and, indeed, had no wish to alter their course or to assume an oversight role. Rather, the RAP team wanted to ensure that all efforts were recognized and coordinated so as to minimize duplication of effort and unnecessary wastage of resources.

12.2.3 Clear Designation of Responsibilities, Timetables, and Anticipated Costs

To a large extent, implementation means program delivery: concrete, observable actions, paid for by a public or private agency. Program delivery in turn implies an administrative structure for program development, delivery, and oversight. Actual plan implementation therefore begins with assignment of responsibility and timetables for specific actions. An obvious place to begin in making these assignments is in jurisdictional responsibilities, such as those defined under national or state/provincial constitutions. If a jurisdiction has the authority to enact legislation, and has in fact done so, then that jurisdiction can legitimately be expected to carry out activities consistent with that legislation and its enforcement.

Sometimes the authors of the plan review jurisdictional responsibilities and propose an assignment of tasks. This proposal is then debated in a suitable forum, often in political councils or Cabinets, and the details of responsibilities gradually hammered out. Partnership agreements and cost-sharing arrangements may be useful mechanisms for achieving agreement about responsibilities. Similarly, technical factors may dictate the most probable costs and timetables for major project actions, but these may later be amended in discussion with political bodies, either to defer or accelerate actions to accommodate political and economic needs, or to share or trade off responsibilities as appropriate for the implementing agencies. Here again, as in so many stages of the watershed planning process, consensus is essential. A consensus built on a decision to delay a costly project may be a good approach if all parties are in agreement and the project is actually (if eventually) built. A prescriptive arrangement assigning responsibility to an agency without its concurrence may result in repeated delays and, ultimately, in failure to build the project at all.

12.2.4 Effective Water Management Laws, Regulations, and Policies

As an extension of its constitutional responsibilities, a jurisdiction usually has a number of tools available for water management. Often, these will have been employed in the past and can be extended or reapplied for current management purposes. Figure 12.1 shows how demand management tools can be used to reduce pressure on water resources quantity and quality (as opposed to supply management tools that would increase supply to meet unrestricted demand). Appropriate demand management tools include a range of legal, paralegal, and nonlegal measures, such as:

- Legal instruments, including extraction permits or licenses, ambient water quality standards, effluent discharge standards, and similar instruments, and the fines and other sanctions associated with noncompliance
- Economic instruments, including effluent charges, grants and subsidies for pollution abatement activities, taxes, and regulations that create markets affecting water and emission rights
- Voluntary instruments, such as nonbinding memoranda of understanding between a government and an industrial or industrial association, to reduce the volume or improve the quality of waste emissions, to avoid or eliminate the use of certain chemicals, or to effect similar actions

There are advantages and disadvantages to each of these instruments. Some generate revenues that can be used for water management purposes. Others, like voluntary programs, generate no revenues but are inexpensive to administer and therefore may be cost-effective even if less prescriptive than legislative tools. In the past, regulatory agencies have been reluctant to adopt economic instruments, whether because of a fear that they will be seen as licenses to pollute or otherwise misuse water, or because they are cumbersome, or perhaps just different, to administer, as compared with traditional legislative instruments like permits. (Economic instruments are, in fact, usually based in statutes or regulations; that is, a legal instrument is used to create a framework within which economic instruments like taxes and effluent charges can operate.)

Probably the most common implementation obstacle in terms of laws and policies is enforcement, especially in less developed countries. Implementation of programs that involve or depend on legal instruments requires that those affected by the legislation be monitored regularly by the responsible agency to determine whether they are in compliance with legal requirements. And there must be sufficient political will and resources directed to enforcing regulations, including prosecution or other sanctions against offenders, in the event of noncompliance. McLelland (1987) and other authors have observed that most countries now have in place a general legislative framework for environmental protection, but enforcement of that framework is rarely satisfactory. Problems that may arise in enforcement include:

- Insufficient staff for compliance monitoring
- Inadequate staff training
- Exceptions and variances to legal requirements granted where improvements would be costly or difficult or would adversely affect an export market
- Graft and corruption (e.g., bribery) to avoid compliance with legal requirements
- Lack of public understanding of the issue and of cause-effect relationships

- High costs of monitoring (e.g., costs of labor, laboratory analysis of wastewater samples, sample and data storage and management, etc.)
- Lack of incentives to comply with legislation

12.2.5 Ongoing Tracking of the Degree and Success of Implementation

Despite the best intentions of all concerned, some intended projects will not be implemented at the scope or within the time period originally envisioned. It is important that implementation progress be tracked regularly and consistently, both to reassure political bodies that progress is being made and to encourage continued participation by all concerned. Implementation tracking also allows screening of individual projects as they are put into place, to make sure that their design and performance are consistent with the goals of the watershed plan.

Regular reports on the progress of implementation also allow the plan to be updated to take advantage of new ideas and emerging technologies that may not have been apparent when the plan was originally completed.

Finally, regular monitoring of project implementation is important to determine the condition and operability of structures and devices that have already been installed. It is not uncommon, especially in less developed countries where other infrastructure may be lacking, community awareness may be low, and parts difficult to obtain, that installed equipment quickly falls into disrepair and may even be stripped for other uses. McLelland (1987) reports a survey of 589 villages in Maharashtra, India, where 36% of hand pumps were found to have failed, either from lack of proper installation or because they had not been adequately maintained. Schramm (1980) cites the example of the state of Zacatecas, Mexico, where in 1978, 50% of all rural water supply works were not functioning and a further 20% were only partially operational. Only 30% of all rural water supply works were working properly. In the same state, only 60% of the nominally irrigated land actually produced a crop in 1976.

In these cultures it may especially important to employ what is sometimes termed "appropriate technology"—that is, technology that is known and understood in the community and that has a high likelihood of continued success over many years.

12.2.6 Ongoing Monitoring and Reporting of Progress

The goal of the watershed plan is, ultimately, to move the watershed toward a more desirable condition. This progress will cost money, probably much of it from public coffers, and the implementing agency must therefore be accountable to the public for its expenditures. Part of this accountability is being able to demonstrate progress toward the goals of the plan.

Regular reporting of progress allows the effectiveness of individual management actions to be evaluated. Recall that, prior to implementation, water managers must forecast the performance of each option and select those that appear most promising. Postimplementation monitoring allows each action's true performance to be evaluated and compared against predictions. If a given action is not found to be as effective as anticipated, it can be replaced or altered as appropriate.

Routine surveillance and monitoring should build on the baseline data collected during the planning stage and should allow water managers to:

- Evaluate changes in watershed conditions because of, or independent of, plan-related actions
- Assess compliance with statutory, regulatory, and voluntary requirements
- Reveal areas where continued or enhanced remediation is required

As a general rule, all monitoring data should be available to the public, either in annual reports or on file in a central but easily accessible location. Some jurisdictions are now employing computer technology to make data available to the public. As one example, Geographic Information Systems (GIS) databases can be displayed using simpler display software so that a user can just "point and click" on a map to bring up more detailed tables, graphs, or maps. In another instance, a kiosk with weatherproof protection housed a computer terminal on a busy street corner. Passersby were able to view monitoring data from continuous sensors in an effluent discharge stream. As discussed in Section 12.2.2, it is essential that all project data be adequately stored and maintained so that they can be retrieved and manipulated as necessary for project analysis. In multijurisdictional and international basins this can be a challenging task, because data collection and laboratory analysis practices may differ from jurisdiction to jurisdiction. Section 10.4.2 described the problems encountered in trying to match Canadian and U.S. GIS records on agricultural practices, slope, soil type, and similar factors in the Great Lakes Basin.

12.2.7   Ongoing Public Education and Communication Programs

Ideally, the watershed plan represents a social consensus about the condition of the watershed. Some management actions, such as control of household hazardous wastes and responsible use of storm drains, require participation throughout the community. Other, more costly, actions, such as sewage treatment plant upgrading or flood control structures, may involve tax increases or other economic impacts on the community. Ongoing public education and communication (which, as noted earlier, can be linked to other external programs) can consolidate and enhance the social consensus achieved in the planning process.

McLelland (1987) notes that national papers on developing countries, prepared for the United Nations, list the following topics in *decreasing* order of importance: industry, agriculture, health, energy, natural resources, transportation, human settlements, communications, and *environment*. Yet, as she points out, public awareness of, and response to, environmental issues is a well-documented phenomenon, stimulated by increasing media coverage and by local and regional environmental crises that have focused public attention.

In less developed countries, literacy rates are often lower than in economically advanced countries, and this may pose a challenge to public education. In Uttar Pradesh, India, and in many other nations, appropriate sewage disposal and safe-birthing information has been successfully distributed, using pictorial representations such as sketches of mothers washing children's hands following defecation. More difficult in some developing countries is the widespread belief that governments are omnipotent and will be able to identify and correct any problems that may arise.

Sometimes indirect public education results from unrelated activities, with benefits that are not anticipated when the primary project is designed. Jacobson and Robles (1992) report that careful tour guide training has been an important vehicle for public education about environmental protection in Costa Rica. As ecotourism continues to increase in that country, it has become especially important to ensure that ecological reserves can support the educational needs of visitors while protecting the reserves' natural resources. This goal can be fulfilled by careful tour guide training, taking into account resource management requirements, visitor needs, and local economic conditions. Jacobson and Robles have demonstrated that a good tour guide training program, developed in collaboration with local communities, scientists, park managers, and the hotel industry, can help mitigate negative tourism impacts and provide environmental education to adults and children in the local community, in addition to enhancing visitor experience and providing local economic benefits.

12.2.8 Periodic Review and Revision of the Plan

To be an effective framework for watershed management, the plan must be reviewed and evaluated periodically, preferably at predetermined intervals. There are many reasons for this. Probably the most important of them is that the watershed itself is dynamic: its population is growing or shrinking, its water resources are improving or deteriorating, land uses and resource utilization patterns are changing, and societal views of these changes are shifting as they are influenced by internal and external factors. The plan must be reviewed and revised as information about the watershed is updated.

Periodic review also allows water managers to take advantage of newer water control and pollution abatement technologies, which may offer advantages of cost or performance over older approaches. Twenty years ago, for example, channel improvements such as dredging, bank stabilization with riprap or gabion baskets, and similar measurements were common; today there is a trend toward natural channels and vegetative rather than structural measures.

Computer simulation methods are also continually improved, especially as computer technology changes. When the Grand River Basin Water Management

Plan was published in Ontario in the early 1980s, the plan was based on 20-year computer projections run on an IBM mainframe computer at considerable expense. Fifteen years later the Grand River Simulation Model, a purpose-built Fortran model of the river system, could be run on any basic desktop computer, faster and more efficiently than was possible in the early years. Now, as the plan is under revision, the Grand River Conservation Authority is taking advantage of improvements in computing technology and recent research advances to update portions of the Grand River Simulation Model, to improve its simulation of algal and rooted aquatic plant growth, and to link it to a regional groundwater model.

Some planning exercises have remained static, one-time efforts. But a regular schedule of plan review encourages water managers to use the plan, and perhaps the option-testing framework on which the plan is based, as a foundation for regular dialogue about water resources within the river basin. Indeed, as the Royal Commission on the Future of the Toronto Waterfront (1992) puts it, a good implementation framework, including periodic review, will truly integrate environmental matters, provide a fair and consistent process, and ensure that information, evaluation, and decision making are shared and accessible by all parties in the watershed. In the long run, it believes, this will lead to greater planning efficiency and may shorten the time required for subsequent studies and agency approvals.

12.2.9 Adequate Funding

It almost goes without saying that implementation will fail if there is not enough money to pay for projects, infrastructure, and an appropriate administrative system for the plan. As McLelland (1987) has pointed out, the risks of failure because of economic constraints, and the implications of failure for a fragile economy, are more extreme in less developed countries than in economically advanced countries. Teclaff and Teclaff (1987) observe that control of transmedia, transboundary pollution hinges on the availability of money and manpower. These authors further note that, in fact, the existence of an administrative structure (they point to binational water management commissions and other similar institutions) may obscure the fact that insufficient money is available to actually carry out projects. Even large and powerful organizations may run short of funds. Teclaff and Teclaff cite the example of the Commission of the European Economic Community, which completed a major study in 1979 on the underground water resources of the Community but was unable to print and disseminate the study's 10 country reports and 152 maps because of insufficient funds.

The Canada-United States International Joint Commission (IJC) regularly reports on how the scarcity of research funding and manpower limits the two countries' ability to manage water resources effectively in the Great Lakes Basin. While funds may be available within each country for short-term local projects, funding for oversight agencies like the IJC is seldom adequate for the management, or even oversight, of an entire watershed.

12.3 WHY IMPLEMENTATION SOMETIMES FAILS

Goodman and Edwards (1992) review the history of integrated water resources planning in developed and less developed countries and provide some insight into why implementation sometimes fails. They note that a single framework for effective integrated water resources planning is not possible for all countries, or even for all regions within a single country, because these entities differ as to:

- Natural resources
- Population distribution and styles of living
- Economy
- Political, institutional, and legal structures

Planned programs may therefore fail if they are not successful in matching projects to watershed conditions, particularly if they:

- Incorporate projects that are too ambitious to be achieved with available financial and other resources, especially in less developed countries
- Incorporate projects that do not meet the expressed goals of the plan
- Fail to guide development properly
- Underestimate or ignore environmental impacts of the plan or individual projects
- Underestimate or ignore social impacts, such as disruption of community cohesion, displacement of groups or individuals, or loss of key social structures that provide a focus for community activities
- Overestimate the efficiency or availability of existing institutional frameworks to support implementation, or fail to create an organization with adequate staffing and responsibilities to ensure that projects are sustained beyond construction and early operation stages
- Lack good leadership
- Lack adequate staff training
- Disregard or underestimate the legal implications of the plan
- Lack an adequate infrastructure of facilities and services such as roads, marketing organizations, and other components
- Fail to recognize national, regional, local, and individual costs and benefits of the project in formulating and analyzing management options and selecting priorities for action

Goodman and Edwards (1992) emphasize the special needs of less developed countries in integrated watershed planning. In those countries, sophisticated methodologies, such as the computer simulation methods described in

Chapter 7 of this book, and the considerations of risk and uncertainty described in Chapter 8, require data resources that are simply not available and will be too costly to acquire and/or require interpretation by an experienced analyst who understands the applications and limitations of both data and methodology. These requirements may be beyond the capability of planners in less developed countries. The simpler techniques described in Chapter 6 of this book may be helpful in such cases, but even access to necessary literature may be limited in some countries. In particular, the problems of uncertainty described in Chapter 8 have received increasing attention in water resources planning in economically advanced countries. Less developed countries are particularly ill equipped to forecast the outcomes of projects under risk and uncertainty, or to deal with the economic implications of failure of a development project, because of scarce capital and trained labor and because the infrastructure for basic services may be incomplete. If a development project fails, the consequence may not be simple inconvenience, as would be the case in an economically advanced country, but rather a failure to provide basic needs and services to communities.

Wilkes (1975) describes implementation issues in the management of the Rhine River basin; many of these issues will also arise in developing countries. A number of obstacles are identified:

1. The need to go outside the basin boundaries for water, and the paucity of international arrangements to ensure adequate water supply for all users.

2. The failure to integrate pollution control with control of water quality, including sufficient water purification to ensure that water flowing beyond political boundaries is of adequate quality for consumption in neighboring jurisdictions. This problem becomes more acute when protection of future water supplies and future populations are considered.

3. The danger of diverting attention from supraregional problems as a result of decentralization of water management responsibilities. By this, Wilkes means that although there are many benefits to regional management of water resources, there may also be a danger in that only problems that are exclusively regional or communal in dimension may be "managed," while supraregional problems, such as transboundary water supply, may fall outside a regional agency's mandate.

4. The lack of automatic links between planning for "growth" and for water supply, including the coordination of land use permitting with permits for water withdrawals. When water suppliers have no real say in where development will go, shortages will occur and water management problems will increase.

5. Failure to give officials duties to ensure lowered use of water, especially tools that allow officials to obtain leverage over existing users to force efficient water use, and tools that permit controls on new and additional water extractions.

6. The risk of ignoring solutions developed outside the river basin through

excessive focus on management within the basin. Solutions developed by nations and regions outside the watershed may have value in the watershed but may be overlooked or discarded in favor of those that are based on direct experience within the basin.

7. Failure to keep political influence out of technical judgments, especially in relation to increased water use, future development, and political good faith in avoiding unwarranted or irresponsible water use.

8. Lack of planning for out-of-basin needs and demands, including trends in international development and other forces.

Although Wilkes was writing more than 20 years ago, his observations remain important and current today. For example, Ontario's Planning Act, which governs land use and zoning, contains no explicit or implicit requirement that development proposals must consider environmental impacts (cf. Wilkes' obstacle 4 in the preceding list). Several years ago a land-use planning commission was established in Ontario, headed by former Toronto mayor John Sewell, and charged with reexamining the Planning Act to improve its responsiveness to environmental concerns. The commission duly reported its recommendations, which included a general statement of principle that land use planning must be based on sound and sustainable environmental practices, and suitable amendments to the Act were drafted. With a change of provincial government in 1995, however, the New Democratic Party was replaced by the Conservative Party and the proposed changes to the Planning Act were swiftly discarded. There is currently no requirement in Ontario to incorporate environmental considerations in land use planning or zoning approvals.

According to Schramm (1980), one common error in making the jump from planning to implementation is to assume that sound economic criteria applied in planning analysis will guarantee economic feasibility in the real world. As discussed in Chapter 8, the choice of a discount rate, or minimum acceptable rate of return, is essentially an arbitrary decision made by the planner. Depending on whether public or private investments are involved (or a mixture of the two), different economic criteria and different market rates will apply. Schramm notes that:

> Many private investments and activities, confidently predicted on the basis of the original benefit-cost analysis, simply do not materialize. The overall projects turn out to be costly failures, unless government takes the further step to subsidize the private investments as well.

This idea—that a planning prediction is simply that, and not a fact—again seems so obvious as to be trivial. And yet, time and again, planners produce economic forecasts, computer simulations, and population projections that, although essentially guesses, are interpreted and used as fact. As Schramm notes, when the underlying assumptions for those guesses are incorrect or inaccurate, grave errors can be made in planning estimates. This may be the particular value of the Adaptive Environmental Assessment and Management approach, as described by Holling (1978) and Grayson et al. (1994) (see Section 3.4.2): the product of the planning exercise is not a set of "facts," but rather a community consensus as to the desired plan of action and its probable effectiveness.

12.4 CASE STUDIES

This book has presented many examples of watershed management techniques in use around the world. In this section, several case studies are examined and compared in an attempt to determine the factors that encouraged success or caused failure.

12.4.1 *The Tennessee Valley Authority, United States* The Tennessee Valley Authority (TVA) is often cited as the best-known attempt at integrated watershed management (McDonald and Kay 1988). Indeed, through its sheer size and duration, it is believed to have influenced water management strategies throughout the world.

The TVA, originally envisioned as an electric power development project, was begun in the early 1930s. It grew out of an earlier development for a nitrate plant to supply war munitions. This plant was begun in 1917 but decommissioned in 1919 before it was fully operational. The plant was to have required significant quantities of power, to be provided by two thermal generating plants and, later, by hydroelectric power from the Wilson Dam development. It is sometimes referred to as the Muscle Shoals facility, in reference to a nearby river feature. Throughout the 1920s and early 1930s, the facility was the subject of hot debate. Some people wanted to use it to produce nitrate fertilizer, but the plant was suited only to the outdated cynamid process, which required large quantities of electric power. If the plant had been converted to this use, it would have demanded all the power supplied by the hydroelectric plant, denying domestic customers access to that power. A variety of other public and private interests were also involved in this debate, which lingered into the beginning of the Depression. President Franklin D. Roosevelt saw the Muscle Shoals plant as an opportunity for wider economic development of the basin, not just as a local project. He envisioned an electric power development at the center of a suite of projects, including navigational improvements and flood controls in the Tennessee River, as well as soil erosion control, reforestation, improvements in agricultural land use, and increased and diversified industrial development in the watershed.

A central feature of the TVA, consistent with Roosevelt's aims for the project, was the "Force Account," which allowed construction in the river and valley to be carried out by TVA itself, employing local laborers, rather than by construction companies from outside the basin. The TVA provided welfare and education to its employees and their families and was therefore fundamental in the economic recovery of the area.

TVA activities continued well into the 1940s with the development of additional electric power generation capacity, using a series of thermal plants. Construction projects were funded by bond sales and revenues from the sale of power and other sources. Federal subsidies are still received for environmental protection, agricultural extension in the valley, and various other purposes, but these funds are small relative to the gross revenues from TVA projects. Today the combination of inexpensive electricity, easily available, trained workers, and a pleasant environment contribute to the area's appeal as a site for industrial development. The project remains a landmark of successful public ownership and multiproject development.

The Danube River Basin

The Danube has been called one of the most international rivers in the world (Linnerooth 1990), with 10 European countries bordering the 2,850 km-long river. The river has more than 300 tributaries, and its basin supports a population of more than 70 million people. The various countries bordering on the Danube collaborate on management of the shared resource through the nonbinding Danube Declaration, which requires cooperation (Linnerooth 1990):

- Between eight countries spanning Eastern and Western Europe,
- In the absence of effective and enforceable international legal rules,
- In the absence of a basinwide planning or decision agency,
- Between numerous national and international authorities with diverse, conflicting interests,
- On problems for which the geopolitics of the "upstream" and "downstream" countries creates disincentives for cooperative behavior,
- On issues characterized by serious scientific gaps and uncertainties, and
- In an atmosphere of increasing concern about the long-term effects of toxic pollutants and acute awareness that pollutants cross national boundaries.

Today the most pressing issues in the Danube are deteriorating water quality and conflicting water use demands, particularly those related to the generation of electric power. Several proposals for hydroelectric development and the construction of barrages are under discussion, but agreement has been difficult to achieve because of the inequitable distribution of project costs and benefits (most opportunities for hydroelectric development exist in the headwaters of the river, in Western Europe, but the environmental costs of development will be felt throughout the river basin). Specific issues now under discussion in the Danube basin include:

- Maintenance of river flow for electric power generation and for waste disposal.

- Maintenance and expansion of navigable waterways (the river experiences heavy winter ice and high spring floods and has large shallow stretches and hazardous rapids).

- Maintenance of an adequate and secure water supply for irrigation, industrial uses, and other purposes.

- Protection of water quality for potable water, irrigation, fishing, recreation, tourism, and nature preservation (water quality has traditionally been measured using dissolved oxygen, pH, bacteria counts, temperature, and hardness; more recently toxic substances have been added to the list).

- Flood prevention.

- Preservation of the river and its watershed for recreation, tourism, and nature preservation.

Linnerooth (1990) emphasizes the importance of cooperation through bilateral, stepwise negotiations. In view of the weak framework for international water law and institutions, co-basin nations may cooperate through building up a reservoir of goodwill upon which to draw in future negotiations. Compensation arrangements ("side payments") may assist in reaching agreement, as may straightforward payment by one country to have another country clean up pollution.

Progress in the Danube is hampered by the lack of an existing mechanism for multilateral, integrated decision making. A Danube Commission exists for the resolution of navigation issues on the river but has not, to date, been involved in wider management debates. Although this Commission might have been expanded to incorporate a wider mandate, this move was blocked by regional interests that preferred to limit the powers of the central agency. Nevertheless, the countries signatory to the Danube Declaration have expressed their willingness to work toward improvement using narrowly focused agreements between pairs or small clusters of countries. This means that, rather than employing integrated decision making at the watershed scale, water management in the Danube will proceed through a series of locally bounded and bilateral bargaining sessions. As a first step, the signatories to the Danube Declaration are wrestling with the problem of what constitutes adequate water quality and what testing methods are to be preferred in collecting water quality data. Linnerooth argues that as discussions move into consideration of management measures, there may be value in the use of computer simulation techniques as a means of communicating and displaying complex information and as a way of providing "neutral" evidence. Linnerooth terms this approach "mutual learning" and emphasizes the value of shared problem definition, information collection, and data analysis in the promotion of a shared goal for the watershed.

The Rhine River Basin

Like the Danube, the Rhine is one of Europe's largest rivers, its watershed supporting a population of more than 50 million people. The Rhine catchment area encompasses nine European nations and one of the most important industrialized regions in the world. Other important water uses on the Rhine include fishing, navigation, waste disposal, and a variety of recreational uses. As a result of these activities, the Rhine has become heavily polluted with mercury, arsenic, cadmium, lead, copper, zinc, chromium, and chlorides (Kiss 1985).

Two organizations are expressly charged with integrated management of the Rhine River and its basin. The Central Commission for the Navigation of the Rhine dates from 1815 and has as its main goal the provision of free navigation on the river and equal treatment for all ships. The Commission, through its navigation-related activities, has some responsibility for pollution control on the Rhine, including the transportation of dangerous goods, but its main function is to provide good navigational facilities throughout the river. The International Commission for the Protection of the Rhine Against Pollution was created in 1963. Its primary task is to prepare and carry out research on the nature, importance, and origin of pollution on the Rhine, on behalf of several member nations. The Commission conducts a range of such activities and reports its findings annually to member nations. It also collaborates and cooperates with other agencies interested in the protection of natural waters.

Studies by the International Commission have clearly shown that most of the pollution on the Rhine originates from industrial sources and from discharges from major urban centers. A Rhine Convention on Chemical Pollution emphasizes integrated management of river water quality and reminds member nations that Rhine River water is used for human consumption, for livestock watering, and for the support of wild animal populations. The Convention also stresses the need to support other beneficial water uses such as fishing, recreation, irrigation, and industrial water supply. Notwithstanding these provisions, surface water discharges are subject primarily to approval by individual governments, but not by any multilateral agency. Discharges are supposed to comply with emission standards established by the International Commission. The Convention also includes provisions for notification of chemical spills and for the resolution of water use conflicts (arbitration is the preferred approach).

Recent legal decisions in the Rhine basin have shown that pollution control will require international cooperation. Kiss (1985) notes that progress on water quality management will not occur solely through the application of liability rules, which do not work well at the international level because concerned governments did not wish to raise the issue of international responsibility of industrial polluters. Despite slow progress on water management and reluctance to acknowledge international liability, Kiss states that the drafting of international rules and the constant institutional cooperation between concerned nations offer promising approaches to effective management of international basins like the Rhine. Intergovernmental institutions are necessary to foster permanent, or at least stable, cooperation, but their resources must be extended by linkages to individual member states and sharing of resources across nations. It may require political action to ensure that the necessary resources are diverted and that shared environmental resources are protected.

The Upper Wye Catchment, United Kingdom

The River Wye is one of the United Kingdom's most important rivers, draining the south-central part of Wales and portions of western England. The basin has a total area of more than 4,000 km², but for planning purposes the total area has been divided into two subbasins: the Upper Wye catchment (the River Wye and its lakes and tributaries down to the city of Hay on Wye) and the Lower Wye catchment (the rest of the basin down to the mouth of the river at its confluence with the River Severn, near the Bristol Channel). The Wye passes through varied country, including hilly upland areas, rolling agricultural lands, and industrial towns. Its complex land uses and the international extent of its basin make it essential to plan future uses of the basin carefully. In June 1993, the National Rivers Authority (NRA) began a comprehensive catchment planning exercise for the upper portion of the River Wye basin to ensure that natural waters are protected and, where possible, improved for the benefit of future generations. The NRA approached the plan in several steps (NRA 1993a, 1993b):

1. Identifying catchment uses
2. Setting appropriate water use and water quality targets
3. Determining the current "state of the catchment"
4. Identifying specific issues and management options
5. Developing a process for public consultation

Catchment uses in the Upper Wye are diverse. Agriculture is predominant, including sheep and dairy farming, crop production, and moorlands. There is a small amount of mining activity and a larger forestry industry in upland areas. A good network of roads and a major railway line serve the larger centers of Builth Wells (population 2,040), Llandrindod Wells (population 4,943), Llanwrtyd Wells, Rhayader, Talgarth, and a number of smaller villages. Several of the urban centers are tourist attractions (a result of their historical development as spa towns), so there is a seasonal influx of population during the summer months. Normal population density is about 18 people per square kilometer. The area supports a wide variety of plants, mammals, fish, and birds. Some areas of the catchment have been designated as Environmentally Sensitive Areas, and the River Wye itself is a "Site of Special Scientific Interest" under national law. The area also contains part of a national park. Water supply and sewage treatment in the area is provided by two companies, Welsh Water (Dwr Cymru) and Severn Trent Water. The activities of these companies, including the volume of their water abstractions and the quality of their finished water, are overseen by Her Majesty's Inspectorate of Pollution.

The Upper Wye is a major spawning and nursery area for salmon and may be, according to the NRA, the best salmon fishery in England and Wales. During the spawning season illegal fishing of salmon occurs, with the effect of reducing fish stocks directly, through taking of adult fish, and indirectly, by reducing the spawning population. There is a lively market for salmon from the river, whether caught legally or illegally, and this market may tend to encourage poaching (illegal fishing). In recent years the salmon catch has declined in the Upper Wye, suggesting that the numbers of salmon in the stream have also declined. If real, this decline may be a result of various factors, including illegal fishing, legal fishing, acidification of waters causing depleted phytoplankton and zooplankton stocks, forestry practices, avian predators, physical barriers to migration, and weather conditions. Other important fish species include brown trout and rainbow trout (Both of which are stocked), eels, shad, and a variety of "coarse" (as compared to sport) fish, including carp, chub, dace, pike, and grayling. Commercial fishing is limited to an eel trap maintained by the NRA on a tributary of the Wye.

Water abstractions are regulated by the NRA under a permit system. Small abstractions, such as residential wells, do not require a permit. The area's groundwater resources are an important resource for Dwr Cymru, the local water company. Abstracted water is mostly returned to the catchment via discharges of sewage treatment plant effluent to rivers and streams. Surface water is also abstracted for drinking water supply, primarily from reservoirs, some of which are used for local electricity generation (possible only when the reservoirs are full). Some increase in potable water demand is expected as the area develops further, but this increase is expected to be small. Reservoir water is also released to the river during low flow periods under an agreement between Dwr Cymru and the NRA. Extensive "drawdown" in the reservoirs has been a matter of concern to ecologists because of its potential to affect aquatic biota. Agricultural abstractions occur throughout the watershed for general agricultural use (including livestock watering), spray irrigation, and fish farms. About 60% of general-use water is returned to the river after use, almost all of the fish-farm water (which can be large in volume), but virtually none of the spray-irrigation water. Industrial water use is limited in the Upper Wye because of the area's largely rural character. Two abstraction licenses have been granted for sand- and gravel-washing operations and related activities. A third license allows groundwater abstraction for industrial use. Future growth in demand is hard to predict but will likely be less than 1% per year.

There are 39 sewage treatment works operating in the Upper Wye catchment, most of them owned by Dwr Cymru. All are monitored by the NRA. In urban areas, there are several stormwater overflows (drainage from streets and roofs), but the impact of these is thought to be negligible. There are very few industrial effluent dischargers in the basin. Filter backwash is discharged from Dwr Cymru's water treatment plants, and there is a small amount of discharge of effluent from fish farms, quarries, and sawmills. Spillage of chemicals including chlorine (used in water and wastewater disinfection) is always a possibility.

Three landfill sites are located in the basin, but none is located close to a stream and so they are not thought to pose any threat to water resources.

The appearance of the river is important to residents and others who visit the area. Recreational activities in the basin include walking, boating (primarily white-water canoeing), sailing and water skiing on the lakes, and a very limited amount of swimming (which is discouraged by the NRA).

The water management issues identified by NRA, and the options for their resolution, were as follows:

1. *Acidification*, resulting from the deposition of acid precipitation (caused by the burning of fossil fuels) on acid-sensitive soils and "soft" (low-hardness) waters. Acidification impairs the survival of fish eggs and young fish, reduces aquatic species diversity, and thus may affect carnivorous species like otters and fish-eating birds. It may also increase the need for treatment of drinking water. Options may include restricting emissions of acid-causing gases from smokestacks, choosing species other than conifers for reforestation, and possibly adding lime (calcium carbonate) to raise pH in natural waters.

2. *Impaired fisheries*, resulting from low dissolved oxygen in some areas. The sources of—and thus the solutions for—this issue are not presently clear. The problem may be a result of pollution from industrial and/or agricultural activities.

3. *Blue-green algae in lakes and ponds*. Algae blooms create water quality problems by using large quantities of oxygen at night, through respiration, and through the oxygen-demanding decay of dead plant tissue. Here again, there is little information available on the causes of the problem, although they are likely related to enrichment of natural waters with nutrients, especially phosphorus and nitrogen, which may enter water through a variety of human activities in the form of sewage discharge, agricultural runoff, and industrial effluents.

4. *Low flows in summer months*. Portions of the river exhibit very low flows during the summer, the time when the demand for spray irrigation is greatest. Water used in spray irrigation is not returned to the river, so this type of abstraction can have a significant negative effect on river flows. Solutions include storage of more water in reservoirs through the winter months so that more is available for release in summer. Farmers can also construct on-farm reservoirs or grow crops that are less dependent on spray irrigation.

5. *Lack of knowledge about the environmental impacts of abstraction.* Current understanding of the impacts of water abstractions on groundwater supplies and surface water systems is very weak. Additional research in this area is needed.

6. *Lack of knowledge about the impacts of development of base flows.* Some people believe that land drainage and land use changes have altered the

natural flow regime in the river, making it "flashier"—faster to respond to rainfall—and lower in its base flows than in the natural condition. Again, more research into these possible impacts is needed to clarify the extent of hydrologic changes following development.

7. *Need to protect and enhance the wildlife resource.* Human activities affect wildlife habitat in many ways. Protective measures may include consideration of wildlife requirements in approving abstraction licenses, land drainage permits, discharge permits, and land use planning applications. River-edge vegetation should be protected where sheep and cattle trample banks as they enter the river to drink. The impact of recreational activities on wildlife should be evaluated carefully, with a view to limiting certain activities when and where necessary.

8. *Decline in salmon and brown trout stocks.* Over the past 20 years, the NRA has observed a decline in the number of spring salmon and brown trout caught on the Wye. The reasons for this decline are not clear but are likely to be complex, including acidification, habitat degradation, physical barriers to migration, and overfishing. Controls or improvements in all these areas are possible and should result in increased fish stocks. The impact of avian predators can be evaluated through separate research.

9. *Flooding at Builth Wells and Llanelwedd.* Flooding of a main road through these communities occurs in floods with a return period of greater than 1 in 4 years. Thirteen homes and 20 commercial properties are affected by this flooding. Flood defenses have been considered but were rejected as too costly and environmentally undesirable.

The preceding list identifies some areas where existing uses match or conflict with existing quality in the Upper Wye catchment. It also gives some indication of where problems are most urgent, where additional information is required, and where concerns are probably minor. It does not, however, give a good picture of future land uses and pressures on catchment water resources.

For the Upper Wye, the NRA chose to focus the planning process on a draft catchment management plan. This plan, which was prepared by the NRA alone, then became the basis for consultation between the NRA and all those with interests in the catchment. In the Upper Wye process, participants were encouraged to comment on the issues and options identified in the plan, suggest alternative options for resolving identified issues, and raise additional issues not identified in the plan. Following the consultation period, the NRA considered comments for incorporation into a revised final plan that will then form the basis for the NRA's actions within the catchment and for the agency's interaction with other organizations.

No information is currently available about the success of this planning initiative, although the NRA expects that it will influence not only the agency's own action plans and statements of policy but also those of developers, planning authorities, and others involved in day-to-day management of the catchment. Individual project actions are still being identified, and implementation will follow over the next decade or so.

The Grand River Basin Water Management Study, Canada

The Grand River is one of the largest rivers in the Province of Ontario, with a huge river basin encompassing mixed residential, urban, and large areas of intensive agricultural land use. The watershed contains five major cities, which together account for at least 10% of the province's population. The river flows from the Dundalk Highlands, the highest point in Ontario, through rolling countryside and glacial outwash plains, into Lake Erie. Until the last few decades the river supported a cold-water trout fishery, but in recent years only less desirable warm-water species have been present. Swimming, boating, and a variety of other recreational uses are popular in the river basin, which includes a number of park and conservation areas. River water is used for potable water supplies, for irrigation and livestock watering, for industrial process and cooling waters, and for many other purposes.

Although the river experiences inputs of a variety of contaminants, including oxygen-demanding materials, suspended solids, heavy metals, and industrial organic compounds, the major focus of public concern has, in recent years, been the growth of rooted aquatic plants and algae (nuisance species) resulting from excess nutrient loadings to the river. This plant growth is also a major obstacle to the restoration of the fishery because it uses large quantities of dissolved oxygen at night, when photosynthesis is suppressed. Some reaches regularly exhibit nighttime dissolved oxygen levels below those necessary to support cold-water fish species. The main cause of this problem is thought to be phosphorus, which enters the stream in sewage treatment plant discharges and in agricultural runoff.

The impetus for watershed planning in the Grand River Basin was, and continues to be, urban development pressure. The basin supports many industries, and population is growing quickly as industrial development proceeds. When the original Grand River Basin Water Management Study was conducted, the Ontario Ministry of the Environment placed limits on further development in the basin pending expansion and improvement of sewage treatment facilities. A separate set of concerns centered on persistent flooding and high flood damages in some communities along the river.

The Grand River Basin Water Management Study was initiated by the Grand River Conservation Authority (GRCA), a basin management agency that operates at the provincial level with the cooperation of local municipalities. Funding for the study was obtained from the Ontario Ministry of the Environment, which formally cosponsored many of the plan activities and all of its reports. As discussed in Chapter 3, the plan employed a two-tiered advisory committee structure. The senior advisory committee consisted of local decision makers, including local politicians and senior government bureaucrats. A second tier of technical committees provided specialized advice on a range of topics from hydrology to economics. Both levels included community representatives. The advisory committees were charged with developing a scope of work for the basin management plan, beginning with problem definition. Technical committees and subcommittees discussed issues such as data acquisition and data quality, nature and adequacy of computer simulation efforts, and similar matters.

The Grand River study employed a comprehensive continuous computer simulation model of the lower river basin (see Figure 7.7), where most development was occurring. The model was developed by Ministry of the Environment engineers and scientists, in collaboration with GRCA, specifically for the planning initiative and was not adapted from any existing model. It incorporated separate subroutines to generate point- and nonpoint-source flows, simulate the growth of several species of aquatic plants and algae, and predict water levels, dissolved oxygen concentration, and the concentration of several other water quality constituents, including phosphorus.

The model proved cumbersome to calibrate and validate (see Chapter 7), and the complex output from the simulation runs were difficult to translate into lay language. Eventually the planning team developed an index of water quality that incorporated data on several key parameters, and this index was used as one decision criterion. Other criteria included flood frequency and magnitude, and percentage of time in which dissolved oxygen levels dipped below target levels.

The study team developed and tested a wide range of urban, agricultural, and in-stream management measures, including major structures such as a dam and channel improvements. A range of sewage treatment plant options were also considered.

The planners employed a series of regional consultations and direct involvement of community members in decision making to evaluate and screen management strategies. The process was time-consuming and costly, requiring about six years and more than $4 million, in addition to "in kind" contributions such as staff time and computing facilities contributed by the provincial government. Several good plans were proposed, each with clear advantages and disadvantages. In the end, the committees chose a medium-cost alternative that included sewage works upgrades at several locations and a variety of other channel improvements, but no dam. These improvements were all in place by the mid-1980s, and local municipalities continue to collaborate with GRCA as the plan, and simulation model, are continually updated to reflect changing conditions and technology.

12.5 LESSONS LEARNED

Section 1.2 presented some elements of a successful integrated watershed management effort. In summary, these are:

1. Adequate expertise for multiple-objective planning and evaluation procedures, especially in economic, social, and environmental areas

2. Adequate resources of time and money for planning and implementation
3. Consideration of a wide range of alternatives to solve observed problems
4. A flexible, adaptable plan, reviewed and amended at regular intervals
5. Representation of all parties affected by the plan and its implementation
6. Sufficient authority to enforce conformity of execution with construction and operating plans

Examination of the case studies presented earlier in this chapter and through-out this book reveals the truth of these requirements. The development and implementation of workable solutions is not a process that has a single correct outcome. Rather, it is an ongoing process of dialogue with the community, learning about needs, teaching about options, and building consensus about an ideal watershed condition and the best way to get there.

The central feature of this process is choice: humans use and affect water resources and have many choices available as to how and when that use occurs. Integrated watershed planning means developing a social consensus about best choices. Figures 12.2 and 12.3 illustrate mass flows of energy, raw materials, and waste under two different operating philosophies, the first a high-use high-waste system, and the second a conservative, low-waste approach.

Sustainable management of water resources requires water users to make conscious choices that may sometimes reduce personal benefits in favor of community or intergenerational benefits. The continuing theme throughout this book is that those choices cannot be made solely on the basis of scientific evidence.

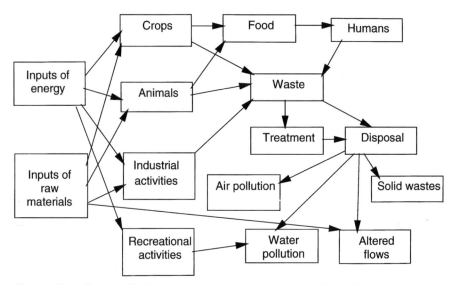

Figure 12.2 *Forces affecting water quantity and water quality under high-use, high-waste management systems.*

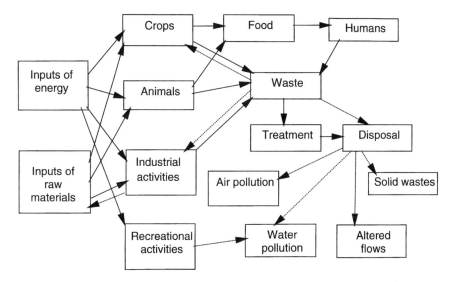

Note: Dashed lines indicate reduced use, reduced impact, or reuse/recycle.

Figure 12.3   *Forces affecting water quantity and water quality under conservative, low-waste management systems.*

Indeed, the science is not very clear, or may be contradictory, on key questions affecting water management. In addition to scientific evidence, sustainable watershed management requires community understanding and support, which in turn will generate political will and, thus, economic and human resources to make changes. Although those resources are now scarcer than they have been for many decades, public awareness of and concern about water resources issues are now higher than ever before. Integrated watershed management will therefore depend on the formation of partnerships between governments and the public, across disciplines and international borders, and among water users with different interests and values. This is a huge challenge, but one for which the payoff will be sustainable management of water resources for our own and future generations.

REFERENCES

Caponera, Dante A. 1985. Patterns of cooperation in international water law: Principles and institutions. *Natural Resources Journal* 25:563–587.

Dubosc, A. 1992. *The French Water Agencies.* Paris, France: Ministère de l'Environnement.

Goodman, A. S., and K. A. Edwards. 1992. Integrated water resources planning. *Natural Resources Forum* 16(1):65–70.

Grand River Conservation Authority. 1982. *The Grand River Basin Water Management*

Study. Summary Report. Cambridge, Ont.: Grand River Conservation Authority, in conjunction with the Ontario Ministry of the Environment.

Grayson, R. B., J. M. Dooland, and T. Blake. 1994. Application of AEAM (Adaptive Environmental Assessment and Management) to water quality in the Latrobe River catchment. *J. Envir. Management* 41:245–258.

Holling, C. S. 1978. *Adaptive Environment Assessment and Management*. Chichester, United Kingdom: John Wiley & Sons.

Jacobson, S., and R. Robles. 1992. Profile: Ecotourism, sustainable development, and conservation education: Development of a tour guide training program in Tortuguero, Costa Rica. *Environmental Management* 16(6):701–713.

Kiss, Alexandre. 1985. The protection of the Rhine against pollution. *Natural Resources Journal* 25:613–637.

Koudstaal, Rob, Frank R. Rijsberman, and Hubert Savenije. 1992. Water and sustainable development. *Natural Resources Forum* 16(4):277–290.

Lee, Terence. 1992. Water management since the adoption of the Mar del Plata Action Plan: Lessons for the 1990s. *Natural Resources Forum* 16(3):202–211.

Linnerooth, Joanne. 1990. The Danube River Basin: Negotiating settlements to transboundary environmental issues. *Natural Resources Journal* 30:629–660.

McDonald, Adrian T. and David Kay. 1988. *Water Resources Issues and Strategies*. Harlow: Longman Scientific and Technical.

McLelland, Nina. 1987. Improved efficiency in the management of water quality. *Natural Resources Forum* 11(1):49–57.

National Rivers Authority. 1993a. *National Rivers Authority Strategy* (eight-part series encompassing water quality, water resources, flood defense, fisheries, conservation, recreation, navigation, research, and development). Bristol, United Kingdom: National Rivers Authority Corporate Planning Branch.

———. 1993b. *Upper Wye Catchment Management Plan Consultation Report*. Cardiff, Wales: National Rivers Authority, Welsh Region.

Ontario Ministry of the Environment and Energy. 1993. *Metropolitan Toronto Stage II Remedial Action Plan: Clean Waters, Clear Choices*. Toronto: Ministry of the Environment and Energy, Central Region.

Royal Commission on the Future of the Toronto Waterfront. 1992. *Regeneration: Toronto's Waterfront and the Sustainable City: Final Report*. Toronto: Queen's Printer.

Schramm, Gunter. 1980. Integrated river basin planning in a holistic universe. *Natural Resources Journal* 20: 787–805.

Teclaff, Ludwik A., and Eileen Teclaff. 1987. International control of cross-media pollution: An ecosystem approach. *Natural Resources Journal* 27:21–53.

Viessman, Warren Jr. 1990. Water management issues for the nineties. *Water Resources Bulletin* 26(6):883–891.

Wilkes, Daniel. 1975. Water supply regulation. In *Regional Management of the Rhine*, edited by Chatham House Study Group. London: Chatham House.

Young, Mike. 1992. Sustainable investment: The economic challenge. In *Environment and Development*, edited by UNESCO. Publication No. 166, Vol. 42(2). London: Taylor and Francis.

Index